Introduction to the Physics of Waves

Balancing concise mathematical analysis with the real-world examples and practical applications that inspire students, this textbook provides a clear and approachable introduction to the physics of waves.

The author shows through a broad approach how wave phenomena can be observed in a variety of physical situations and explains how their characteristics are linked to specific physical rules, from Maxwell's equations to Newton's laws of motion. Building on the logic and simple physics behind each phenomenon, the book draws on everyday, practical applications of wave phenomena, ranging from electromagnetism to oceanography, helping to engage students and connect core theory with practice. Mathematical derivations are kept brief and textual commentary provides a non-mathematical perspective. Optional sections provide more examples together with higher-level analyses and discussion.

This textbook introduces the physics of wave phenomena in a refreshingly approachable way, making it ideal for first- and second-year undergraduate students in the physical sciences.

Tim Freegarde is a Senior Lecturer in Physics at the University of Southampton, where his research explores the use of light to trap, cool and manipulate atoms and particles. He has taught wave-related subjects to physics undergraduates of all levels for over 15 years.

Introduction to the Physics of Waves

TIM FREEGARDE

University of Southampton

CAMBRIDGE
UNIVERSITY PRESS

CAMBRIDGE
UNIVERSITY PRESS

University Printing House, Cambridge CB2 8BS, United Kingdom

Published in the United States of America by Cambridge University Press, New York

Cambridge University Press is part of the University of Cambridge.

It furthers the University's mission by disseminating knowledge in the pursuit of education, learning and research at the highest international levels of excellence.

www.cambridge.org
Information on this title: www.cambridge.org/9780521197571

First published 2013

A catalogue record for this publication is available from the British Library

Library of Congress Cataloguing in Publication data
Freegarde, Tim, 1965–
Introduction to the physics of waves / Tim Freegarde, University of Southampton.
pages   cm
Includes bibliographical references and index.
ISBN 978-0-521-19757-1
1. Waves – Textbooks.   2. Wave-motion, Theory of – Textbooks.   I. Title.
QC157.F74   2012
531'.1133 – dc23   2012025066

ISBN 978-0-521-19757-1 Hardback
ISBN 978-0-521-14716-3 Paperback

Additional resources for this publication at www.cambridge.org/waves

Contents

Preface   *page* ix
Acknowledgements   x
Symbols   xi

1 The essence of wave motion   1
   1.1   Introduction   1
   1.2   A local view of wave propagation   2
   1.3   Cause and effect   4
   1.4   Examples of wave disturbance   9
   Exercises   10

2 Wave equations and their solution   12
   2.1   Wave equations   12
   2.2   Waves on long strings   17
   Exercises   22

3 Further wave equations   23
   3.1   Waves along a coaxial cable   23
   3.2   Electromagnetic waves   28
   3.3   Ocean waves   30
   3.4   Capillary waves   37
   3.5   Gravity waves in compressible fluids 40
   3.6   Waves and weather   44
   Exercises   46

4 Sinusoidal waveforms   47
   4.1   Sinusoidal solutions   47
   4.2   Energy of a wave motion   50
   4.3   The tsunami   53
   4.4   Normal modes, standing waves and orthogonality 57
   Exercises   62

5 Complex wavefunctions   63
   5.1   Complex harmonic waves   63
   5.2   Dispersion in dissipative systems   65

5.3   Phasors and geometric series   67
Exercises   69

6 **Huygens wave propagation**   71
6.1   Huygens' model of wave propagation   71
6.2   Propagation in free space   71
6.3   Reflection at an interface   77
6.4   Refraction at an interface   78
6.5   Fermat's principle of least time   81
Exercises   82

7 **Geometrical optics**   85
7.1   Ray optics   85
7.2   Refraction at a spherical surface   86
7.3   The thin lens   90
7.4   Fermat's principle in imaging   92
Exercises   93

8 **Interference**   96
8.1   Wave propagation around obstructions   96
8.2   Young's double-slit experiment   97
8.3   Wavefront dividers   100
8.4   The Michelson interferometer   101
Exercises   104

9 **Fraunhofer diffraction**   106
9.1   More wave propagation around obstructions   106
9.2   Diffraction by a single slit   107
9.3   Babinet's principle   111
9.4   The diffraction grating   112
9.5   Wavefront reconstruction and holography   119
9.6   Definition of Fraunhofer diffraction   121
9.7   The resolution of an imaging system   123
Exercises   123

10 **Longitudinal waves**   125
10.1   Further examples of wave propagation   125
10.2   Sound waves in an elastic medium   125
10.3   Thermal waves   129
Exercises   132

11 **Continuity conditions**   134
11.1   Wave propagation in changing media   134
11.2   The frayed guitar string   134

11.3 General continuity conditions and characteristic impedance 140

11.4 Reflection and transmission by multiple interfaces 143

11.5 Total internal reflection 146

11.6 Frustrated total internal reflection 152

11.7 Applications of internal reflection and evanescent fields 154

11.8 Evanescent-wave confusions and conundrums 155

Exercises 156

12 Boundary conditions 158

12.1 The imposition of external constraints 158

12.2 The guitar and other stringed musical instruments 159

12.3 Organ pipes and wind instruments 161

12.4 Boundary conditions in other systems 165

12.5 Driven boundaries 167

12.6 Cyclic boundary conditions 167

Exercises 169

13 Linearity and superpositions 171

13.1 Wave motions in linear systems 171

13.2 Linearity and the superposition principle 172

13.3 Wavepackets 173

13.4 Dispersion and the group velocity 176

Exercises 179

14 Fourier series and transforms 180

14.1 Fourier synthesis and analysis 180

14.2 Fourier series and the analysis of a periodic function 182

14.3 Alternative forms of the Fourier transform 186

14.4 Mathematical justification of Fourier's principle 188

14.5 The spectrum 192

14.6 Orthogonality, power calculations and spectral intensities 193

14.7 Fourier analysis of dispersive propagation 195

14.8 The convolution of waveforms 197

14.9 Fourier analysis of Fraunhofer diffraction 201

14.10 Fourier-transform spectroscopy 203

Exercises 204

15 Waves in three dimensions 207

15.1 Waves in multiple dimensions 207

15.2 Wave equations in two and three dimensions 207

15.3 Plane waves and the wavevector 209

15.4 Fourier transforms in two and three dimensions 211

15.5 Diffraction in three dimensions 212

15.6   Wave radiation in three dimensions   216
15.7   Polarization   219

16 Operators for wave motions   225
16.1   The mathematical operator   225
16.2   Operators for frequency and wavenumber   226
16.3   The expectation value: the mean value of an observable   227
16.4   The uncertainty: the standard deviation of an observable   229
16.5   Operator analysis of a Gaussian wavepacket   230
16.6   Complex electrical impedances   232
Exercises   233

17 Uncertainty and quantum mechanics   235
17.1   The bandwidth theorem   235
17.2   Wave–particle duality   237
17.3   The quantum-mechanical wavefunction   238
17.4   Measurement of the quantum wavefunction   241
Exercises   245

18 Waves from moving sources   247
18.1   Waves from slowly moving sources   247
18.2   Waves from quickly moving sources   252
18.3   The wake of a ship under way   256
Exercises   261

19 Radiation from moving charges   263
19.1   Solution of the electromagnetic wave equation   264
19.2   Retarded electromagnetic potentials   268
19.3   Retarded electromagnetic fields   275
19.4   Radiation from moving charges   279
Exercises   281

Appendix: Vector mathematics   283
A.1   Cartesian coordinates   283
A.2   Spherical polar coordinates   284

References   286
Index   291

Preface

When we revised our Southampton undergraduate programme to draw into a single course the wave phenomena hitherto distributed among optics, electromagnetism, thermal physics, quantum mechanics and solid-state physics, there seemed to be no single text to recommend. This book is an expanded version of the lecture notes that resulted, and its aim, beyond covering wave physics in its own right, is to introduce the common phenomena and analytical methods that are encountered in these individual fields as well as in the disciplines such as oceanography from which we have always drawn a further audience.

There were nonetheless some excellent textbooks for individual aspects. Coulson's classic [15] provides a concise and elegant mathematical introduction; French's once ubiquitous volume [29] is admirable for its clarity and brevity; and Crawford's brilliant and popularly acclaimed approach through everyday examples [16] suffers only from being long out-of-print. Many other texts are highly satisfactory in the areas they cover, and references to their recent editions are included throughout the following chapters.

One privilege for the author of any new volume is to have a new range of scientific and technological examples upon which to draw. Oscillations in the circulations of the oceans have been quite recently recognized; the extraction of power from ocean waves and tides is only now emerging as practical and necessary; the electronic control of holographic arrays has been possible for just a few years; and the quantum mechanics of coherent systems now underpins major research fields and devices that not long ago appeared impossible.

Yet a particular delight in the physics of waves is that it is readily encountered in so many everyday phenomena – from mountain clouds to surface ripples and from sound and music to the colours of feathers and oil films – to which Pretor-Pinney's brilliantly approachable guide [72] is a joy that takes few scientific liberties. Many seminal studies were therefore performed centuries ago by scientists who are now venerated, and it is a miracle of our age that their original writings are often available, via the Internet, without having to don white gloves in the vaults of a national library. For those like me to whom it brings a particular thrill to read these classic works, I have given a number of references. While scholars of old were often handicapped by the mathematical limits of their eras, their writings generally prove much more profound and insightful than we are commonly led to believe.

Acknowledgements

It is a great pleasure to express my heartfelt thanks to those who have helped me to compile this book. My Southampton colleagues benevolently allowed me to divert them when I repeatedly demonstrated that only by attempting to explain something does one discover the limits of one's own understanding; the students and post-docs in my research group provided a good-humoured sounding-board; and it was Rob Eason who suggested that my lecture notes might make a textbook in the first place. My editors John Fowler and Claire Poole at Cambridge University Press offered invaluable technical guidance and bore my repeated deadline-busting with great forbearance; Steven Holt's meticulous copy-editing immeasurably improved the 'final' manuscript; and Sehar Tahir of Aptara/CUP's T$_E$Xline provided help and advice with exceptional speed and efficiency.

My gliding friends Dave Aknai and Alison Randle enthusiastically offered photographs to illustrate mountain waves and boat wakes. For permission to reproduce other photographs I am grateful to Walter Spaeth of ARTside, Arthur Barlow and James Kennedy of New College, Oxford, John Teufel of NIST, and the companies Holoeye, OpenHydro, Pelamis and Raytheon. Other figures are reproduced with permission from the journal *Nature*, and from the US Navy and Air Force, JPL, Minnesota Sea Grant, NASA and NOAA, all of which generously place their materials in the public domain.

Final thanks go to my parents and Deirdre for their constant encouragement and deliveries of tea to the distant garret, and in particular to Dad for painstakingly proof-reading a late draft of the entire manuscript.

Symbols

The following symbols are used throughout this book for the variables and constants given below. Physical constants and scalar variables are set in *italics*; vectors are in **bold**. Operators and unit vectors are indicated by a circumflex (e.g. $\hat{\mathbf{k}}$). The accepted values of physical and mathematical constants are given, and SI units are indicated where appropriate.

a	amplitude, coefficient, distance (slit width)		m
b	amplitude, coefficient		
c	speed of light	$2.997\,924\,58 \times 10^{8}$	m s^{-1}
	specific heat capacity		J K^{-1} kg^{-1}
d	distance (between slits)		m
d	differential, infinitesimal increment		
∂	partial differential		
e	elementary charge	$1.602\,176 \times 10^{-19}$	C
	Jones vector or component		
e	Euler's number	$2.718\,282$	
f	function,		
	frequency,		Hz
	focal length		m
g	acceleration due to gravity		m s^{-2}
h	height, elevation		m
	Planck's constant	$6.626\,069 \times 10^{-34}$	J s
$\hbar$	Planck's constant$/(2\pi)$	$1.054\,572 \times 10^{-34}$	J s
i	index		
i	imaginary unit ($\sqrt{-1}$)		
$\hat{\mathbf{i}}$	unit vector along the x axis		
j	index		
$\hat{\mathbf{j}}$	unit vector along the y axis		
k	index,		
	wavenumber, wavevector		rad m^{-1}
k_{B}	Boltzmann's constant	$1.380\,650 \times 10^{-23}$	J K^{-1}
$\hat{\mathbf{k}}$	unit vector along the z axis		
l	index,		
	length		m

m	index,		
	mass		kg
n	index, number of moles		
p	pressure,		Pa
	momentum,		kg m s^{-1}
	dipole moment,		C m
p	image–lens distance		m
q	index,		
	charge		C
r	radius, radial coordinate, position vector		m
	amplitude reflectivity		
s	coordinate variable along arbitrary path,		m
	object–lens distance		m
t	time,		s
	amplitude transmission		
u	transformed position, e.g. $x - vt$		
v	velocity		m s^{-1}
v_{p}	phase velocity		m s^{-1}
v_{g}	group velocity		m s^{-1}
w	grating slit width		m
x	coordinate variable		m
y	coordinate variable		m
z	coordinate variable		m
A	cross-sectional area		m^2
$\mathbf{A}$	vector potential		V s m^{-1}
B	magnetic flux density		T
C	capacitance,		F
	capacitance per unit length,		F m^{-1}
	specific heat capacity		J K^{-1} kg^{-1}
D	electric displacement field		C m^{-2}
E	electric field strength,		N C^{-1} $\equiv$ V M^{-1}
	modulus of elasticity		Pa
$\mathcal{E}$	energy		J
F	force		N
G	gravitational constant	$6.674\,28 \times 10^{-11}$	m^3 kg^{-1} s^{-2}
H	magnetic field strength		A m^{-1}
$\hat{\mathcal{H}}$	Hamiltonian operator		J
I	current		A
$\mathcal{I}$	intensity		W m^{-2}
J	current density		A m^{-2}
$\mathcal{K}$	kinetic energy		J
L	inductance,		H
	inductance per unit length		H m^{-1}

M	mass per unit length		kg m^{-1}
N	number of grating rulings		
P	pressure		Pa
	degree of polarization		
$\mathcal{P}$	power		W
Q	total charge		C
Q	heat		J
R	radius of curvature		m
	gas constant	8.314 472	J K^{-1} mol^{-1}
S	surface (area) of integration		m^2
	Stokes vector or parameter		
$\mathcal{S}$	entropy		J K^{-1}
T	period		s
$T(t)$	function of single variable t		
$\mathcal{U}$	potential energy		J
V	voltage, scalar potential		V
	volume		m^3
W	tension		N
$X(x)$	function of single variable x		
$Y(y)$	function of single variable y		
$Z(z)$	function of single variable z		
Z	impedance		Ω
Z_0	impedance of free space	376.730	Ω
α	angle		rad
γ	friction coefficient,		kg s^{-1}
	specific heat ratio, Lorentz factor		
$\delta(x)$	Dirac delta function		
δ_{mn}	Kronecker delta function		
δx	small finite increment of x		
ϵ	energy density		J m^{-3}
ε_0	permittivity of free space	$8.854\,188 \times 10^{-12}$	F m^{-1}
ζ	vertical displacement		m
η	refractive index		
ϑ	angle, phase angle		rad
κ	thermal conductivity,		W m^{-2} K^{-1}
	radiation enhancement factor		
λ	wavelength		m
μ_0	permeability of free space	$4\pi \times 10^{-7}$	H m^{-1}
ν	frequency		Hz
$\bar{\nu}$	spectroscopists' wavenumber		m^{-1}
ξ	horizontal displacement		m
π	Archimedes' constant	3.141 593	

ρ	density (mass per unit length)	kg m^{-1}
	density (mass per unit area)	kg m^{-2}
	density (mass per unit volume)	kg m^{-3}
σ	conductivity,	A m^{-2} V^{-1}
	standard deviation	
τ	period, characteristic time, duration	s
φ	angle, phase angle	rad
ψ	wavefunction,	
	angle, phase angle	rad
ω	angular frequency	rad s^{-1}
Δ	finite increment or change; uncertainty	
Δ^2	variance	
∇	vector differential	
Φ	magnetic flux	V s
Σ	sum	
Θ	temperature	K, °C
Ψ	wavefunction superposition	
Re	real part	
Im	imaginary part (a real value)	
Arg	argument	
|...|	modulus	
$\langle...\rangle$	mean, expectation value	
[...]	retarded value	
$*$	convolution	

1 The essence of wave motion

1.1 Introduction

The physics of waves is too often presented only in a few rather straightforward and sometimes uninspiring contexts: the motion of strings, sound, light and so on. Students may be led to regard the topic with disdain; and they may be left with some crucial misconceptions, such as that all waves are sinusoidal. Wave physics may hence be considered an old-fashioned field with little relevance to the more modern, exciting areas of quantum physics, nanotechnology and cosmology. Yet there is plenty to find interesting just in classical and modern optics and the physics of musical instruments; and wave phenomena prove to be central to most of the fascinating and newly emerging branches of both fundamental and applied physics.

Most aspects of physics may be viewed from two perspectives: one, named after Lagrange, addresses particles, while the other, due to Euler, considers fields. We may, for example, establish the electromagnetic properties of matter by considering the Coulomb forces among all the constituent charged particles; or we may describe the material's bulk response to a field and tackle the problem that way. This duality pervades most areas of physics and, while one of the alternatives often proves vastly more convenient than the other, the two are ultimately quite consistent, equivalent viewpoints.

When we extend our analysis to dynamic systems, the particle approach becomes a form of 'kinetics' or ballistics, and changes in the field description are manifest as waves. So, when we consider the physics of waves, we're really addressing the general subject of time-dependent field theory. Stringed and wind instruments and so on are merely particularly easy examples to visualize.

In this book, we address the physics of waves in a fairly thorough fashion, but illustrate each fundamental concept with practical examples such as sonar, imaging optics, water ripples, radar and so on. As we proceed, we develop a range of techniques that have much broader application, and allow us to acquire the concepts and methods behind many other areas – the most important of which is probably that of quantum mechanics, where particle–wave duality is clearly, impressively and sometimes confusingly apparent.

1.2 A local view of wave propagation

You have doubtless already met countless examples of what are termed *waves*, from the surface waves in ripple tanks to the vibrations of a guitar string or the air column in an organ pipe, to radio waves and light. You may have been led to believe that each of these is a periodic, and perhaps even sinusoidal, travelling disturbance. But what of other, more commonplace examples? Should physicists consider ocean waves, tidal waves, bow waves and shock waves? Are journalists justified in writing of a *wave of fear* or *wave of protest*?

We shall consider many examples of wave propagation, and manifestations of wave phenomena, in the following chapters; and our understanding will stem as much from these accumulated examples as from succinct summarizing statements. We shall see for example that waves need not be periodic, that the propagated properties may sometimes be neither transverse nor longitudinal, and that it is not only for simple physical systems that a wave-propagation description can have some validity. Common principles of propagation allow very different systems to show the same characteristic phenomena of refraction, reflection, dispersion, superposition, interference and diffraction. We shall thus come to understand features of wave motion both in an abstract, generic sense and through the specific manifestations in particular examples.

It is nonetheless helpful to begin by defining what we mean by a wave, and we shall in this chapter consider three aspects. First, we shall consider what happens at a local level to allow a wave to propagate through a given region, and we shall see that important concepts are the point-to-point propagation of a physical effect, and the time lag as the effect travels from one point to another. Secondly, we shall consider the forces of interaction between two charges or masses, and shall see that waves occur when, for whatever reason, there is a finite speed of propagation between them. Finally, we shall briefly consider the nature of the disturbance propagated by the wave.

Numerous examples will illustrate these characteristics throughout this book, and a couple of specific examples of the origins of wave behaviour are considered in this chapter. To identify the defining characteristics of a wave, and illustrate the mechanisms by which it propagates, however, we start with a rather everyday, and somewhat unscientific, example.

Suggested reading

Suggestions for further reading are indicated throughout the following chapters by the sketch of an open book in the margin. For recent editions of common textbooks, abbreviated to the author's surname, specific pages or sections are given. Full details of all cited texts may be found towards the end of this book.

Fig. 1.1
The Mexican wave *la ola*, reportedly seen in North America since the 1960s, gained international popularity after the World Cup in Mexico in 1986.   ⓒ Walter Spaeth, ARTside.de

1.2.1 *La ola*

We begin in Mexico City, where the Azteca Stadium was the centre of the 1968 Olympic Games and, in the lulls between events, television viewers across the world were delighted to watch a new phenomenon. Restless spectators joined in as their playful neighbours stood and waved briefly in synchronism, and the original motion spread as a ripple around the crowd. The *Mexican wave* (Fig. 1.1) had (reportedly, at least) made its international debut.

I. Farkas *et al.* [20].

The Mexican wave is found typically to travel at 12 m s^{-1} and have a width of around 15 seats. Only 30 or so spectators are needed to start a full wave.

A crowd of human spectators is, of course, a poorly defined, wildly nonlinear and inhomogeneous medium, yet the Mexican wave clearly illustrates the crucial characteristics of wave motion. The **disturbance** at any point (in this case, the vertical displacement of a particular spectator's hands, say) is a response to the action of a neighbour in raising or lowering his/her own hands; and the response is slightly delayed, through either inertia or simply the time lag in perception. These properties prove to be completely general.

1.2.2 Microscopic definition of a wave

A **wave**, then, is

> a collective bulk disturbance in which what happens at any given position is a delayed response to the disturbance at adjacent points.

The progress of a disturbance from one point to its neighbours, and thence to their neighbours and so on, is known as **propagation**; and the wave propagates through a **medium**, which may, for example, be water, air or glass, a guitar string or drum skin, an electrical cable, a crowd of spectators, or pure vacuum. The medium showing the wave motion need not necessarily be linear or homogeneous, and the wave does not need to be sinusoidal or even periodic.

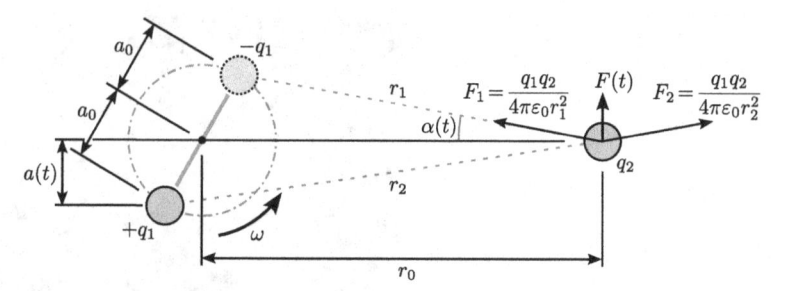

Fig. 1.2 Coulomb interaction between the charges $+q_1$ and $-q_1$ of a rotating dipole and a test charge q_2.

We shall see that it is often helpful to consider such cases, but they are specific examples rather than the only forms allowed.

Positions within the medium are described by *coordinates* (usually in one, two or three dimensions, depending upon whether the medium is a string, drum skin or ocean), and at each position the wave is described by one or more further variables – such as the fluid pressure, electric field strength or elevation of a spectator's arms – that quantify the disturbance. These variables, to which we shall generally refer explicitly, are all **wavefunctions**. Mathematically, each wavefunction depends upon, or is defined by, the coordinates of the position at which it is determined, together with a further variable: *time*.

1.3 Cause and effect

One of the most common examples of waves is the electromagnetic wave, manifest as light, radio waves and so on. We're often tempted to adopt quite a local picture of electromagnetic wave propagation, and consider a lonely photon traversing the Universe, or some region of space in which electric and magnetic fields oscillate in synchronism for no obvious reason. But it can be helpful to consider as well the sources of waves, and the effects that they subsequently have as they are detected: for electromagnetic waves, the overall process is simply that moving charges influence other charges through a retarded Coulomb interaction. All waves have a cause and effect, and their explicit inclusion can help to clarify otherwise mysterious phenomena.

1.3.1 Electromagnetic waves

Consider the situation in Fig. 1.2, in which two equal but opposite charges $\pm q_1$, separated by a distance $2a_0$, form a *dipole* that rotates slowly, with angular frequency ω, about its centre. This is our 'transmitter' or wave source. Some distance away, our 'receiver' (or detector) is a single charge q_2. Both charges of the dipole will exert a Coulomb force on q_2; in each case, the magnitude will be $q_1 q_2/(4\pi\varepsilon_0 r^2)$, where $r = r_1, r_2$ is the distance between the test charge and the dipole charge; and the direction, whether attractive or repulsive, will be along the line joining the charge centres. If $a_0 \ll r_0$, where r_0 is the distance between the test charge and the dipole centre, then the horizontal component of the force due to $+q_1$ will cancel out that due to $-q_1$. We're therefore left with just the vertical components, which will add. We shall work out shortly the magnitude of this force, but for now it suffices to understand how it arises.

C.-A. de Coulomb [14]

As the dipole rotates slowly about its centre, the vertical – or *transverse* – force on q_2 will grow and fade, and the oscillatory motion of the charges in the dipole is thereby transmitted to the other charge. The rotating dipole here represents the flow of charges in the antenna of a radio transmitter and, at a different scale, the variation of the charge distribution in the electron 'cloud' of a radiating atom. So far, then, we've considered nothing more than the electrostatic Coulomb interaction between separated charges.

For the rotating dipole of Fig. 1.2 $a(t)$ may be written explicitly in the form $a(t) = a_0 \cos(\omega t + \varphi)$, where a_0, ω and φ characterize the specific example. But we do not need to be so specific, and therefore write in terms of a general *function* $a(t)$ that indicates only that the vertical distance a depends upon the time t.

The function $a(t)$ should not be confused with the simple product at.

The transverse force component is found by multiplying the individual Coulomb forces F_1 and F_2 by $\sin\alpha(t)$, where $\sin\alpha(t) \approx a(t)/r_0$ and $a(t)$ is the vertical component of a_0 at time t, and adding them to yield the net transverse force $F(t)$ experienced by q_2. With a little working, and assuming $a_0 \ll r_{1,2}$ and $r_{1,2} \approx r_0$, this thus turns out to be

$$F(t) = \frac{2q_1 q_2}{4\pi\varepsilon_0 r_0^3} a(t). \tag{1.1}$$

Rather than characterize the effect of the dipole by the *force* it exerts upon a specific charge q_2, we consider the strength of the electric **field**, which is simply the force exerted upon the test charge divided by the value of the test charge q_2 itself. We may therefore write

$$E(t) = \frac{F(t)}{q_2} = \frac{2q_1}{4\pi\varepsilon_0 r_0^3} a(t). \tag{1.2}$$

We now increase the speed at which the 'transmitter' dipole rotates, and the theory of relativity imposes a time lag or *retardation* on the force caused by the dipole: any force caused by the charges of the dipole is delayed by a time r_0/c, so that q_2 sees the dipole as it was a time r_0/c earlier. When we take this into account, our expression for the electric field experienced by q_2 at a time t becomes

$$E(t) = \frac{2q_1}{4\pi\varepsilon_0 r_0^3} a\left(t - \frac{r_0}{c}\right) \tag{1.3}$$

and we thus have our wave: a function that depends upon what was happening at another position some time previously.

Feynman [22] Chapters 1 and 2

This is how all waves propagate. We can add a medium – other charges between q_1 and q_2 – so that the disturbance can also travel from q_1 to q_3 (say) and then from q_3 to q_2, and this turns out to be the origin of the refractive index of a material and so on. But the basic principle is that what happens at one point in space has an influence upon what will happen elsewhere, with a time lag because of the finite speed of propagation. Waves, then, are what happens to a system of forces when we move things around quickly enough for delays to be significant.

You may have spotted that the description here of electromagnetic radiation isn't complete: it gives an oscillating electric field, but no accompanying magnetic field; and the electric field decreases with the cube of the distance from the dipole, which, although correct at small distances, does not account for the generally more important field that varies as r_0^{-1} and allows the radiation of energy. This is because, while we introduced the time lag required by the theory of relativity, we omitted to perform the Lorentz transformations that take into account the speeds of the moving charges, and to include the variation in retardation as the individual dipole charges move towards and away from the test charge.

We shall see in Chapter 19 that a full calculation introduces important corrections to the electric field. Some can be simply written in terms of the motions of the dipole charges, and correspond to radiated electric fields that vary with the inverse or inverse square of the distance. Other terms turn out to be non-zero only when q_2 is also moving; these yield the radiated magnetic field components. Magnetism, it turns out, is simply the relativistic correction to the electrostatic Coulomb force.

Billingham & King [7]
pp. 210–212
D. H. Frisch & L. Wilets [30]

1.3.2 Macroscopic definition of a wave

The example of electric dipole radiation allows us to offer a second definition of a wave, as

> a time-dependent feature in the field of an interacting body, due to the finite speed of propagation of a causal effect.

While this is more of a *description* of a wave than an indication of how it arises, it is both a useful summary of how waves are manifest and a reminder that they always ultimately emanate from some form of source.

At a more abstract level, waves may be regarded as dynamic solutions to a time-dependent field theory. Just as simple systems of interacting particles, which in static equilibrium would be stationary, more generally show oscillatory motions, so the steady-state fields of stationary charges or masses turn into waves in the more general dynamic case.

1.3.3 Gravitational waves

Just as electromagnetic waves result from the retardation of an oscillating electrostatic force, so variations in a gravitational force can be manifest in gravitational waves. The distances involved are, naturally, cosmological.

Figure 1.3 shows a rather hypothetical scenario in which we have the ability to move planets. One planet, of mass m_1, can be moved in an oscillatory or rotary motion of amplitude a_0, and we are interested in the effect of this

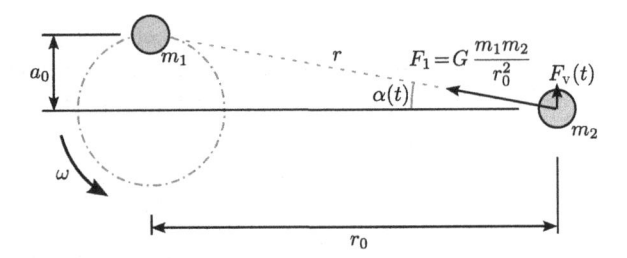

Fig. 1.3   Gravitational interaction between a moving mass m_1 and a test mass m_2.

A much-considered source of gravitational waves is a coalescing binary star. In a typical example, neutron stars, each of 1.4 solar masses, are separated by a few tens of kilometres, and rotate about their common centre of mass several times a second. Coalescence occurs within a few minutes because the stars lose energy and angular momentum through gravitational radiation with spiral wavefronts as illustrated below:

Such mind-boggling systems are the favoured sources to be observed by extremely sensitive (and expensive) gravitational-wave detectors currently being built at various sites world-wide and planned for space.

For details of laser interferometric gravitational-wave detection see [89–91].

motion upon a test planet of mass m_2. We shall concern ourselves only with the vertical component of the force experienced by m_2, for the large horizontal component will be largely constant. As shown, the gravitational force is given by the usual formula to be inversely proportional to the square of the distance r_0 between the masses. We shall assume that $a_0 \ll r_0$.

The vertical component of the gravitational force is given by

$$F_{\mathrm{v}}(t) = F \sin \alpha(t) = F \frac{a(t)}{r_0} \qquad (1.4)$$

and leads to an apparent gravitational field at m_2 with a vertical component g_{v} given by $m_2 g_{\mathrm{v}} = F_{\mathrm{v}}(t)$, and hence

$$g_{\mathrm{v}} = G \frac{m_1}{r_0^3} a(t). \qquad (1.5)$$

Because of the finite speed of gravitational propagation, however, the effect is retarded, and depends not on the current position of planet m_1 but upon that seen by an observer on planet m_2 – just as with the electromagnetic wave:

$$g_{\mathrm{v}}(r, t) = G \frac{m_1}{r_0^3} a \left(t - \frac{r_0}{c} \right). \qquad (1.6)$$

In reality, a motion such as that of m_1 must be due to rotation about another heavenly body, as shown in Fig. 1.4. The second mass has the unfortunate effect of cancelling out, to first order, the gravitational field oscillations produced by the first. The result is that only higher-order, *quadrupole* radiation components may be detected. Their calculation requires a slightly more careful evaluation.

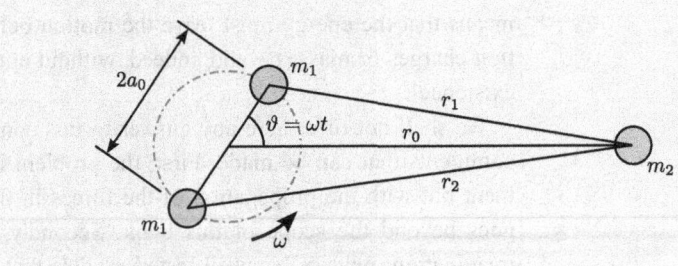

Fig. 1.4   Gravitational interaction between a rotating pair of masses m_1 and a test mass m_2.

By application of the cosine rule, $r_{1,2}^2 = r_0^2 + a_0^2 \pm 2a_0 r_0 \cos \vartheta$, where ϑ defines the orientation of the pair of masses with respect to the test mass, the net vertical component of the gravitational field is found to be

$$g_v = G\frac{m_1 a_0 \sin \vartheta}{r_0^3}\left[\left(1+\frac{a_0^2}{r_0^2}-2\frac{a_0}{r_0}\cos\vartheta\right)^{-1} - \left(1+\frac{a_0^2}{r_0^2}+2\frac{a_0}{r_0}\cos\vartheta\right)^{-1}\right]$$

$$= G\frac{2m_1 a_0^2}{r_0^4}\sin(2\omega t). \tag{1.7}$$

With the caveat that, as for electromagnetic waves, we have here calculated only the short-range component, it is this field that gravitational-wave observatories are intended to detect.

1.3.4 The æther

In describing the propagation of electromagnetic and gravitational forces, we have tiptoed straight through the middle of one of the greatest controversies of the nineteenth century: how can objects influence each other through a vacuum, unless there is an *æther* – an intangible fluid – that connects them? By the early twentieth century, the existence of an æther had been all but disproved: Maxwell and Einstein had formulated elegant descriptions of the propagation of electromagnetic and gravitational fields and waves that made no reference to the presence or properties of any æther; and the results of various experiments – including those of Michelson and Morley [63, 64] – made any theory of an æther increasingly untenable. In the twenty-first century, then, we are generally comfortable that static forces may be transmitted through a vacuum.

But what of waves? The answer here must be the same, yet there are grounds for disquiet that we should mention, even though they will then be dismissed. The first is the concept of point-to-point propagation, which will become the basis of the Huygens description of wave motion: we tend to imagine photons travelling in straight lines through space, yet diffraction suggests that they can take other routes even when there is nothing to deflect them. The second is the question of radiation. The motions of charges and masses can be damped by transferring their energy to other charges or masses, in a process that is extremely clear under Newtonian mechanics; yet the process of radiation means that the energy must leave the motion before it arrives at the destination charges or masses – and, indeed, without apparently even requiring their existence!

We shall not offer here any answer to this conundrum, but there are some comments that can be made. First, the problem lies not with the wave treatment but with the propagation of the forces in themselves – so the question goes beyond the scope of this book. Secondly, the problem is one of our imagination: physics has well-tested models that account very accurately for how the forces propagate; our problem is the more philosophical 'why?' – so, arguably, it's not even a question for physics! Summarized another way: we have very good descriptions of how electromagnetic and gravitational forces work, and the propagation of electromagnetic and gravitational waves is simply a consequence that we derive from those initial descriptions. Neither the forces nor their wave propagation has yet presented any inconsistency, either with itself or with experimental observation. So, if we find it hard to imagine, that's a defect with our imagination. But it shouldn't stop us asking interesting questions!

1.4 Examples of wave disturbance

In the examples above, we have considered two wave motions in which the propagated property or **disturbance** is a force or displacement *perpendicular* to the direction of wave propagation. It is common to refer to these as *transverse* waves, and to distinguish them from *longitudinal* motions that are *parallel* to the direction of wave propagation. To end this chapter we consider the validity of such categorizations for some common examples of wave motion.

Transverse wave motions, including electromagnetic and gravitational waves, are characterized by the propagation or measurement of a transverse physical displacement; the disturbance, in other words, is a vector property directed perpendicular to the direction of wave propagation. As well as transverse waves in vacuum, there are many examples in what may be considered continuous media. The transverse motion of a guitar string, for example, is considered to result from the curvature of the taut string, so that the tension does not pull in exactly opposite directions on opposite sides of any given point. In this category, we might also consider surface water waves, due to the transverse components of surface tension or gravitational forces, and mountain lee waves in the atmosphere. The skin of a drum is a two-dimensional version of the guitar string; and surface acoustic waves used in optoelectronic devices resemble ocean waves with the bulk elasticity of the medium playing the role of gravity.

Longitudinal waves are those in which the physical displacement or disturbance is once again a vector quantity, but is parallel to the direction of propagation. The classic example here is sound, and the motion results because the pressures either side of a given region are not equal. The pulse of our blood flow may be considered a form of sound wave, as may the pulses observed in the exhausts of jet engines.

There is no reason why a vector disturbance should be aligned parallel or perpendicular to the propagation direction, however: it is possible to generate longitudinal electromagnetic and gravitational waves, and when magnetization propagates as a *spin wave* it can have transverse and longitudinal components.

Terrestrial and solar seismology similarly involve manifestations of both longitudinal and transverse wave motion.

The wave disturbance need not be a physical force or displacement; it could be a scalar quantity, or a vector unrelated to the propagation coordinates. Temperature, for example, may propagate as thermal waves; and the spatially dependent concentrations of chemical reactants and products can change as chemical waves travel through a *reaction–diffusion* system such as the human heart, or as a flame front travels through a flammable material. Perhaps the most fascinating example of a propagated property that is not a displacement is the quantum wavefunction – the mysterious quantity that, according to quantum mechanics, contains everything we can know about a particle. Quite what the quantum wavefunction is, and what it means, are profound questions that must be left to other texts; but the nature of its propagation is identical to that of the other wave manifestations that we'll study here.

One could reason that thermal waves propagate by the *longitudinal* flow of heat, or argue that sound waves may be described by the scalar property of pressure. The simple characterization into transverse, longitudinal and scalar waves is therefore not always straightforward. Fortunately, such labels are really only for convenience; there are few physical consequences, and we shall see that in specific cases an examination of the propagation mechanisms resolves any doubt as to the phenomena observed.

So what of the waves of emotion and protest mentioned earlier? A key aspect of wave propagation is that the disturbance should propagate from point to point in a causal fashion, rather than simply reflect the staggered arrival times at adjacent points via independent routes. So, provided that the fear or protest of each person is inspired by the fear or protest of a neighbour – and social scientists can be reasonably clear in identifying such mechanisms – it may indeed be valid to regard the propagation of such properties as a wave. That the medium through which the wave propagates is composed of people who are discrete, nonlinear and to some extent irreproducible does not differ fundamentally from many granular, nonlinear and noise-ridden examples of more classical physical systems. The wave description may not only serve as a useful shorthand to imply the neighbour-to-neighbour-mediated propagation: subject to the natural imprecision of such systems, it may even allow their mathematical simulation and prediction.

Exercises

The following exercises do not directly concern wave propagation but address the mathematical techniques used in subsequent chapters. Substitute a few arbitrary values for a simple numerical check.

1.1   Given that $\sin(A + B) \equiv \sin A \cos B + \cos A \sin B$ and $\cos(A + B) \equiv \cos A \cos B - \sin A \sin B$, express the following in terms of simple sine/cosine functions of the individual variables ($\sin x$, $\cos b$, etc.) alone (e.g. $\cos(2\vartheta) \to \cos^2 \vartheta - \sin^2 \vartheta$):

1. $\sin(X - Y)$ 3. $\sin(2\vartheta)$ 5. $\cos(2a - 3b)$
2. $\cos(X - Y)$ 4. $\sin(3\vartheta)$

1.2   Express the following purely as linear combinations of sine or cosine functions (e.g. $\sin^2 X \to (1 - \cos(2X))/2$):

1. $\cos A \cos B$ 3. $\cos^2(\omega t)$ 5. $\cos(\omega_1 t)\sin(\omega_2 t)$
2. $\sin(at)\sin(bt)$ 4. $\cos(\omega_1 t)\cos(\omega_2 t)$

1.3   Given that $\exp(i\vartheta) \equiv \cos \vartheta + i \sin \vartheta$, where $i^2 = -1$, determine the following, writing your answers in the form $\exp(i\vartheta)$. For each expression, evaluate its real part, imaginary part, modulus and argument. There may in some cases be more than one solution. The terms a, b, α, β and ϑ should all be taken to be real.

1. $(\exp(i\vartheta))^2$ 3. $\exp(i\alpha)\exp(i\beta)$ 5. $a \exp(i\alpha) + b \exp(i\beta)$
2. $\sqrt{\exp(i\vartheta)}$ 4. $\exp(i\alpha)/\exp(i\beta)$

1.4   Recalling where necessary the product and chain rules

$$\frac{\mathrm{d}}{\mathrm{d}x}ab \equiv a\frac{\mathrm{d}b}{\mathrm{d}x} + b\frac{\mathrm{d}a}{\mathrm{d}x}, \qquad \frac{\mathrm{d}}{\mathrm{d}x}f(g) \equiv \frac{\mathrm{d}f}{\mathrm{d}g}\frac{\mathrm{d}g}{\mathrm{d}x},$$

determine the following:

1. $\dfrac{\mathrm{d}}{\mathrm{d}\vartheta}\sin\vartheta$ 7. $\dfrac{\mathrm{d}}{\mathrm{d}x}\sin^2(4x)$ 12. $\dfrac{\mathrm{d}}{\mathrm{d}\beta}\exp(-ia\beta)$

2. $\dfrac{\mathrm{d}}{\mathrm{d}\vartheta}\cos\vartheta$ 8. $\dfrac{\mathrm{d}}{\mathrm{d}x}\cos^2(bx)$ 13. $\dfrac{\mathrm{d}}{\mathrm{d}y}\sin y \cos y$

3. $\dfrac{\mathrm{d}}{\mathrm{d}\vartheta}\sin(3\vartheta)$ 9. $\dfrac{\mathrm{d}}{\mathrm{d}\beta}\exp\beta$ 14. $\dfrac{\mathrm{d}}{\mathrm{d}y}\sin(2y)\cos(3y)$

4. $\dfrac{\mathrm{d}}{\mathrm{d}\vartheta}\sin(a\vartheta)$ 10. $\dfrac{\mathrm{d}}{\mathrm{d}\beta}\exp(a\beta)$ 15. $\dfrac{\mathrm{d}}{\mathrm{d}t}\sin(at)\exp(-it)$

5. $\dfrac{\mathrm{d}}{\mathrm{d}x}\sin^2 x$ 11. $\dfrac{\mathrm{d}}{\mathrm{d}\beta}\exp(i\beta)$ 16. $\dfrac{\mathrm{d}}{\mathrm{d}\vartheta}\cos(b\vartheta)\exp(-ia\vartheta)$

6. $\dfrac{\mathrm{d}}{\mathrm{d}x}\cos^2 x$

1.5   Determine the following:

1. $\int_{x_1}^{x_2} \sin(kx)\,\mathrm{d}x$ 3. $\int_{t_1}^{t_2} \cos(\omega t)\exp(-iat)\,\mathrm{d}t$
2. $\int_{t_1}^{t_2} \exp(-iat)\,\mathrm{d}t$ 4. $\int_0^{\infty} \cos(\omega t)\exp(-\alpha t)\,\mathrm{d}t$

Wave equations and their solution

2.1 Wave equations

Pain [66] pp. 108–112
French [29] pp. 161–167
Crawford [16] pp. 48–56

We have discussed qualitatively in the previous chapter how wave motions result from the delayed response to an adjacent displacement. But how can we quantify the motion? How can we determine the speed of propagation, the energy contained within the wave, whether the wave shape changes as it propagates, and so on? The answer is that we must first quantify the physics of the wave system by writing explicit equations that describe how the adjacent disturbance influences a given point and how the response is delayed.

For any system, we shall see that it is possible to write a **wave equation** that embodies the governing physics in the form of a **partial differential equation** relating derivatives of the wave displacement (wavefunction) with respect to time and position. Although different systems have different wave equations, most are similar in form and can be solved using the same methods.

Our general approach is therefore to begin with the specific physics of the particular situation, derive the corresponding wave equation, and then solve it in a fairly generic way before applying specific conditions to determine a particular solution. We shall in this chapter consider as an example the plucked string of a musical instrument, and our analysis will allow us to determine the string's motion after it has been released. We shall meet many further examples in later chapters; each may be tackled by following the same strategy.

To give a sense of direction to this treatment, however, we shall begin in reverse by considering phenomenologically the sort of mathematical expressions that we shall expect for our wave solutions, and hence the form that we might expect our wave equations to take.

2.1.1 Travelling waves

We have already seen how wave propagation may be viewed as the effect of retardation upon what at short range we might regard as an instantaneous interaction. We shall shortly consider how the propagation characteristics, including the speed and hence retardation, are determined by the specific physics of a situation. It is common in science, though, for our understanding to be led by an attempt to explain observed patterns of behaviour.

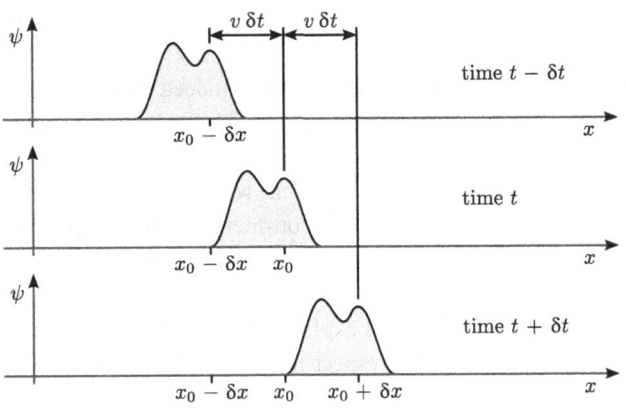

Fig. 2.1 A travelling wave $\psi(x - vt)$ has the same shape at successive times $t - \delta t$, t and $t + \delta t$, but is each time displaced by $\delta x = v\,\delta t$, where v is the wave propagation velocity.

Before outlining the approach that we shall adopt for the rest of this book, we therefore first consider what may be deduced from the commonly observed characteristic of **travelling waves**, illustrated in Fig. 2.1, which is at least approximately shown in a wide variety of wave situations: that a propagating wave will remain essentially unchanged in shape, but simply be displaced according to its propagation velocity v, when compared with that a short time earlier. The wave displacement ψ, which for our sportsground example of a Mexican wave will be some property like the height of the spectator's hands, will at a position x and time t be equal to that at an adjacent position $x - \delta x$ at a slightly earlier time $t - \delta t$, where for our example δx will be the distance to the adjacent spectator and δt the time it takes for the spectator to respond to his or her neighbour's movement. Put mathematically,

$$\psi(x, t) = \psi(x - \delta x, t - \delta t). \tag{2.1}$$

Any point of constant displacement $\psi(x, t) = \psi_0$ therefore travels with a speed v given by $\delta x = v\,\delta t$, as does the wave overall.

If equation (2.1) is true for all points and times, and the ripple of our Mexican wave remains the same shape as it propagates around the stadium, then it follows that the wave displacement ψ at any time t and position x will equal that found at time $t = 0$ a distance $\delta x = vt$ away at a position $x - vt$, and the wave may be expressed as a function of a single variable $u \equiv x - vt$ that corresponds to the position variable x with a time-dependent offset. We may therefore write

Travelling wave

$$\psi(x, t) = \psi(x - vt) \equiv \psi(u). \tag{2.2}$$

This **travelling wave** is the phenomenon whose origins and characteristics we shall now explore.

Although it will be clear in specific examples that the wave disturbance or wavefunction is an electric field, or the distance through which a string is displaced, or the pressure of a fluid, it is sometimes convenient to refer to it with a generic label. For this, we shall use the Greek character ψ, pronounced *psi*.

2.1.2 The Mexican-wave equation

It is possible (and is indeed the case for *nonlinear* systems, as in Exercise 5.8) that the waveform of a travelling wave must be of a particular shape; but it is a general observation that the wave shape is often of little consequence, and that different wave shapes propagate in essentially the same fashion. We shall briefly consider how this suggests the existence of a **wave equation** that embodies the specific physics of wave propagation in any system.

It turns out that we may express the properties of our travelling-wave solutions without explicit definition of their shapes by relating the derivatives of $\psi(x, t)$ with respect to x and t. We may do this from first principles by subtracting $\psi(x, t - \delta t)$ from each side of equation (2.1),

$$\psi(x, t) - \psi(x, t - \delta t) = \psi(x - \delta x, t - \delta t) - \psi(x, t - \delta t). \qquad (2.3)$$

We now divide both sides by δt, substitute $\delta x = v\, \delta t$, and take the limit as δt and δx tend to zero; with a little rearrangement, this gives

$$\lim_{\delta t \to 0} \frac{\psi(x, t) - \psi(x, t - \delta t)}{\delta t} = -v \lim_{\delta x \to 0} \frac{\psi(x, t - \delta t) - \psi(x - \delta x, t - \delta t)}{\delta x}.$$
$$(2.4)$$

The limiting ratios on each side are, by definition, the *partial derivatives* of the wavefunction ψ with respect to position and time, and are related through the propagation velocity v by what we shall come to regard as a *wave equation*

$$\frac{\partial \psi}{\partial t} = -v \frac{\partial \psi}{\partial x}. \qquad (2.5)$$

Note, though, that we have not yet *derived* the wave equation, and that equation (2.1) could equally yield any equation of the form

$$\frac{\partial^n \psi}{\partial t^n} = (-v)^n \frac{\partial^n \psi}{\partial x^n}, \qquad (2.6)$$

with integer n. We might argue that n should be even, because otherwise the equation would need to be modified if the wave propagation direction and hence sign of v were changed, but otherwise all values of n are so far equally possible.

Such considerations guide us to examine the governing mechanisms of the Mexican wave for ways in which such derivatives are related: the temporal derivative $\partial \psi / \partial t$ could represent the action of our muscles in raising our arms or body, and the spatial derivative $\partial \psi / \partial x$ the difference that we observe between our current stance and that of our neighbour. We therefore in general follow this path in reverse. If, for this rather fickle example, we could establish and quantify how we react to the difference that we observe between our own stance and that of our neighbour, then we could derive the governing wave equation and solve it to determine the wave solutions that may result.

This is indeed the approach that we shall take throughout this book. A system exhibiting wave motion may be described by a wave equation relating derivatives (not necessarily the first) with respect to time to derivatives with respect to position; and the wave equation is in each case derived by considering the physical processes by which the system responds to changes at neighbouring points. We shall later encounter the electromagnetic wave equation resulting from Maxwell's equations and Schrödinger's equation for the quantum motion of a free particle. We start, however, by considering the simpler example of a guitar string.

2.1.3 Partial differentiation and partial differential equations

We have already referred in Section 2.1.2 to the **partial derivatives** of a wave displacement or wavefunction. The following reminder of what is meant by such objects may safely be skipped by those already familiar with their use.

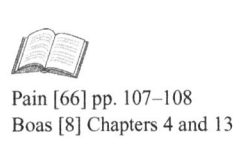

Pain [66] pp. 107–108
Boas [8] Chapters 4 and 13

With wave motions, as with any other form of dynamics, we are interested in how quantities vary with time. The disturbance measured at any point will therefore be a function of time, and we may indicate this explicitly by writing it as $\psi(t)$: this means nothing more than ψ, but is just a way of reminding ourselves of the temporal dependence. If $\psi(t)$ were the vertical, z, coordinate of a mass m subject to gravity, for example, then Newton's equation for the motion of the mass would be simply

$$m\frac{\mathrm{d}^2\psi}{\mathrm{d}t^2} = -mg. \tag{2.7}$$

Our waves involve the motions of a large number of different particles at a large number of different positions. One way of labelling these points (or particles) would be by numbering them and then referring to their displacements as ψ_1, ψ_2 etc. The nth displacement $\psi_n(t)$ would then describe the nth particle at the nth position. The displacement ψ therefore depends upon both t and n. As the number of points increases, this notation becomes increasingly cluttered, and we choose to identify the particular ψ by the position of the point or particle, for example

$$\psi(x, t). \tag{2.8}$$

The symmetry of this notation suggests that x can be treated in just the same way as t, so, to indicate the acceleration of particles at a given point, we specify that the position coordinate is constant. Equation (2.7) then becomes

$$m\left(\frac{\partial^2\psi}{\partial t^2}\right)_x = -mg. \tag{2.9}$$

The *full derivative* $\mathrm{d}\psi$ has been replaced by the *partial derivative* $\partial\psi$, to indicate that we are considering the dependence of the function ψ upon only one of its parameters (the time, t) and we have indicated with the subscript that the other parameter, the position coordinate x, is held fixed.

The function ψ does not need to vary continuously from point to point – it could be a series of discrete values such as the positions of atoms in a lattice – but, in the context of waves, it usually turns out to do so at least to a good approximation. For example, a guitar string might be represented by

$$\psi(x, t) = a \cos(kx)\sin(\omega t) \tag{2.10}$$

so that, at $x = x_0$, the displacement will be $\psi(x_0, t)$, and the acceleration will be

$$\left(\frac{\partial^2 \psi}{\partial t^2}\right)_{x_0} = a \cos(kx_0)(-\omega^2)\sin(\omega t). \tag{2.11}$$

We have therefore differentiated ψ with respect to t, taking $a \cos(kx_0)$ to be constant.

We can also take a snapshot of the guitar string at time $t = t_0$: the displacement will be $\psi(x, t_0)$. The gradient of the string at time t_0 is then found by differentiating with respect to x, this time treating $a \sin(\omega t)$ as a constant:

$$\left(\frac{\partial \psi}{\partial x}\right)_{t_0} = -a \sin(kx)\sin(\omega t). \tag{2.12}$$

One of the clearest ways to picture a function of two variables is as a landscape, whereby the height h of the land above sea level varies with longitude and latitude, or with the x and y coordinates of the grid reference. Suppose, then, that we wish to know how much higher a point 10 m to the northeast of us is. We could reach that point by walking $\sim$7 m to the east and then the same distance to the north. The height gained in the first leg would be approximately

$$\Delta h_1 = 7 \left(\frac{\partial h}{\partial x}\right)_y, \tag{2.13}$$

while the second leg makes a similar contribution,

$$\Delta h_2 = 7 \left(\frac{\partial h}{\partial y}\right)_x, \tag{2.14}$$

where the x axis is taken to run to the east and the y axis to the north. Strictly, the partial derivative for Δh_2 should be evaluated at $x + 7$ rather than at x, and of course the above expressions assume constant gradients, but, as the distances involved are reduced, these become ever better approximations. In the limit of moving infinitesimal distances $\mathrm{d}x$ and $\mathrm{d}y$, we find that the change in height is given exactly by

$$\Delta h = \Delta h_1 + \Delta h_2 = \left(\frac{\partial h}{\partial x}\right)_y \mathrm{d}x + \left(\frac{\partial h}{\partial y}\right)_x \mathrm{d}y. \tag{2.15}$$

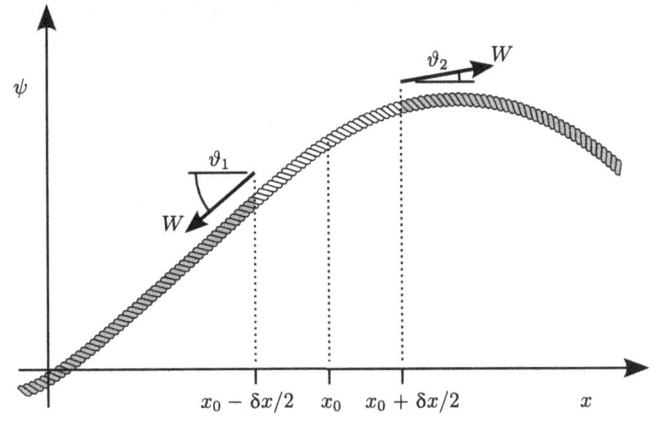

Fig. 2.2 Forces acting on an element of a taut string subject to a tension W.

For simplicity, it is quite common to drop the brackets and fixed variable, so that $(\partial h/\partial x)_y$ becomes simply $\partial h/\partial x$. That this is a partial derivative, with the other variables held constant, is then implicit.

2.2 Waves on long strings

To illustrate how a wave equation may be derived by considering the physics underlying a given situation, and how the wave motion may then be determined, we take the example of the transverse motion of a taut string, such as the string of a musical instrument, a washing line, or the overhead cables of an electrified railway. Specifically, we consider the string of an acoustic guitar.

Figure 2.2 shows a section of a guitar string, taken to be distant from the bridge and fret that mark its ends. We shall neglect gravity and assume that the string tension W N does not vary significantly with small displacements from its resting position, which we shall describe by the wavefunction $\psi(x, t)$.

If the string has a uniform mass per unit length of M (measured for example in kg m^{-1}) then the short element between $x = x_0 - \delta x/2$ and $x = x_0 + \delta x/2$ will have a mass of $M \, \delta x$. The net force upon this element will be the vector sum of the forces acting on its two ends, which have the same magnitude W but in general act at slightly different angles $\vartheta_{1,2}$ to the x axis. If these angles are small, so that $\cos \vartheta_{1,2} \approx 1$, then the longitudinal components of the tension will cancel out, and we need only consider the transverse components.

The net transverse force in the positive ψ direction will be given by $W(\sin \vartheta_2 - \sin \vartheta_1)$, measured here in N, where $\vartheta_{1,2}$ are defined by the gradients of the string displacement $\tan \vartheta = \partial \psi/\partial x$ at each end. If ψ varies slowly with x, so that $\vartheta_{1,2}$ are again small, we may make the approximation $\sin \vartheta \approx \partial y/\partial x$ and hence, applying Newton's second law, write

$$M \, \delta x \left(\frac{\partial^2 \psi}{\partial t^2} \right)_{x_0} = W \left[\left(\frac{\partial \psi}{\partial x} \right)_{x_0 + \delta x/2} - \left(\frac{\partial \psi}{\partial x} \right)_{x_0 - \delta x/2} \right], \qquad (2.16)$$

where $\partial^2 \psi / \partial t^2$ is the acceleration of the string element. Dividing through by δx and taking the limit as δx tends to zero, we arrive at the wave equation

$$M \frac{\partial^2 \psi}{\partial t^2} = W \frac{\partial^2 \psi}{\partial x^2}. \qquad (2.17)$$

While this form emphasizes the origins of the two differential terms in the inertia of the element and the tension to which it is subjected, it can be more convenient to rearrange equation (2.17) slightly to give

$$\frac{\partial^2 \psi}{\partial t^2} = \frac{W}{M} \frac{\partial^2 \psi}{\partial x^2}. \qquad (2.18)$$

Equation governing transverse waves on a string

Validity: shallow waves on an infinitely flexible string with no frictional losses.

This equation completely governs the motion of the string, and we shall find that, provided we know the initial displacement and motion $\psi(x, t = 0)$ and $\partial \psi / \partial x (x, t = 0)$, the future motion of the string is entirely predictable.

2.2.1 Solving the wave equation

use physics/mechanics to write partial differential wave equation for system

↓

insert generic trial form of solution

↓

find parameter values for which trial form is a solution

Weighty volumes have been written on the solution of partial differential equations, and there are indeed wave systems that require solutions of some complexity. Fortunately, unless we venture into particularly technical or esoteric systems, we shall usually meet only a few types of wave equation, all of which are quite straightforward to solve. Indeed, the number of forms of the wave equation is so small that we shall quickly learn to recognize the type of solution that is needed, and solving the wave equation then merely requires the insertion of a general form of that solution and the derivation of values for the free parameters that are involved.

Usually, this simple and pragmatic approach will suffice, but with more complex systems it must be remembered that the success of a particular trial form of wave solution does not necessarily exclude other possibilities. New phenomena have been discovered, and Nobel prizes won, by those who have checked sufficiently carefully their initial assumptions and looked for missing solutions.

2.2.2 Travelling-wave solutions

Pain [66] p. 112

We shall first look for travelling waves of the form $\psi(x, t) = \psi(x - vt)$ given in equation (2.2). Although this is a function of both x and t, it depends upon the single quantity $u \equiv x - vt$; the wave shape therefore does not change with time except for being translated a distance vt along the x axis. We may then differentiate $\psi(x - vt)$ using the chain rule

$$\frac{\partial}{\partial s}\psi(u) = \frac{d\psi}{du}\frac{\partial u}{\partial s}, \tag{2.19}$$

where $s = x, t$. On substituting the travelling wave into equation (2.18), and differentiating using the chain rule as above, we obtain

$$v^2\frac{d^2\psi}{du^2} = \frac{W}{M}\frac{d^2\psi}{du^2}, \tag{2.20}$$

which is valid for any wave shape $\psi(x - vt)$ provided that

$$v^2 = W/m, \tag{2.21}$$

and hence

$$v = \pm\sqrt{W/M}, \tag{2.22}$$

the signs corresponding to forward and backward travelling waves. The wave speed thus increases with the tension, reflecting how strongly a point is coupled to its neighbours, and decreases with the string density, according to the inertia in response.

Our wave equation has the important property of *linearity*: that if $\psi_1(x, t)$ and $\psi_2(x, t)$ are solutions, then so is the general **superposition** $\psi = a\psi_1(x, t) + b\psi_2(x, t)$, where a and b are arbitrary constants. We shall consider linearity at length in Chapter 13. In this example, we have seen that there are two classes of solution, corresponding to the forward- and backward-travelling waves. If we work in terms of the positive root $|v|$ of equation (2.22), our general solution may be written

$$\psi(x, t) = \psi_+(x - |v|t) + \psi_-(x + |v|t), \tag{2.23}$$

where $\psi_+(x, t)$ and $\psi_-(x, t)$ are, respectively, the forward- and backward-going parts. We stress that our analysis so far holds for any waveforms ψ_+ and ψ_-.

Having found our general solution, how do we determine the specific solution for a given system? Suppose that we know the initial waveshape $\psi(x, 0)$ and its velocity $\partial\psi/\partial t(x, 0)$ at $t = 0$. Setting $t = 0$ in equation (2.23) and its temporal derivative gives us two equations:

$$\psi(x, 0) = \psi_+(x) + \psi_-(x), \tag{2.24a}$$

$$\frac{\partial\psi}{\partial t}(x, 0) = \frac{\partial\psi_+(x - |v|t)}{\partial t}(x, 0) + \frac{\partial\psi_-(x + |v|t)}{\partial t}(x, 0). \tag{2.24b}$$

The shapes of our two travelling-wave components ψ_+ and ψ_- must therefore add to give the initial waveform of the string, and their transverse velocities $\partial\psi_\pm/\partial t$ must add for each element of the string to give its initial velocity.

Application of the chain rule to equation (2.24b) then gives

$$\frac{\partial\psi}{\partial t}(x, 0) = -|v|\frac{d\psi_+(u_+)}{du_+}(x, 0) + |v|\frac{d\psi_-(u_-)}{du_-}(x, 0), \tag{2.25}$$

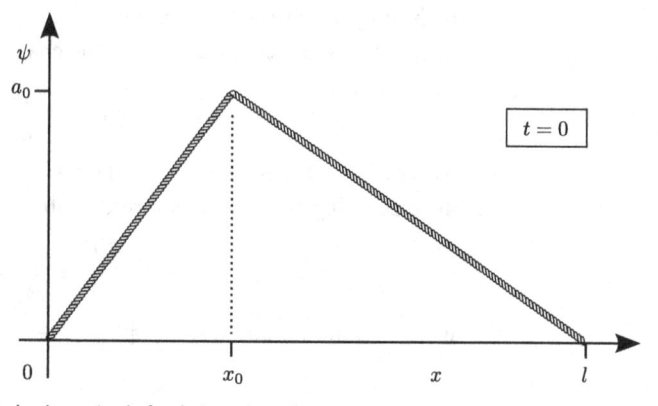

Fig. 2.3   The plucked guitar string before being released.

where $u_+ \equiv x - |v|t$ and $u_- \equiv x + |v|t$ so that, when $t = 0$ and hence $u_+ = u_- = x$, equation (2.25) gives

$$\frac{\partial \psi}{\partial t}(x, 0) = |v| \left\{ \frac{\mathrm{d}\psi_-(x)}{\mathrm{d}x} - \frac{\mathrm{d}\psi_+(x)}{\mathrm{d}x} \right\}. \qquad (2.26)$$

For each point x along the initial waveform, we therefore have two equations (2.24a) and (2.26) to define the two unknown functions ψ_+ and ψ_-, and our initial conditions hence suffice to provide a unique solution.

2.2.3 The plucked guitar string

We may now consider the detailed motion of a plucked guitar string, which we take to be fixed by the bridge and fret at $x = 0$ and $x = l$. The action of plucking the string involves pulling it at a single point x_0 so as to displace it transversely through a distance a_0; the string is then released at time $t = 0$. The initial displacement $\psi(x, 0)$, which we label $\psi_0(x)$, may therefore be taken to be the triangular form shown in Fig. 2.3, and the initial velocity $\partial \psi / \partial t(x, 0)$ may everywhere be taken to be zero. These initial conditions, together with equations (2.24a) and (2.26), fully describe the physics of the situation and allow us to determine the forward- and backward-travelling components $\psi_\pm(x, t)$ and hence the subsequent motion of the string.

For an initially stationary string, equation (2.26) shows that the gradients of $\psi_+(u_+)$ and $\psi_-(u_-)$ are everywhere equal at $t = 0$ when $u_+ = u_- = x$, so

$$\psi_-(x) = \psi_+(x) + c, \qquad (2.27)$$

where c is a constant of integration. On combining equations (2.24a) and (2.27) and assuming the initial form $\psi_0(x)$ we hence obtain

$$2\psi_+(x) + c = \psi_0(x), \qquad (2.28)$$

so that

$$\psi_+(x) = \tfrac{1}{2}(\psi_0(x) - c) \qquad (2.29)$$

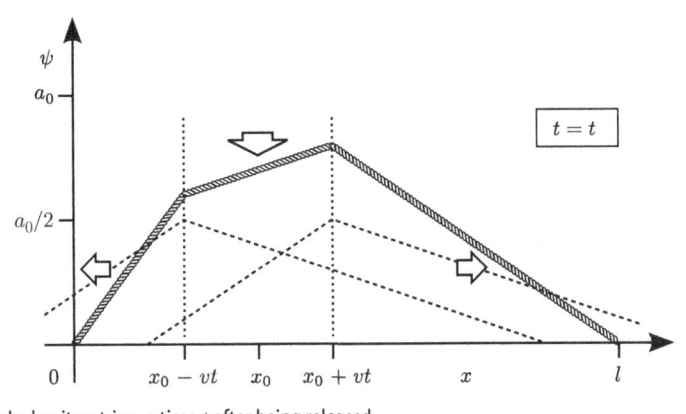

Fig. 2.4 The plucked guitar string, a time t after being released.

and therefore

$$\psi_-(x) = \tfrac{1}{2}(\psi_0(x) + c).$$ (2.30)

The constant of integration hence disappears when the components are added, and may be neglected. Substituting equations (2.29) and (2.30) into equation (2.23) thus gives the waveform at any time t,

$$\psi(x, t) = \tfrac{1}{2}\big[\psi_0(x - |v|t) + \psi_0(x + |v|t)\big].$$ (2.31)

The forward- and backward-travelling components are seen to be identical and each equal to half of the initial displacement; the motion of the string, described by equation (2.31), is found by adding these two components in their respective positions at any later time $t > 0$ when, as shown in Fig. 2.4, they will have moved through distances $\pm vt$. We see that the overall displacement shows a flattening that starts from x_0 where the string was plucked and propagates outwards with the wave speed v. Beyond the flattened region, the displacement remains unchanged – a reminder that information about the release of the string cannot propagate faster than the wave can travel.

This analysis above works well for the initial motion of the string near where it was plucked, but the eagle-eyed will have spotted that it is not a complete treatment and breaks down as the wave approaches the fixed ends of the string. Strictly, the initial condition expressed in equation (2.24a) applied only in the range $0 \leq x \leq l$ for which the initial displacement was defined; beyond this, we can only say that $\psi(0, t) = \psi(l, t) = 0$. This additional condition turns out to be satisfied provided that, for any value of u,

$$\psi_+(u) = -\psi_-(2l - u) = \psi_+(u - 2l),$$ (2.32)

hence defining the components beyond the initial range, and in the example above the two component waves are therefore periodic sawteeth. This is an example of the effect of **boundary conditions**, which we shall examine in detail in Chapter 12.

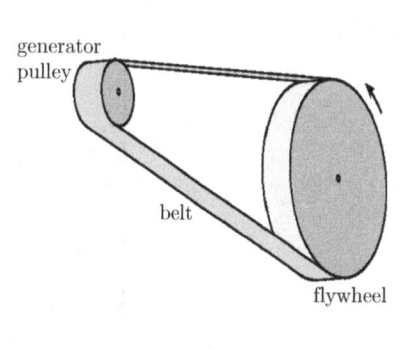

generator
pulley

belt

flywheel

Fig. 2.5   A steam-powered showman's traction engine and the transmission belt used to drive its generator.

Exercises

2.1   Waves on the lowest (C) string on a correctly tuned 'cello propagate with a speed of 90 m s^{-1}. If the string has a mass per unit length of 0.014 kg m^{-1}, find the tension to which it must be subjected.

2.2   Waves on the highest (A) string of the 'cello of Exercise 2.1 must travel at about 3.4 times the speed of waves on the C string. If the string tensions are approximately equal, find the diameter of the A string if it is made of solid steel with a density of 7850 kg m^{-3}.

2.3   Figure 2.5 shows the leather belt used to transmit power from the flywheel of a steam-powered traction engine to the electrical generator at the front. Waves may naturally propagate along the belt, in exactly the same way as they travel along the string of a guitar.

   If the tension in the belt is W, its mass per unit area is ρ, the width of the belt is d, and the belt displacement ψ is expressed as a function of the distance x and time t, sketch a diagram indicating these parameters and derive the wave equation

$$\frac{\partial^2 \psi}{\partial t^2} = \frac{W}{\rho d} \frac{\partial^2 \psi}{\partial x^2}. \tag{2.33}$$

2.4   If $W = 100$ N, $\rho = 10$ kg m^{-2} and $d = 10$ cm, calculate the speed of waves travelling along the pulley belt of Exercise 2.3.

2.5   If the flywheel of Exercises 2.3 and 2.4 has a diameter of 1 m, at what engine speed (in revolutions per second) will the wave speed equal the speed of the belt, allowing a stationary disturbance to build up?

3   Further wave equations

In Chapter 2 we saw how the motion of a guitar string could be established by considering the physical mechanisms that governed it and using them to determine the wave equation for the system. In this chapter we shall see that the same approach can be applied to a wide range of physical systems. We begin with detailed derivations of the wave equations for electromagnetic waves along a coaxial cable and in free space, and then examine ocean waves and ripples on a fluid surface, showing how they may be extended to describe a variety of atmospheric and oceanic phenomena.

In each case, we begin by determining how the disturbance or displacement at any point is affected by that at adjacent points, and how the physical properties of the system determine how quickly it can respond. This allows us to derive a partial differential equation that describes the wave propagation and embodies all the relevant physics. What remains, as before, is the purely mathematical solution of this wave equation.

Although this chapter provides many useful illustrations of the principles outlined in Chapter 2, the reader may safely skip these examples without missing any fundamental principles upon which we shall build later.

3.1   Waves along a coaxial cable

Coaxial cables are used for the distribution of electrical signals, and include microphone cables, television aerial leads and, indeed, most of the connections between audio and video apparatus. They are also used for most analogue signals in laboratory experiments because of their well-defined electrical characteristics and relative immunity to interference.

Coaxial cables typically comprise a thin copper conductor surrounded by a braided copper sleeve with an insulating spacer to keep them from touching. Like any pair of isolated conductors, these act as a capacitor, for an electric field will exist between them whenever one conductor is electrically charged with respect to the other. They also, when connected at the end of the cable by further circuitry, exhibit inductance, so that the current through the conductors takes a finite time to respond to a change in the voltage applied. These physical phenomena, described by the laws of Gauss and Faraday, are sufficient to determine how the waves corresponding to electrical signals propagate along

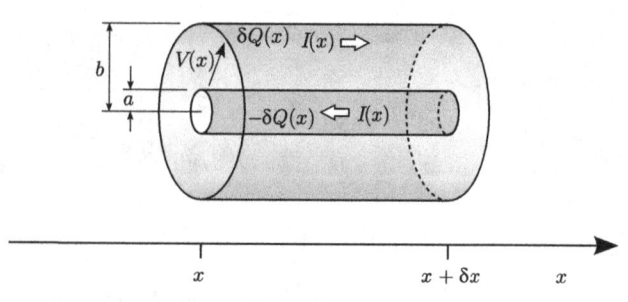

Voltages and currents along a section of coaxial cable.

coaxial cables. We begin by assuming a given inductance and capacitance per unit length, and then calculate these terms explicitly from first principles.

We consider a section of coaxial cable of length δx, aligned along the x-axis as shown in Fig. 3.1, with an inner conductor of radius a and outer conductor of inside radius b. The voltage between the two conductors at any point is written as $V(x)$; equal but opposite charges of $\delta Q(x)$ are assumed to occur on the two conductors; and the current $I(x)$ is taken to flow along the outer conductor and return with equal magnitude along the inner. We assume translational and rotational symmetry (straight cables of circular cross-section), negligible resistance and dimensions that are small compared with the wavelength. To avoid clutter, the time dependence of the voltage and so on will be left implicit.

The principle of charge conservation allows us first to relate changes in the stored charge to gradients in the current, so that any difference between the current flowing into the section and that leaving it must appear as a change in the charge stored:

$$\frac{\partial}{\partial t}\delta Q(x) = I(x) - I(x + \delta x). \tag{3.1}$$

We assume that the cable is characterized by a capacitance per unit length C, which allows us to write the charge in terms of the voltage $V(x)$ and capacitance $C\,\delta x$ of the element:

$$\delta Q(x) = (C\,\delta x)V(x). \tag{3.2}$$

Inserting equation (3.2) into equation (3.1) and taking the limit as $\delta x \to 0$ thus gives

$$C\frac{\partial V}{\partial t} = \lim_{\delta x \to 0}\frac{I(x) - I(x + \delta x)}{\delta x} = -\frac{\partial I}{\partial x}. \tag{3.3}$$

We now consider the rectangular loop ABCD in Fig. 3.2, formed by parallel sections of the inner and outer conductors and two radii at x and $x + \delta x$, through which the current is presumed to induce a magnetic flux $\delta\Phi(x)$. Faraday's law of induction relates this to the potential differences along the radii,

$$\frac{\partial}{\partial t}\delta\Phi(x) = V(x) - V(x + \delta x) \tag{3.4}$$

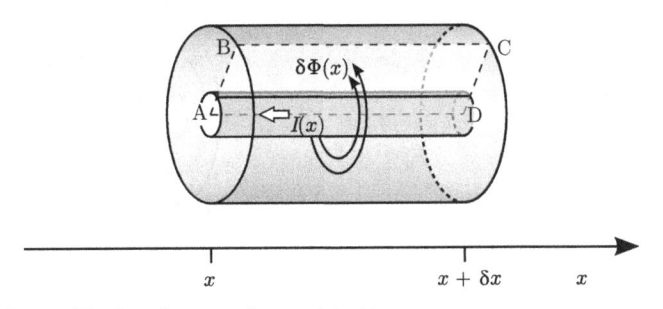

Fig. 3.2 Magnetic flux resulting from the current in a coaxial cable.

and, if we define a self-inductance per unit length L such that

$$\delta\Phi(x) = (L\,\delta x)\,I(x),\tag{3.5}$$

we may combine equations (3.4) and (3.5) and take the limit as $\delta x \to 0$ to give

$$L\frac{\partial I}{\partial t} = \lim_{\delta x \to 0}\frac{V(x) - V(x + \delta x)}{\delta x} = -\frac{\partial V}{\partial x}.\tag{3.6}$$

The manipulations of equations (3.4)–(3.6) thus resemble those of equations (3.1)–(3.3) and, overall, we have two equations relating derivatives of the current and the voltage, which we now combine to give a single differential equation for one variable. On differentiating equation (3.3) with respect to time we obtain

$$C\frac{\partial^2 V}{\partial t^2} = \frac{\partial^2 I}{\partial x\,\partial t},\tag{3.7a}$$

while differentiating equation (3.6) with respect to position gives

$$L\frac{\partial^2 I}{\partial x\,\partial t} = \frac{\partial^2 V}{\partial x^2},\tag{3.7b}$$

allowing elimination of $\partial^2 I/\partial x\,\partial t$ to give

$$\frac{\partial^2 V}{\partial t^2} = \frac{1}{LC}\frac{\partial^2 V}{\partial x^2}.\tag{3.8}$$

Equation for electromagnetic waves along a coaxial cable

Validity: negligible resistance, translational and rotational symmetry, $b \ll \lambda$

Had we instead differentiated equation (3.3) with respect to position and equation (3.6) with respect to time, we would have arrived at an identical expression for the current:

$$\frac{\partial^2 I}{\partial t^2} = \frac{1}{LC}\frac{\partial^2 I}{\partial x^2}.\tag{3.9}$$

Equations (3.8) and (3.9) are equivalent versions of the wave equation for our coaxial cable, and are identical in form to the wave equation (2.18) for guitar strings; the forms of their solutions will therefore be mathematically identical.

3.1.1 Air-spaced coaxial cables

The capacitance and inductance per unit length of a coaxial cable are quite straightforward to obtain if the space between the inner and outer conductors

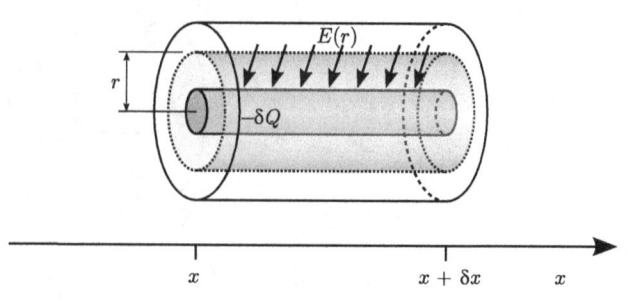

Fig. 3.3 Radial electric field within a coaxial cable.

is a uniform insulating medium. The simplest case of this is a vacuum, from which the more practical example of an air-spaced cable differs only slightly, and the use of a solid dielectric spacer requires merely a simple modification. We again consider sections of the cable, but now take slices that are thin enough for the electric and magnetic fields to be approximately uniform.

To determine first the capacitance per unit length, we consider the slice shown in Fig. 3.3. The charge $-\delta Q$ on the inner conductor results in a radial electric field $E(r)$, which we integrate over the surface of the concentric cylindrical region of space of radius r. Application of Gauss's law then gives

$$E(r) \times 2\pi r \, \delta x = \frac{-\delta Q}{\varepsilon_0}, \qquad (3.10)$$

where ε_0 is the permittivity of free space and, for the cases of air or a solid dielectric, should be multiplied by the relative permittivity of the medium ε_r. It follows that the electric field at a radius r is given by

$$E(r) = \frac{-\delta Q}{2\pi \, \delta x \, \varepsilon_0} \frac{1}{r}. \qquad (3.11)$$

The voltage $V(x)$ is then found by integrating the radial field component between the two conductors:

$$
\begin{aligned}
V(x) &= \int_a^b -E(r)\mathrm{d}r \\
&= \frac{\delta Q}{2\pi \, \delta x \, \varepsilon_0} \int_a^b \frac{1}{r} \, \mathrm{d}r \\
&= \frac{\delta Q}{2\pi \, \delta x \, \varepsilon_0} \big[\ln r \big]_a^b = \frac{\delta Q}{2\pi \, \delta x \, \varepsilon_0} \ln\left(\frac{b}{a}\right),
\end{aligned}
\qquad (3.12)
$$

so that

$$\delta Q = \frac{2\pi \varepsilon_0}{\ln(b/a)} V(x) \delta x. \qquad (3.13)$$

Comparison with equation (3.2) hence yields the capacitance per unit length

Capacitance per unit length

$$C = \frac{2\pi \varepsilon_0}{\ln(b/a)}. \qquad (3.14)$$

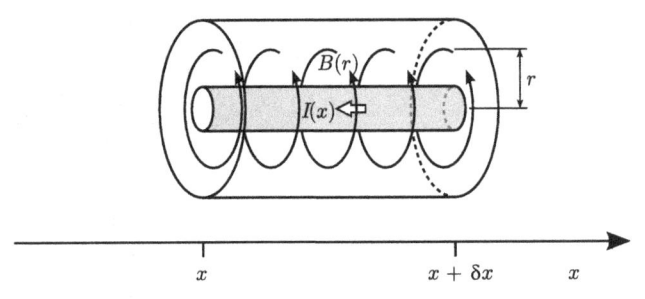

Fig. 3.4   Azimuthal magnetic field within a coaxial cable.

To determine the inductance per unit length, we consider the slice shown in Fig. 3.4. The current I induces an azimuthal magnetic field $B(r)$, which we find by applying Ampère's law to a circle of radius r around the central conductor,

$$\oint B(r)\mathrm{d}s = \mu_0 I, \tag{3.15}$$

where μ_0 is the permeability of free space and for a magnetic material should be multiplied by the relative permeability μ_r. It follows that

$$B(r) = \frac{\mu_0 I}{2\pi r}, \tag{3.16}$$

and hence the flux $\delta\Phi$ through the radial section ABCD in Fig. 3.2 will be

$$
\begin{aligned}
\delta\Phi(x) &= \delta x \int_a^b B(r)\mathrm{d}r \\
&= \delta x \frac{\mu_0 I}{2\pi} \int_a^b \frac{1}{r}\,\mathrm{d}r \\
&= \frac{\mu_0 I}{2\pi} \delta x \ln\left(\frac{b}{a}\right).
\end{aligned}
\tag{3.17}
$$

Comparison with equation (3.5) thus yields the inductance per unit length

$$L = \frac{\mu_0}{2\pi} \ln\left(\frac{b}{a}\right). \tag{3.18}$$

The factor $1/(LC)$ in the wave equation (3.8) is thus given by

$$
\begin{aligned}
\frac{1}{LC} &= \frac{2\pi}{\mu_0 \ln(b/a)} \frac{\ln(b/a)}{2\pi\varepsilon_0} \\
&= \frac{1}{\mu_0\varepsilon_0} = c^2,
\end{aligned}
\tag{3.19}
$$

where c, as we shall see explicitly in Section 3.2, is the speed of light in vacuum. By inserting equation (3.19) into the wave equation (3.8) we may derive the propagation speed as in Section 2.2.2; if $\varepsilon_\mathrm{r} \approx \mu_\mathrm{r} \approx 1$, we find that waves travel along air-spaced coaxial cables with the speed of light.

More generally, when the space between the conductors is filled with a dielectric or magnetic medium, waves propagate along the cable at a speed

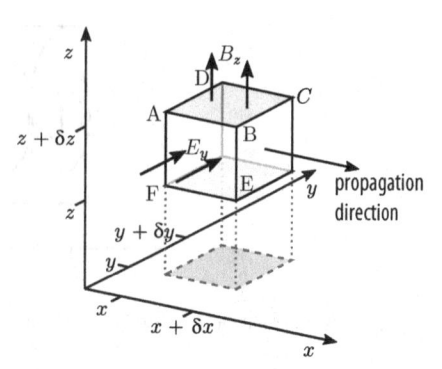

    Element of free space in a region of propagating plane electromagnetic waves.

$c/\sqrt{\varepsilon_r\mu_r}$. For the typical laboratory example of a solid polyethylene spacer with $\varepsilon_r \sim 2.25$, the propagation speed is therefore reduced to around $c/1.5$.

3.2 Electromagnetic waves

Our finding in Section 3.1.1 that the speed of propagation of signals along coaxial cables depends upon the permittivity and permeability of the region between the conductors, and not upon the dimensions of the conductors themselves, suggests that the waves that we have analysed are more general in nature. It transpires that they are examples of freely propagating electromagnetic plane waves, comprising oscillating transverse electric and magnetic fields that again are orthogonal both to each other and to the direction of propagation.

We shall examine the full three-dimensional form of electromagnetic waves in Section 15.3.1, but we may establish the most important characteristics by considering the following example in which plane waves, with translational symmetry perpendicular to the direction of propagation, travel in free space parallel to the x axis, with electric and magnetic fields parallel to the y axis and z axis, respectively.

Our analysis in many ways follows that of the coaxial cables of Section 3.1, but in the absence of the conductors our integration paths are more arbitrarily chosen. We simply consider an elemental cuboid of space, with sides of length δx, δy and δz, as shown in Fig. 3.5. We again derive the wave equation from the fundamental electromagnetism equations, but this time we concentrate upon changing fields rather than moving charges and currents.

We begin with Faraday's law for a magnetic field B_z in the z direction,

$$\oint \mathbf{E} \cdot d\mathbf{s} = - \int \frac{\partial B_z}{\partial t} \, dS \tag{3.20}$$

where **s** describes the perimeter of a surface element of area $dS \equiv |\mathbf{dS}|$, with the vector **dS** lying normal to the element, and **E** is the associated electric field. We apply this to the loop ABCD, yielding

$$\int_A^B \mathbf{E} \cdot \mathbf{ds} + \int_B^C \mathbf{E} \cdot \mathbf{ds} + \int_C^D \mathbf{E} \cdot \mathbf{ds} + \int_D^A \mathbf{E} \cdot \mathbf{ds} = -\frac{\partial B_z}{\partial t} \, \delta x \, \delta y. \qquad (3.21)$$

In terms of the x, y and z components of the fields E_x etc., this becomes

$$\int_x^{x+\delta x} E_x(y) \mathrm{d}x + \int_y^{y+\delta y} E_y(x) \mathrm{d}y + \int_{x+\delta x}^x E_x(y+\delta y) \mathrm{d}x + \int_{y+\delta y}^y E_y(x+\delta x) \mathrm{d}y$$
$$= -\frac{\partial B_z}{\partial t} \, \delta x \, \delta y \qquad (3.22)$$

so that, since in our example $E_x = 0$,

$$\left[E_y(x+\delta x) - E_y(x) \right] \delta y = \frac{\partial B_z}{\partial t} \, \delta x \, \delta y \qquad (3.23)$$

and hence, on dividing by $\delta x \, \delta y$ and taking the limit $\delta x \to 0$, $\delta y \to 0$, we obtain

$$\frac{\partial E_y}{\partial x} = \frac{\partial B_z}{\partial t}. \qquad (3.24)$$

We now similarly apply Ampère's law, which describes the magnetic field **B** that results when a current of density J_y flows in the y direction or an electric field in the same direction changes to produce a displacement current dE_y/dt,

$$\oint \frac{1}{\mu_0} \mathbf{B} \cdot \mathbf{ds} = \int \left(J_y + \varepsilon_0 \frac{\partial E_y}{\partial t} \right) \mathrm{d}S. \qquad (3.25)$$

This we apply to the loop ABEF, giving, in the same fashion as above,

$$\frac{\partial B_z}{\partial x} = \mu_0 \varepsilon_0 \frac{\partial E_y}{\partial t}. \qquad (3.26)$$

Differentiating equation (3.24) with respect to x and equation (3.26) with respect to t allows the magnetic field to be eliminated, giving the electromagnetic wave equation

Equation for electromagnetic waves in vacuum

Validity: no free charges

$$\frac{\partial^2 E_y}{\partial x^2} = \mu_0 \varepsilon_0 \frac{\partial^2 E_y}{\partial t^2}. \qquad (3.27)$$

This is essentially the same as equation (3.8), where the electric field is the voltage divided by the distance between the conductors. Differentiating these expressions the other way around yields the same wave equation for the magnetic field B_z. Electromagnetic waves in vacuum therefore again propagate with speed $c \equiv 1/\sqrt{\mu_0 \varepsilon_0}$.

The same working shows that, in the presence of a uniform dielectric or magnetic medium with relative permittivity ε_r and relative permeability μ_r, the speed of propagation will once more be $c/\sqrt{\varepsilon_r \mu_r}$.

3.3 Ocean waves

Billingham [7] pp. 109–123
Main [60] Chapter 13
Feynman [23] Vol. I, p. 51-4

Another fine example of the physics behind everyday wave propagation concerns the motion within and on the surface of a body of water. These waves can vary from rapid ripples on the surface of a pond to the spectacular breakers when ocean swell reaches the shore, and in general require a fully three-dimensional nonlinear treatment for their analysis.[1] The underlying physics is nevertheless straightforward, and its analysis is quite accessible if we confine ourselves to shallow waves and the limiting regimes in which the wave motion is dominated by a single process.

In the following sections, we therefore consider separately the details of low- and high-frequency waves in deep and shallow water, before sketching out how they may be combined for a general solution for shallow waves in intermediate cases. We then extend our treatment to compressible fluids in order to explain some consequences for our weather and climate.

3.3.1 Shallow-water gravity waves

We first restrict ourselves to shallow, long-wavelength waves that result from gravitational restoring forces, and whose kinetic energy lies principally in the horizontal motion of the water. These include shallow-water ocean waves, or *swell*, which are generated initially by the action of wind over the surface of the sea and can travel for many hundreds of kilometres. By *shallow* here we refer to the depth of the water in comparison with the horizontal scale of the wave motion; in many cases even the deep ocean is shallow in this respect.

We address principally the one-dimensional motion of the oceans along the propagation direction; while some vertical motion must occur, the vertical velocities and energies for shallow-water waves prove to be small. We assume a level seabed and translational symmetry along the horizontal y axis, which is equivalent to requiring roughly linear wavefronts but proves not to be a significant limitation. We may therefore consider a water surface whose height h above a given datum varies only along the x axis, as shown in Fig. 3.6.

We then imagine dividing the ocean into thin vertical slices, which might, for example, be identified by dropping sheets of a thin plastic film from the surface to the seabed. We arrange these sheets to be uniformly spaced with a separation δx when the sea is calm and the water surface is level, so that the volume of water contained between each pair of sheets is the same. When the surface is perturbed by a wave motion, we find that the variation in water height within any individual slice is small, and that we may approximate the slice to one of constant height, thus overall forming the stepped profile shown in Fig. 3.6.

---

[1] Water waves, according to Feynman [23], while '... *easily seen by everyone ... are the worst possible example, because ... they have all the complications that waves can have*'.

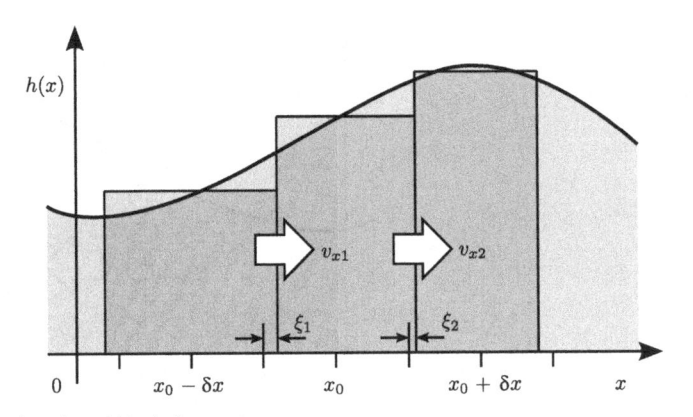

Horizontal motion within shallow-water waves.

We assume that the sheets remain flat as the wave progresses; this proves to be an alternative definition of the 'shallow-water' regime. Since no water can flow through the sheets, the variation in height $h(x)$ must be accompanied by a horizontal motion of the sheets, so that the volume of water remains the same. That is, if the two sheets around x are displaced by ξ_1 and ξ_2, and we consider a column of water of width δy in the y direction, then

$$h(x)(\delta x + \xi_2 - \xi_1)\delta y = \text{constant.} \tag{3.28}$$

On differentiating this product with respect to time, we hence obtain

$$\left(\delta x + \xi_2 - \xi_1\right)\delta y \frac{\partial h}{\partial t} + h\,\delta y \left(\frac{\partial \xi_2}{\partial t} - \frac{\partial \xi_1}{\partial t}\right) = 0. \tag{3.29}$$

We now neglect $\xi_{1,2}$ in comparison with δx and variations of h in comparison with the average h_0, divide by δx, and cancel out the common factor δy, to give

$$\frac{\partial h}{\partial t} = -h_0 \frac{d\xi_2/dt - d\xi_1/dt}{\delta x}. \tag{3.30}$$

The derivatives $d\xi_1/dt$ and $d\xi_2/dt$ are simply the horizontal velocities of the dividing sheets, shown in Fig. 3.6 as v_{x1} and v_{x2}. On taking the limit as the slice thickness $\delta x \to 0$, we thus obtain

$$\frac{\partial h}{\partial t} = -h_0 \lim_{\delta x \to 0} \frac{v_{x2} - v_{x1}}{\delta x} \to -h_0 \frac{\partial v_x}{\partial x}. \tag{3.31}$$

We thus have a differential equation relating the temporal derivative of one characteristic of the wave, the water height h, to the spatial derivative of another, the horizontal velocity v_x.

To derive a second differential equation relating these two properties, we consider the difference in hydrostatic pressure resulting from the variation in height of the water surface. Consider the column of water around x_0 whose velocity in the x direction we again represent by v_x, as shown in Fig. 3.7. At any

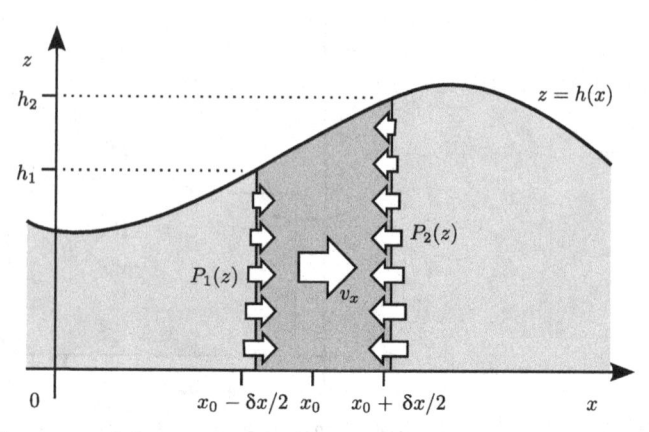

Fig. 3.7   Hydrostatic pressure variations accompanying shallow-water waves.

height z, there will be a difference in hydrostatic pressure across the column due to the difference in height of water above that level,

$$P_1(z) - P_2(z) = (h_1 - h_2)\rho g, \tag{3.32}$$

where ρ is the density of the water and g the acceleration due to gravity. As a result, a net horizontal force acts upon the column of water, resulting in its acceleration $(\partial v_x / \partial t)$. For a layer of depth δz, we may therefore write

$$(h_2 - h_1)\rho g \, \delta z \, \delta y = \rho \, \delta x \, \delta y \, \delta z \, \frac{\partial v_x}{\partial t}. \tag{3.33}$$

On cancelling out the common factors, dividing by δx and taking the limit $\delta x \to 0$, we thus reach our second differential equation:

$$\frac{\partial v_x}{\partial t} = \lim_{\delta x \to 0} \frac{h_2 - h_1}{\delta x} g \to -g \frac{\partial h}{\partial x}. \tag{3.34}$$

We now combine equations (3.31) and (3.34), which both relate h and v_x (we may choose to eliminate either), to give the equation for shallow ocean waves,

$$\frac{\partial^2 h}{\partial t^2} = -h_0 \frac{\partial^2 h}{\partial x \, \partial t} = g h_0 \frac{\partial^2 h}{\partial x^2}. \tag{3.35}$$

Equation for shallow-water gravity waves
Validity: incompressible fluid of negligible viscosity and surface tension, water shallow in comparison with horizontal scale of wave motion

On solving this wave equation as in Section 2.2.2, we see that shallow ocean waves travel with a speed v that varies with the square-root of the water depth,

$$v = \sqrt{g h_0}. \tag{3.36}$$

Since these shallow-water waves result from the gravitational restoring force upon a displaced column of water, they are sometimes known as *gravity waves*.

3.3.2 Velocities in water wave motion

It is important to distinguish between the different velocities that we encounter in our analysis. We have derived in equation (3.36) a value for the propagation speed v of the wave motion: for a given depth of water and gravitational acceleration, this is a constant. As the *wave* propagates steadily forwards, however, individual regions of *water* will move back and forth and up and down: v_x is the horizontal velocity of the water at a given position, and $\partial h/\partial t$ is the vertical velocity of water near the surface, at any time.

It is instructive to consider in more detail the motion of a given pocket of water whose undisturbed position is a height z_0 above the datum, which is then displaced by the wave through a vertical distance ζ to an instantaneous height $z_0 + \zeta$; for water at the surface, this gives $z_0 = h_0$ and $z_0 + \zeta = h$. We now adapt equation (3.28) to the column of water that extends from the seabed up to this pocket, and hence write

$$(z_0 + \zeta)(\delta x + \xi_2 - \xi_1)\delta y = \text{constant} \tag{3.37}$$

and hence, proceeding as before,

$$\frac{\partial \zeta}{\partial t} = -z_0 \lim_{\delta x \to 0} \frac{v_{x2} - v_{x1}}{\delta x} \to -z_0 \frac{\partial v_x}{\partial x}. \tag{3.38}$$

We thus see that the vertical velocity $v_z \equiv \partial \zeta/\partial t$ depends linearly upon the unperturbed vertical position z_0, and is given simply by

$$v_z = \left(\frac{z_0}{h_0}\right) \frac{\partial h}{\partial t}, \tag{3.39}$$

where $h(x, t)$ is the instantaneous height of the water surface, by which we characterize the wave motion overall. For a travelling wave $h(x, t) = f(u)$, where $u \equiv x - vt$, this gives

$$v_z = -v \left(\frac{z_0}{h_0}\right) \frac{\mathrm{d}f}{\mathrm{d}u} = -\frac{v z_0}{h_0} \frac{\partial h}{\partial x}, \tag{3.40}$$

which, using equation (3.34) and the wave speed from equation (3.36), gives

$$v_z = \frac{v z_0}{g h_0} \frac{\partial v_x}{\partial t} = \frac{z_0}{v} \frac{\partial v_x}{\partial t}. \tag{3.41}$$

The instantaneous vertical velocity is therefore at all times proportional to the horizontal acceleration.

This relationship between the horizontal and vertical velocities at any time determines the path traced out by a pocket of water as the wave passes. An interesting case is when the waves are sinusoidal and hence of the form $h(u) = h_1 \sin(ku)$. Substituting this into equations (3.39) and (3.41) and integrating gives $v_x = (v h_1/h_0)\sin(ku)$ and $v_z = [-v z_0 h_1/(k h_0)]\cos(ku)$, where we have set the constant of integration to zero to correspond to the absence of an underlying flow of the ocean. Integrating these expressions gives the position coordinates of the pocket of water relative to its undisturbed position,

$$\xi = \frac{1}{k} \frac{h_1}{h_0} \cos(ku), \tag{3.42a}$$

$$\zeta = z_0 \frac{h_1}{h_0} \sin(ku). \tag{3.42b}$$

While the horizontal displacement ξ therefore follows a sinusoidal motion with depth-independent amplitude $h_1/(kh_0)$, the sinusoidal variation of the vertical displacement ζ is delayed by a quarter of a cycle and increases in amplitude with the distance z_0 above the seabed. The water therefore describes elliptical paths whose vertical range is a factor kz_0 times the horizontal range.

Since we have limited our treatment so far to water that is shallow in comparison with the wavelength, the factor kz_0 will be small; as previously asserted, the kinetic energy therefore lies principally in the horizontal motion of the water. For typical ocean depths of 100–10 000 m, this regime is limited to wave periods that are rather greater than 20–200 s, respectively; it is interesting, and perhaps significant, that the typical period of ocean swell lies at the limit of this regime.

3.3.3　Deep-water gravity waves

For deeper water, or wave shapes that are smaller in scale (harmonic waves of higher frequencies whose wavelengths are comparable to or shorter than the water depth), our initial assumptions prove invalid, and the vertical motion must be taken into account in our description of the underlying physics: a significant fraction of the kinetic energy will be associated with the vertical motion, and Newton's second law must be applied in this direction as well. A full analysis should also account for the elasticity of and friction within a viscous fluid, the vortices of rotational flow and so on, and will not be attempted here, but we may obtain some key features by again assuming the water to be incompressible – so that a given mass always occupies the same volume – and by assuming its viscosity to be zero.

Main [60] Chapter 13
Billingham [7] Chapter 4
Crawford [16] Section 7.3

For an incompressible fluid whose horizontal and vertical displacements are given by ξ and ζ, we may modify equation (3.37) so that, for an element of unperturbed width δx and height δz starting at (x_0, z_0),

$$(\delta x + \xi_2 - \xi_1)(\delta z + \zeta_2 - \zeta_1) = \text{constant.} \tag{3.43}$$

On differentiating with respect to time, and assuming $\xi,\ \zeta \ll \delta x,\ \delta z$, we hence obtain

$$\delta z\big[v_x(x + \delta x) - v_x(x)\big] + \delta x\big[v_z(z + \delta z) - v_z(z)\big], \tag{3.44}$$

so that, on dividing by $\delta x\,\delta z$ and taking the limits $\delta x,\ \delta z \to 0$, we obtain

$$\frac{\partial v_x}{\partial x} + \frac{\partial v_z}{\partial z} = 0. \tag{3.45}$$

In three dimensions, this equation may be written

$$\nabla \cdot \mathbf{v} = 0.$$

The absence of viscosity proves to result in *irrotational* flow, so that the integral of the velocity around any path is zero:

$$\oint \mathbf{v} \cdot \mathbf{ds} = 0. \tag{3.46}$$

Applied around our element of width δx and height δz, this gives

$$v_x(z)\delta x + v_z(x + \delta x)\delta z - v_x(z + \delta z)\delta x - v_z(x)\delta z = 0, \tag{3.47}$$

In three dimensions, this equation may be written

$$\nabla \times \mathbf{v} = 0.$$

so that, on dividing again by $\delta x\,\delta z$ and taking the limits $\delta x,\ \delta z \to 0$, we obtain

$$\frac{\partial v_x}{\partial z} - \frac{\partial v_z}{\partial x} = 0. \tag{3.48}$$

Equations (3.45) and (3.48) take the place of equation (3.31); similarly, it is necessary to re-write equation (3.34) to account for vertical as well as horizontal motion, usually resulting in what are known as the Bernoulli or Navier–Stokes equations. While a full solution of these governing equations is somewhat complex and will not be treated here, the form of sinusoidal solutions – with a depth-dependent amplitude – may be explored by substituting into equations (3.48) and (3.45) the trial solutions

$$\xi = \xi_0(z_0)\cos(kx - \omega t), \tag{3.49a}$$

$$\zeta = \zeta_0(z_0)\sin(kx - \omega t), \tag{3.49b}$$

giving

$$\frac{\mathrm{d}\xi_0}{\mathrm{d}z_0} = k\zeta_0, \tag{3.50a}$$

$$\frac{\mathrm{d}\zeta_0}{\mathrm{d}z_0} = k\xi_0. \tag{3.50b}$$

Differentiating equation (3.50a) and substituting from equation (3.50b) yields a separate equation for ξ_0:

$$\frac{\mathrm{d}^2\xi_0}{\mathrm{d}z_0^2} = k^2\xi_0. \tag{3.51}$$

Equation (3.51) may now be solved to determine $\xi_0(z_0)$, and equation (3.50b) then used to derive $\zeta_0(z_0)$, to give the full solution for a given situation. We shall see in Section 12.4.4 that, on taking into account the effect of the seabed, we obtain the overall solutions

$$\xi = \xi_0(z_0)\cos(ku), \tag{3.52a}$$

$$\zeta = \zeta_0(z_0)\sin(ku) = \xi_0(z_0)\tanh(kz_0)\sin(ku), \tag{3.52b}$$

where $u \equiv x - vt$ and

The hyperbolic functions here result from the *boundary conditions*, due to the seabed, which we examine in more detail in Section 12.4.4.

$$\xi_0(z_0) = \frac{\cosh(kz_0)}{\sinh(kh_0)}h_1, \tag{3.52c}$$

$$\zeta_0(z_0) = \frac{\sinh(kz_0)}{\sinh(kh_0)}h_1, \tag{3.52d}$$

with the wave velocity given by

$$v = \sqrt{\frac{g}{k}\tanh(kh_0)}. \tag{3.53}$$

The Open-Centre Turbine tidal-stream device on a test rig at the European Marine Energy Centre.
Commercial deployment on the seabed will use a gravity base, rather than the pile structure seen here.
© OpenHydro

These equations correctly reduce to our previous results in the limit of small kh_0, when $\tanh(kh_0) \to kh_0$.

In the deep-water limit of large kh_0, when $\tanh(kh_0) \to 1$,

Wave velocity in deep water

$$v \to \sqrt{\frac{g}{k}} \equiv \sqrt{\frac{g\lambda}{2\pi}}. \qquad (3.54)$$

We shall see in Section 18.3 that this strong dependence upon the wavelength λ accounts for the intriguing form of the Kelvin ship wake. The magnitudes ξ_0 and ζ_0 of the horizontal and vertical motions tend to the same amplitude, $h_1 \exp[-k(h_0 - z_0)]$, and the motion at any height becomes circular.

The amplitude of the motion decreases exponentially with depth, falling by a factor of 500 with each extra wavelength of depth. Submarines and semi-submersible oil platforms, whose buoyancy chambers typically lie tens of metres below the water surface, thus experience less wave motion than do surface vessels in the same conditions. The exponential dependence of the wave amplitude upon depth means that deep-water gravity waves may be regarded as *evanescent waves*, which we address in more detail in Chapter 11.

3.3.4 Ocean wave power

Our analysis of gravity waves in the ocean shows that the horizontal motion is often much greater than, and never less than, the vertical motion. The horizontal flow of the ocean would therefore appear to be a more promising source of hydroelectric power, especially in shallow water, than the vertical motion of the

A 750-kW prototype wave-energy converter off the coast of the Orkney Isles. © Pelamis Wave Power

surface. This is particularly the case with tidal flows, which may be regarded as very-large-scale shallow-water waves (or as the constant of integration that we omitted in deriving equations (3.42) and (3.52)). Tidal movement can therefore be as strong in the calmer regions near the ocean floor where higher-frequency waves are attenuated. Tidal-stream devices such as that shown in Fig. 3.8 can hence operate effectively when deployed on the seabed, where they are clear of surface shipping, unaffected by storms, and not dependent upon trailing anchor chains or electrical cables.

Practical considerations, however, sometimes favour using the vertical motion. Tidal strengths vary greatly with location, and the motion of deep-water waves decreases rapidly with depth. The power of the ocean swell is therefore greatest at the surface, where it may be readily captured from the relative motion of adjacent, buoyant structures as shown in Fig. 3.9. Nonetheless, there are hydroelectric schemes that are based upon all combinations of horizontal and vertical motion, floating and anchored, using both rotary turbines and reciprocating actions. Cost, reliability, efficiency, seabed geography, weather patterns and many environmental factors all play important rôles.

3.4 Capillary waves

The exponential decrease in the amplitude of ocean gravity waves with distance below the surface means that, for sufficiently high wave frequencies, only a thin layer near the surface of the water plays a significant part in the wave motion. If this layer is sufficiently thin – for water, around a centimetre – surface tension can perturb or dominate the wave-propagation mechanism. Such ripples are known as *capillary waves*. We may derive the behaviour of such waves by initially considering their propagation across a puddle or other shallow stretch of water.

3.4.1 Capillary waves in shallow water

Our analysis of capillary waves combines part of our derivation of shallow-water waves (Section 3.3.1) with that of waves on a stretched string (Section 2.2). Equation (3.31) remains valid, but the pressure is now found, as in equation (2.16), by considering components of the surface tension σ (per unit length within the surface normal to the force) to give, in addition to the hydrostatic pressure $(h_0 - z)\rho g$, a pressure component $P(x)$ defined by

$$P(x_0)\delta x\,\delta y = \sigma\,\delta y\left[\left(\frac{\partial h}{\partial x}\right)_{x_0-\delta x/2} - \left(\frac{\partial h}{\partial x}\right)_{x_0+\delta x/2}\right]. \qquad (3.55)$$

Hence, taking the limit as $\delta x \to 0$,

$$P(x_0)\delta y = -\sigma\,\delta y\,\frac{\partial^2 h}{\partial x^2}. \qquad (3.56)$$

As in equation (3.33), the difference in pressure across any element of width δx results in the acceleration of the element, so that

$$\rho\,\delta x\,\delta y\,\delta z\,\frac{\partial v_x}{\partial t} = \left[P\left(x_0 - \frac{\delta x}{2}\right) - P\left(x_0 + \frac{\delta x}{2}\right)\right]\delta y\,\delta z$$

$$= -\delta x\,\delta y\,\delta z\,\frac{\partial P}{\partial x}$$

$$= \delta x\,\delta y\,\delta z\,\sigma\,\frac{\partial^3 h}{\partial x^3}, \qquad (3.57)$$

where the final step uses the spatial derivative of equation (3.56); hence

$$\rho\,\frac{\partial v_x}{\partial t} = \sigma\,\frac{\partial^3 h}{\partial x^3}. \qquad (3.58)$$

We now differentiate equations (3.31) and (3.58) with respect to t and x, respectively:

$$\frac{\partial^2 h}{\partial t^2} = -h_0\,\frac{\partial^2 v_x}{\partial x\,\partial t}, \qquad (3.59)$$

$$\rho\,\frac{\partial^2 v_x}{\partial x\,\partial t} = \sigma\,\frac{\partial^4 h}{\partial x^4}, \qquad (3.60)$$

and combine the results to give the wave equation

$$\frac{\partial^2 h}{\partial t^2} = -\frac{h_0\sigma}{\rho}\,\frac{\partial^4 h}{\partial x^4}. \qquad (3.61)$$

Equation for shallow-water capillary waves

Validity: high frequencies, negligible viscosity, water shallow in comparison with horizontal scale of wave motion

Note that this wave equation is different in nature from those that we have previously encountered in this chapter, in that the second derivative with respect to time is related to the *fourth* with respect to position. It will therefore have different solutions, which we must find explicitly. It turns out that travelling waves of the form given in equation (2.2) will not generally be solutions of this equation, although travelling *sinusoidal* waves will. We address sinusoidal wave motions in the next chapter.

3.4.2 Capillary waves in deeper water

As with the ocean waves of Section 3.3, we find that, when the water depth is comparable to or greater than the wavelength, the amplitude of the motion is reduced with depth and our initial assumptions are incorrect: the vertical motion of the water must be considered in more detail. In the deep-water limit, the wave equation for the vertical displacement of the water becomes

Equation for deep-water capillary waves
Validity: high frequencies, negligible viscosity, water deep in comparison with horizontal scale of wave motion

$$\frac{\partial^2 \zeta}{\partial t^2} = -\frac{\sigma}{\rho}\frac{\partial^3 \zeta}{\partial x^2\,\partial z_0}. \tag{3.62}$$

This may be solved in terms of sinusoidal waves of the form found in Section 3.3.3, and it is instructive to substitute such solutions into the wave equation (3.62) above. In the deep-water limit, equations (3.52b) and (3.52d) may be combined to give

$$\zeta = h_1 \exp[-k(h_0 - z_0)]\sin(ku), \tag{3.63}$$

so that, for this particular case,

$$\frac{\partial^3 \zeta}{\partial x^2\,\partial z_0} = k\frac{\partial^2 \zeta}{\partial x^2}. \tag{3.64}$$

Equation (3.62) may therefore be written

$$\frac{\partial^2 \zeta}{\partial t^2} = -\frac{\sigma k}{\rho}\frac{\partial^2 \zeta}{\partial x^2}. \tag{3.65}$$

This resembles the wave equation for a taut string encountered in Section 2.2, with the term $\sigma k/\rho$ corresponding to the term W/M that defines the wave-propagation speed in equation (2.18). The string tension W is thus replaced by the surface tension (per unit transverse distance) ρ, and the string mass per unit length M is replaced by the mass per unit surface area ρ/k of a surface layer of effective thickness $1/k$.

3.4.3 Capillary–gravity waves

It is relatively straightforward, although rather more lengthy, to combine the capillary and gravitational effects into the same analysis: we simply add the two restoring forces to give, for waves in shallow water,

Equation for shallow-water capillary–gravity waves
Validity: high frequencies, negligible viscosity, water shallow in comparison with horizontal scale of wave motion

$$\frac{\partial^2 h}{\partial t^2} = gh_0\frac{\partial^2 h}{\partial x^2} - \frac{h_0\sigma}{\rho}\frac{\partial^4 h}{\partial x^4}. \tag{3.66}$$

Substitution of a sinusoidal waveform $h(x,t) = h_1\sin(ku)$, where $u \equiv x - vt$, shows that the wave speed v is given by

$$v = \sqrt{h_0\left(g + \frac{\sigma}{\rho}k^2\right)}. \tag{3.67}$$

Billingham [7] Chapter 4

The transition from capillary- to gravity-wave regimes thus occurs when $g \sim \sigma k^2 / \rho$ so that, for water, capillary effects dominate for wavelengths below a few millimetres. For deeper water, equation (3.67) gains a factor $\sqrt{\tanh(kh_0)/k}$.

Besides neglecting viscosity and rotational flow, we have throughout restricted our treatment to small waves that remain in the *linear* regime; our treatment does not therefore extend to the short, steep waves known as 'chop', or to the 'breakers' much loved by surfers. We have furthermore assumed water to be an incompressible fluid, so that any horizontal contraction must be accompanied by a vertical expansion and vice versa. We shall see overleaf and in Chapter 10 that, when compressibility is taken into account, further wave motions, such as sound and mountain lee waves, are also possible.

3.4.4 Pouring oil on troubled waters

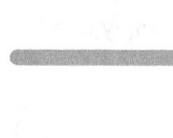

Pliny the Elder [69]
B. Franklin [27]

It has been observed since the days of Pliny the Elder, and was investigated experimentally by Benjamin Franklin, that a little oil on the surface of the sea calms the water beneath it. Trace leaks of fuel or lubrication oil cause a ship to leave a calm area of water in its wake, which is visible from afar even though the ship may seem tiny; and naval commanders have deliberately released oil to calm an otherwise violent sea. The exact mechanisms continue to be debated, but the origins lie in the effect of the oil upon capillary waves on the water surface.

Ocean waves begin as turbulence in the air above. Small-scale variations in pressure distort the sea surface, and these distortions in turn deflect the wind that blows across it, increasing the turbulence and further distorting the surface. Ripples of capillary waves result, and are then themselves amplified by the wind to grow into larger and deeper gravity waves of great force and energy.

The properties of capillary waves, as we have seen, depend upon the surface tension of the fluid, which can be substantially modified by a layer of oil or surfactant, even if it spreads to only a single molecule in thickness. By affecting the ease with which millimetre-sized ripples can be formed, the oil prevents the formation of far more substantial, and sometimes dangerous, ocean waves.

3.5 Gravity waves in compressible fluids

Houghton [45] Chapter 8

The Earth's atmosphere, like its oceans, is a fluid, and exhibits various wave motions that are analogous to those shown by water waves. Besides being a factor of a thousand or more less dense, the atmosphere differs from the oceans in two significant respects. First, it has no distinct surface and instead shows a roughly exponential decrease in density with altitude. Secondly – and the cause of the first difference – it is far from incompressible. As a result, we cannot even for slow variations consider the atmosphere to be approximated by vertical slices characterized at any instant by a single horizontal velocity, and it is necessary to divide the fluid into elements in at least two dimensions. This is also true for a full analysis of deep-water ocean waves.

If, as in Section 3.3.3, the horizontal and vertical displacements are ξ and ζ, the volume δV of an element of fluid with unperturbed dimensions δx, δy and δz will be (neglecting motion in the y direction)

$$\delta V = (\delta x + \xi_2 - \xi_1)\delta y(\delta z + \zeta_2 - \zeta_1). \tag{3.68}$$

For atmospheric waves, and compressible fluids in general, the volume occupied by a given quantity of fluid can vary, and equation (3.28) must be replaced by an *equation of state* relating the volume to the temperature Θ and pressure P of the fluid. For example, for δn moles of an ideal gas,

Ideal-gas equation

$$P\,\delta V = \delta n\, R\Theta, \tag{3.69}$$

where R is the *ideal-gas constant*. Equation (3.28) thus becomes

$$P(\delta x + \xi_2 - \xi_1)\delta y(\delta z + \zeta_2 - \zeta_1) = \delta n\, R\Theta. \tag{3.70}$$

We must therefore take into account the pressure and temperature of the atmosphere at each position, and the thermodynamic constraints upon the compression and expansion of the element. It may commonly be assumed that no heat flows into or out of the element, for example; the process is then said to be *adiabatic*, in which case it is a basic thermodynamic result that the pressure and volume of an ideal gas are related by

$$P\,\delta V^{\gamma} = \text{constant}, \tag{3.71}$$

where $\gamma \equiv c_P/c_V$ is the ratio of the specific heats at constant pressure and constant volume. Equivalently, on substituting from equation (3.69), the pressure and temperature of the element will be related by

$$P^{1-\gamma}\Theta^{\gamma} = \text{constant}. \tag{3.72}$$

Proceeding in this fashion, we may derive a new set of equations, in place of those of our earlier treatment, to take into account the compressibility of the fluid, together with major practical characteristics such as a decrease in temperature, and often an increase of wind speed, with altitude.

3.5.1 Mountain-lee waves

The most visibly spectacular examples of atmospheric gravity waves are those found when, as shown in Fig. 3.10, a brisk wind blows over a range of mountains. Under the right conditions, as illustrated in Fig. 3.11, beautiful bars of *lenticular* cloud form where the air rises, expands and cools enough for the moisture contained to begin to condense. These strong, smooth vertical currents are used by glider pilots to reach altitudes of up to 50 000 feet (15 km) – well above the cruising height of current airliners. Happily, mountain-lee waves are

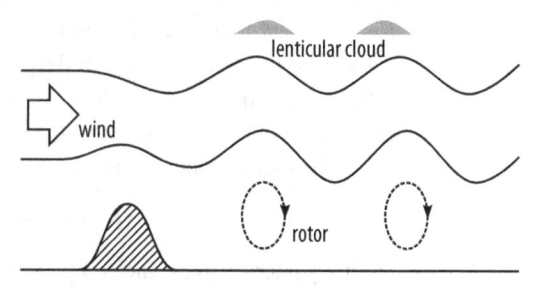

Fig. 3.10
Atmospheric gravity waves in the lee of a mountain. Glider pilots find strong, smooth lift around the leading edges of the lenticular clouds formed in the elevated air; uncomfortable rotor forms near the ground as the flow becomes turbulent. Streamlines show the wave displacements at different heights.

Fig. 3.11
The lenticular cloud of a mountain-lee wave, photographed from a glider (sailplane) that can remain aloft by flying in the rising air.   © David Aknai

reasonably well approximated by a simple model that corresponds to neglecting both vertical structure and horizontal flow and considering the effect of the fluid compressibility upon the motion.

We consider a pocket of air, initially at the same temperature Θ_0 as the surrounding atmosphere at that altitude, and displace it vertically through a distance ζ so that it undergoes a reduction in pressure of $\delta P = -\zeta \rho g$, where ρ is the local air density and g the acceleration due to gravity. We may assume that little heat will flow between the pocket of air and its surroundings during this displacement, and that it therefore expands adiabatically according to equation

(3.72), which we may differentiate with respect to pressure to obtain

$$(1 - \gamma)P^{-\gamma}\Theta^{\gamma} + P^{1-\gamma}\gamma\Theta^{\gamma-1}\left(\frac{\partial\Theta}{\partial P}\right)_S = 0, \tag{3.73}$$

where the subscript indicates that the process is adiabatic and therefore constrained to a constant entropy S. Rearrangement yields

$$\left(\frac{\partial\Theta}{\partial P}\right)_S = \frac{(\gamma-1)P^{-\gamma}\Theta^{\gamma}}{\gamma P^{1-\gamma}\Theta^{\gamma-1}} = \frac{\gamma-1}{\gamma}\frac{\Theta}{P}. \tag{3.74}$$

Since the variation of hydrostatic pressure with height is given by

$$\frac{dP}{dz} = -\rho g, \tag{3.75}$$

we hence obtain the *dry adiabatic lapse rate* – the rate at which a thermally isolated pocket of dry air will cool as it ascends,

$$\begin{aligned}\left(\frac{d\Theta}{dz}\right)_S &= \frac{dP}{dz}\left(\frac{\partial\Theta}{\partial P}\right)_S = -\rho g\frac{\gamma-1}{\gamma}\frac{\Theta}{P}\\ &= -\frac{M\,\delta n}{\delta V}g\frac{\gamma-1}{\gamma}\frac{\Theta}{P}\\ &= -\frac{\gamma-1}{\gamma}\frac{Mg}{R},\end{aligned} \tag{3.76}$$

where the final result is obtained by using the ideal-gas equation (3.69) and writing the density in terms of the molecular mass M as $\rho = M\,\delta n/\delta V$. For the Earth's troposphere the lapse rate is about -0.01 K m^{-1}.

Our pocket of air of mass $M\,\delta n$ will, at its new temperature $\Theta_0 + \zeta(d\Theta/dz)_S$ and at the slightly lower pressure P_1, occupy a new volume δV_1 given by

$$\frac{P_1\,\delta V_1}{R} = \delta n\left[\Theta_0 + \zeta\left(\frac{d\Theta}{dz}\right)_S\right]. \tag{3.77}$$

At this altitude, the ambient temperature will be $\Theta_0 + \zeta\,d\Theta_0/dz$, where $d\Theta_0/dz$ is known as the *environmental lapse rate* and depends upon the particular conditions at that time and place. The number of moles $\delta n'$ of ambient air occupying a volume δV_1 will therefore be given by

$$\frac{P_1\,\delta V_1}{R} = \delta n'\left[\Theta_0 + \zeta\frac{d\Theta_0}{dz}\right]. \tag{3.78}$$

It follows that the net upthrust on the pocket – the difference between its weight and that of the air displaced – will be

$$M\,\delta n\frac{d^2\zeta}{dt^2} = Mg(\delta n' - \delta n) = Mg\,\delta n\left[1 - \frac{\Theta_0 + \zeta\,d\Theta_0/dz}{\Theta_0 + \zeta(d\Theta/dz)_S}\right]. \tag{3.79}$$

On performing a binomial expansion to first order and cancelling out common terms, this gives, for small ζ,

$$\frac{d^2\zeta}{dt^2} = g\left[\left(\frac{d\Theta}{dz}\right)_S - \left(\frac{d\Theta_0}{dz}\right)\right]\zeta. \tag{3.80}$$

The pocket of air therefore performs simple harmonic motion about its initial position at the *Brunt–Väisälä frequency* given in terms of the angular frequency

Brunt–Väisälä frequency

$$\omega_{\mathrm{B}}^2 = g\left[\left(\frac{\mathrm{d}\Theta_0}{\mathrm{d}z}\right) - \left(\frac{\mathrm{d}\Theta}{\mathrm{d}z}\right)_{\!S}\right].$$ (3.81)

For a typical environmental lapse rate $\mathrm{d}\Theta_0/\mathrm{d}z$ of -0.0065 K m^{-1}, this gives an oscillation frequency $\omega_{\mathrm{B}}/(2\pi)$, seen by an observer moving with the pocket, of about 0.1 min^{-1}. If the wind speed is, for example, 15 m s^{-1}, the oscillation wavelength over the ground will be about 9 km.

3.6 Waves and weather

W. Thomson (Lord Kelvin) [84]
C.-G. A. Rossby [74]

The fluid layers that form the Earth's atmosphere and oceans support a variety of further wave motions that play key rôles in our weather and climate. *Kelvin waves* are related to the gravity waves considered in Sections 3.3 and 3.4, but are confined in the lateral direction by *Coriolis* forces that push either against a shoreline or against the opposing Coriolis force the opposite side of the equator. *Rossby waves* occur when the Coriolis force results in an otherwise unconstrained lateral flow until the fluid displacement creates a hydrostatic restoring force. These lateral motions can occur both in the atmosphere and in the ocean, and indeed can be associated with distinct layers within these two fluids.

The phenomenon known as the *El Niño Southern Oscillation* (ENSO), illustrated in Fig. 3.12, is a complex interplay of different wave phenomena in the southern Pacific ocean and the atmosphere above it. Generally, there is an ocean circulation between South America and the Antipodes, whereby the prevailing *trade winds* push water near the ocean surface from east to west, during which journey it is warmed by the Sun; cold water from below rises in its place off the western coast of South America. The atmosphere above thus tends to be cool and clear in the east, but warm and moist in the west, resulting in high pressures off South America and low pressures (and typhoons etc.) towards New Zealand and Australia. The combination of atmospheric pressure difference and ocean flow causes the western ocean to lie about half a metre higher, and be some 8 °C warmer, than it does in the east.

During the *el Niño* phase, named after the infant Jesus because of its tendency to arrive around Christmas, Kelvin waves around the equator transport warm water from west to east, warming the waters off Peru (Fig. 3.12(a)) and causing atmospheric instability and a lower atmospheric pressure that enhances the eastward flow of warm water and in turn favours a dry, high-pressure climate to the west – a reversal of the usual situation. South American fishermen

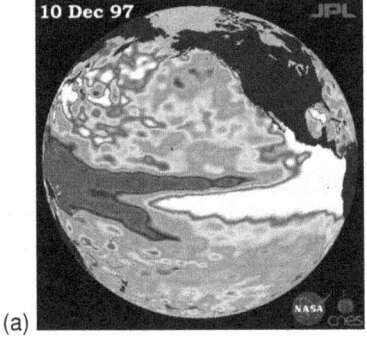
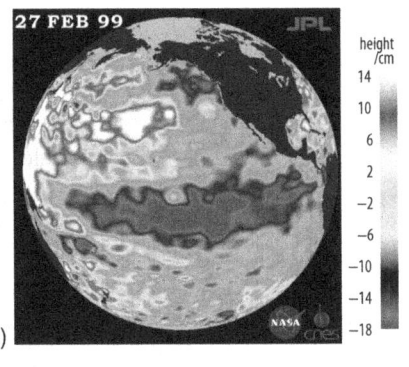

Fig. 3.12   (a) El Niño and (b) la Niña are apparent as variations of the Pacific sea-surface height from its mean level, as revealed by instruments on the TOPEX/Poseidon satellite.   Courtesy of NASA/JPL-Caltech

Fig. 3.13   Equatorial waves are apparent as hook-shaped anomalies from the mean sea-surface temperatures in the eastern Pacific.   Courtesy of NOAA/NCDC

notice warmer, less favourable, waters, while Australia experiences periods of drought.

The equatorial Kelvin waves of el Niño are thought to excite Rossby waves that travel westwards towards, and be reflected and converted to Kelvin waves by, the coast of south-east Asia. When the reflected wave returns to the south Pacific, it disrupts the Kelvin wave, and the climate returns to normal. El Niño conditions, which occur somewhat irregularly every 3–7 years, typically last a year or more, and influence the climate across the globe. Intervening periods of unusually cold water temperatures around the Pacific equator, as illustrated in Figs. 3.12(b) and 3.13, are known as *la Niña*.

Rossby waves in the atmosphere are responsible for the variations in the path of the *jetstream* and, when the fluctuations become large, for the formation of the cyclones and anticyclones (low- and high-pressure systems) that determine our weather on a day-to-day scale.

Exercises

3.1 A coaxial cable comprises an inner conductor 1 mm in diameter that is held within a 5-mm-diameter outer conductor by a spacer of solid PTFE with a relative permittivity (dielectric constant) of 2.05 and a relative permeability of unity. Find the capacitance and inductance per unit length and the time taken for a signal to propagate along a 1-m length of the cable.

3.2 A bath 1.5 m in length is filled with water to a depth of 0.3 m. Find the velocity of shallow-water gravity waves and hence the time it takes for a wave to propagate from one end of the bath to the other and back again.

3.3 The historic sailing vessel *Leader* is 20.8 m long at her waterline. She will oscillate in a 'pitching' motion (i.e. about a horizontal, sideways axis, bow rising while the stern falls and vice versa) with a period of 4 s. The pitching motion may be excited at sea if she encounters waves at her resonance frequency, and the effect will be largest when the wavelength of such waves is about two boat lengths. Assuming the speed of ocean waves to be given by $v = \sqrt{gh_0}$, where g is the acceleration due to gravity and h_0 the depth of the water, find the depth of water in which the pitching motion will be most readily excited (and hence the motion will be most uncomfortable when at anchor).

3.4 How will the 'most uncomfortable depth', determined in Exercise 3.3, vary if the vessel is moving through the water, for example with a speed v_L and on a course at an angle ϑ to the direction in which the waves are travelling?

3.5 For capillary–gravity waves in deeper water, equation (3.67) becomes

$$v = \sqrt{h_0 \left(\frac{g}{k} + \frac{\sigma}{\rho} k \right) \tanh(kh_0)} \qquad (3.82)$$

so that the deep-water limit ($\tanh(kh_0) \approx 1$) gives

$$v = \sqrt{h_0 \left(\frac{g}{k} + \frac{\sigma}{\rho} k \right)}. \qquad (3.83)$$

Show that the wave speed v takes a minimum value when

$$k = \sqrt{\frac{g\rho}{\sigma}} \qquad (3.84)$$

and hence find the corresponding wavelength $\lambda = 2\pi/k$ for water given that $\sigma = 0.078\,\mathrm{N\,m^{-1}}$, $\rho = 1000\,\mathrm{kg\,m^{-1}}$ and $g = 9.81\,\mathrm{m\,s^{-1}}$. This is the wavelength of ripples that persist longest as they propagate away after, for example, a pebble has been dropped into a pond.

4 Sinusoidal waveforms

4.1 Sinusoidal solutions

Pain [66] pp. 109–152
French [29] pp. 201–212

We have already seen in Sections 3.3 and 3.4 that it can be convenient to consider sinusoidal solutions to a wave equation, for their simple transformations under differentiation often allow the differential equations to be reduced to mere algebraic expressions that relate the constant parameters characterizing the wave motion. Sinusoidal waves are certainly not the only wave motions that are possible, but they often prove to be an extremely important and fundamental class of solution. We shall see shortly that they allow us to determine *standing-wave* motions that determine the operation of many musical instruments, and these musical examples and analogies will help us at various points throughout this book. We shall also see that sinusoidal waves can be the only real travelling-wave solutions in *dispersive* systems; and, indeed, that the language of wavelength and frequency pervades into *dissipative* systems for which pure sinusoidal waves are not possible. Perhaps most importantly, we shall see in Chapter 14 that, for linear systems at least, they form a complete set of waveforms that can be combined to represent any possible wave motion.

There are also plenty of common systems for which the observed waves do indeed appear to be approximately sinusoidal, or at least periodic and composed of a few distinct sinusoidal components. As the electromagnetic and gravitational wave examples of Chapter 1 illustrate, many wave sources are periodic. Of these, many in turn are sinusoidal, for simple harmonic motion occurs not only for the individual coordinates of a rigid rotator but also whenever a body moves with sufficiently low energy in a fixed potential well. Even when the source motion is anharmonic, the propagation of the wave through a dissipative medium can often leave only a single dominant sinusoidal component. The cavity magnetron of a microwave oven, the oscillating electron of a radiating atom and the resonant air column of a flute therefore all radiate waves that to a good approximation may be regarded as sinusoidal waves.

This chapter introduces the notation and mathematical representation of sinusoidal waves, their propagation velocity and the energy and power that they carry, before examining their crucial characteristic of orthogonality, which allows us to determine many of the properties of a complex wave motion by simply adding the individual contributions of its sinusoidal components.

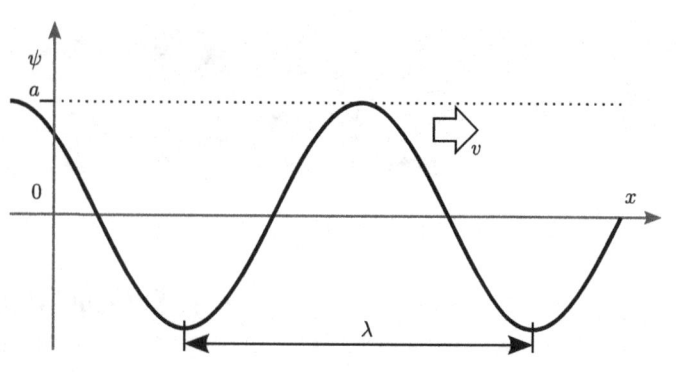

Fig. 4.1 A snapshot of a sinusoidal wave of wavelength λ and amplitude a.

4.1.1 Sinusoidal travelling waves

We have already in Chapter 2 met travelling waves of the form $\psi(x, t) = \psi(u)$, where $u \equiv x - vt$. We shall here consider the specific sinusoidal example

$$\psi(x, t) = a \sin(ku + \varphi), \tag{4.1}$$

where we have allowed for an offset φ. Such waveforms, shown in Fig. 4.1, prove to be solutions to most of the non-dissipative linear wave equations that we meet, with the notable exception of the Schrödinger wave equation for a quantum-mechanical particle; they extend, oscillating for ever, for all values of x and t, and are interesting because they may be characterized by single values of the wavenumber k and hence of the wave angular frequency ω, which is defined by

$$\omega = vk. \tag{4.2}$$

Such a wave motion may be expressed in a variety of ways, for example as

$$\psi(x, t) = a \sin(kx - \omega t + \varphi), \tag{4.3a}$$

$$\psi(x, t) = a \sin\left[2\pi\left(\frac{x}{\lambda} - ft\right) + \varphi\right], \tag{4.3b}$$

$$\psi(x, t) = a \sin\left[2\pi\left(x\bar{\nu} - \frac{t}{\tau}\right) + \varphi\right], \tag{4.3c}$$

It is quite common to omit the word 'angular' and allow this to be implied by use of the symbol ω.

where the parameters are the **amplitude** a, **wavelength** λ, **wavenumber** $k = 2\pi/\lambda$, **spectroscopists' wavenumber** $\bar{\nu} = 1/\lambda$, **frequency** f, **angular frequency** $\omega = 2\pi f$ and oscillation **period** τ. By comparison with equation (4.1), we see that the **wave speed** v relates these properties through

$$v = f\lambda = \frac{\omega}{k}. \tag{4.4}$$

The variable φ here is the initial **phase** of the wave motion at the origin, and describes its offset with respect to the pure sine function ($\varphi = 0$). The phase is a *relative* property, for we could equally have written our wave motion in terms of the cosine function or measured relative to a different origin or reference time; there is no universally accepted absolute phase.

4.1.2 Phase velocity

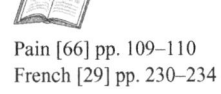

Pain [66] pp. 109–110
French [29] pp. 230–234

In equations (4.1) and (4.3), we introduced the phase φ of the wave motion, which more generally may be regarded as the argument $k(x - vt) + \varphi$ of the sine function at any position x and time t. The phase is therefore relevant only when we express the waveform as a sinusoidal motion, and it is a relative property that allows comparison with other pure sine or cosine functions. More significantly, though, the phase allows us to define two fundamental properties of wave motions, namely the wavefront and the phase velocity.

A **wavefront** is a point, line or surface of constant phase and hence, for the example of equation (4.1), satisfies

$$k(x - vt) + \varphi = \text{constant.} \tag{4.5}$$

This could describe the crest of a wave, a point where the disturbance is zero, or any other point that moves with the wave so as to maintain at all times the same value for the wave displacement ψ. Equation (4.5) can easily be rearranged to give an explicit equation of motion for the wavefront:

$$x = vt + (\text{constant} - \varphi)k. \tag{4.6}$$

The wavefront therefore moves with a constant velocity v, which, because it describes the motion of a point of constant phase, is known as the **phase velocity** of the wave; we shall generally henceforth denote it v_p.

4.1.3 Dispersion

Unlike the wave equation (2.18) for the guitar string of Section 2.2, equation (3.61) for shallow-water capillary waves contains partial derivatives of differing orders. If we repeat the approach of Section 2.2.2 by substituting a general travelling wave of the form of equation (2.2) and applying the chain rule of equation (2.19), we obtain

$$v_\text{p}^2 \frac{\mathrm{d}^2 \psi}{\mathrm{d}u^2} = \frac{h_0 \sigma}{\rho} k^2 \frac{\mathrm{d}^4 \psi}{\mathrm{d}u^4}, \tag{4.7}$$

and there is therefore no common derivative of the wavefunction that can simply be cancelled out like the term $\mathrm{d}^2 \psi / \mathrm{d}u^2$ of equation (2.20). A travelling capillary wave cannot therefore in general propagate without changing its shape – a phenomenon known as **dispersion**.

Sinusoidal waves do, however, prove to be solutions of equation (3.61), for the second and fourth derivatives of a sine function are also sine waves. However, whereas the phase velocities of waves supported by the guitar string of Section 2.2 and of electromagnetic waves in the cables of Section 3.1 proved

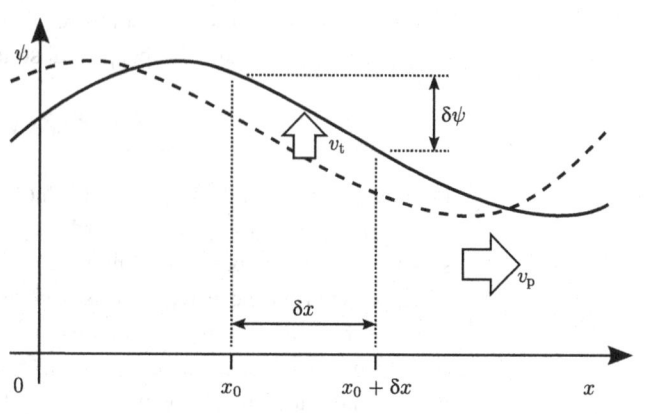

Fig. 4.2   A travelling wave propagating along a section of a guitar string.

to be constant, being given, respectively, by the ratio of the string's tension to its mass per unit length and by the permeability and permittivity of the dielectric spacer, substitution of the general sinusoidal travelling wave of equation (4.1) into the capillary wave equation (3.61) yields

$$v_{\mathrm{p}}^2 = -\frac{h_0\,\sigma}{\rho}k^2. \tag{4.8}$$

The phase velocity v_{p} hence depends upon the wavenumber k of the sinusoidal solution – as we saw in Section 3.3 to be the case for the capillary–gravity and deep-water ocean waves whose wavenumber-dependent phase velocities are given by equations (3.53) and (3.67). This is an alternative signature of the phenomenon of dispersion. Only if they are sinusoidal (and hence k is single-valued) will travelling waves have a definite phase velocity and be solutions to the wave equation.

We shall consider dispersion further in Chapters 5 and 13, and in Chapter 18 we shall see that it is the cause of the remarkable *Kelvin wedge* in the wake of a ship or other watercraft.

4.2   Energy of a wave motion

We know from everyday experience that a wave motion, be it sound, light or the displacement waves of the ocean, conveys energy. For the guitar string of Section 2.2, it is clear that an element of the string undergoing a wave motion may have a higher energy than when in its relaxed state, by virtue of both the increased potential energy of the displaced string and the kinetic energy associated with its motion. In this section, we shall determine these two components of the wave energy and the rate at which it travels; these specific results illustrate some general properties of the energy and power conveyed as a wave propagates.

Figure 4.2 shows a section of a guitar string undergoing a travelling-wave motion $\psi(x, t) = \psi(u)$, where $u \equiv x - v_p t$. To simplify matters, we shall assume that the string tension is not affected by the presence of the wave, and that the wave amplitude is small in comparison with its wavelength.

The string tension may in principle indeed be affected by the wave: this is an example of a longitudinal wave, which we shall consider further in Chapter 10. Usually, the propagation speed of longitudinal waves in guitar strings is considerably higher than of transverse waves, and any longitudinal wave excited will therefore occur at higher, inaudible frequencies.

We first calculate the kinetic energy associated with the transverse motion of a short section of the string of natural length δx and position x_0. The mass of this element will be $\delta m = M \delta x$, where M is the string's mass per unit length, and its transverse velocity v_t will be given by $v_t = \partial \psi / \partial t$. The kinetic energy $\mathcal{K}$ will hence be given by

$$\mathcal{K} = \frac{1}{2} \delta m \, v_t^2 = \frac{M \, \delta x}{2} \left(\frac{\partial \psi}{\partial t} \right)^2. \tag{4.9}$$

Note that, if there is no longitudinal motion of the string, then, regardless of how it is stretched, each point will lie directly above or below its rest position; the string therefore maintains a constant mass for each unit of its length when projected onto the x axis.

To calculate the potential energy of the string element, we must find the work done in stretching the element from its free length δx against the tension W, and for this we must first find the amount by which the string has been extended. The stretched length of the string element, on applying Pythagoras' theorem, will be given by

$$\sqrt{\delta x^2 + \delta \psi^2} = \sqrt{\delta x^2 + \left(\delta x \, \frac{\partial \psi}{\partial x} \right)^2} \tag{4.10}$$

and thus the extension beyond its natural length will be equal to

$$\sqrt{\delta x^2 + \left(\delta x \, \frac{\partial \psi}{\partial x} \right)^2} - \delta x = \delta x \left(\sqrt{1 + \left(\frac{\partial \psi}{\partial x} \right)^2} - 1 \right)$$

$$\approx \frac{1}{2} \delta x \left(\frac{\partial \psi}{\partial x} \right)^2, \tag{4.11}$$

where the final step involves replacing the square root by a binomial expansion and retaining only the leading power in δx. The potential energy $\mathcal{U}$ will be the extension given in equation (4.11) multiplied by the string tension W,

$$\mathcal{U} = \frac{W \, \delta x}{2} \left(\frac{\partial \psi}{\partial x} \right)^2. \tag{4.12}$$

By substituting $\psi(x, t) = \psi(u)$, where $u \equiv x - v_p t$, we find that

$$\frac{\partial \psi}{\partial x} = \frac{d\psi}{du} \frac{\partial u}{\partial x} = \frac{d\psi}{du} \tag{4.13a}$$

and

$$\frac{\partial \psi}{\partial t} = \frac{\mathrm{d}\psi}{\mathrm{d}u}\frac{\partial u}{\partial t} = -v_{\mathrm{p}}\frac{\mathrm{d}\psi}{\mathrm{d}u}. \tag{4.13b}$$

Hence, since $v_{\mathrm{p}}^2 = W/M$, the potential and kinetic energies of the string element given by equations (4.9) and (4.12) are equal at any point, and the total energy, which scales with the square of the wave amplitude, is

$$\mathcal{E} = \mathcal{K} + \mathcal{U} = W\left(\frac{\partial \psi}{\partial x}\right)^2 \delta x = W\left(\frac{\mathrm{d}\psi}{\mathrm{d}u}\right)^2 \delta x. \tag{4.14}$$

This yields the total *energy density* ϵ (the energy per unit string length),

$$\epsilon \equiv \frac{\mathcal{E}}{\delta x} = W\left(\frac{\partial \psi}{\partial x}\right)^2 = W\left(\frac{\mathrm{d}\psi}{\mathrm{d}u}\right)^2. \tag{4.15}$$

Although we have here examined the specific example of transverse waves on a taut string, it turns out that very similar calculations may be performed in most other cases. That the two contributions to the total energy density are equal is an example of the *virial theorem*, which states that the average values of the kinetic and potential energies of any dynamic system will be related; although in the above example we found this equality to apply also at any time and position, this proves to be a special case.

If the potential energy of a system of particles with coordinates r_i depends upon terms like r_{ij}^N where $r_{ij} \equiv |\mathbf{r}_i - \mathbf{r}_j|$, the virial theorem states that the mean kinetic energy will be $N/2$ times the mean potential energy. For our continuous string, r_{ij} corresponds to the relative displacement of adjacent segments, $(\partial \psi/\partial x)\delta x$ and hence, from equation (4.12), $N = 2$; this will be true for any system undergoing small displacements about its equilibrium configuration.

4.2.1 Power transmitted by a wave motion

We have seen above that the energy density associated with any travelling-wave motion, given by equation (4.15), depends purely upon the property $\mathrm{d}\psi/\mathrm{d}u$, which is associated with a given point on the travelling waveform. The energy may therefore be considered to travel with the propagating wave and hence move with the wave speed v_{p}. The power at a given point, which is just the energy passing that point in unit time, will thus be the product of the energy density and the wave speed,

$$\mathcal{P} = W v_{\mathrm{p}}\left(\frac{\mathrm{d}\psi}{\mathrm{d}u}\right)^2, \tag{4.16}$$

and is therefore proportional to the square of the wave amplitude. While this result is again for our specific example of waves on a taut string, we shall find that equivalent expressions may be obtained in other cases. For example, the instantaneous electromagnetic energy density in a non-conducting medium of relative permittivity and permeability ε_{r} and μ_{r} is given in terms of the electric field strength E by

$$\epsilon = \varepsilon_0 \varepsilon_{\mathrm{r}} E^2, \tag{4.17}$$

and hence, with a wave propagation speed $v_{\mathrm{p}} = \sqrt{1/(\varepsilon_0 \varepsilon_{\mathrm{r}} \mu_0 \mu_{\mathrm{r}})}$, the intensity will be

$$I = \sqrt{\varepsilon_0 \varepsilon_{\mathrm{r}}/(\mu_0 \mu_{\mathrm{r}})} E^2. \tag{4.18}$$

We must be careful to remember that, if there are several waves moving simultaneously in different directions, the net power will involve a vector sum taking into account the different wave velocities.

4.3 The tsunami

A spectacular example of the significance of the energy and power conveyed by a wave motion is provided when shallow ocean waves of the type we considered in Section 3.3 run into shelving coastal waters and approach the shore. We shall address the *refraction* of such waves in Chapter 6; here, however, we consider how waves that appear modest out at sea can wreak destructive power when they reach the shore.

We shall throughout the following sections assume the waves to be long in comparison with the depth h_0 of the water through which they travel, and we shall characterize them by their instantaneous height h (where $h - h_0 \ll h_0$) above the seabed. Just as in Section 4.2, where we determined the kinetic and potential energies in terms of the string displacement ψ, here we may determine the energy components in terms of the wave height h.

4.3.1 Energy density of ocean waves

We first calculate the potential energy $\delta \mathcal{U}$ within the wave. This is simply the gravitational potential energy, which we calculate relative to that of the flat ocean of height h_0.

There is, however, a small catch, in that a naïve calculation will find that the trough of a wave has a *lower* potential energy than the equilibrium value, because the column of water to which it corresponds contains less fluid. We must therefore take into account that the total volume of the ocean is unaffected by the wave, and that the rise and fall in the height of a given column of water as the wave passes must be matched by a flow into or out of that column from elsewhere; any change from a level surface therefore involves raising water to a greater height and potential energy, irrespective of whether we are locally forming a crest or a trough. Alternatively, if a crest is formed by raising water from the level surface into the top of the column of water, the raising of water from below the equilibrium level to leave a trough is equivalent to raising a negative amount of water from the level surface through a negative distance. There is, in other words, an asymmetry about the equilibrium surface level in that there is normally air above but water below. The end result is that any perturbation of the level surface increases the total potential energy.

We consider a column of water with horizontal dimensions δx and δy, and height h above the seabed $z = 0$, and determine the potential energy associated with the formation of such a column. If $h > h_0$ then we must move water from a height h_0 to form the raised column. We divide the column vertically into slices of thickness δz and hence mass $\rho\,\delta x\,\delta y\,\delta z$, each of which requires an energy $\rho\,\delta x\,\delta y\,\delta z\,g(z - h_0)$ for it to be raised into position, where z is the distance of the slice above the seabed. The total potential energy of the column is therefore

$$
\begin{aligned}
\delta\mathcal{U} &= \sum_{z=h_0}^{h} \rho\,\delta x\,\delta y\,\delta z\,g(z - h_0) \\
&\rightarrow \int_{h_0}^{h} \rho\,\delta x\,\delta y\,g(z - h_0)\mathrm{d}z \\
&= \frac{\rho g h^2}{2}\,\delta x\,\delta y,
\end{aligned}
\tag{4.19}
$$

where the integral form is obtained by slicing the column ever more thinly to approach the limit $\delta z \rightarrow 0$. If $h < h_0$, the limits of integration and sign of $(z - h_0)$ will both be reversed, and the same expression results.

For the same column with horizontal dimensions δx and δy, the kinetic energy $\delta\mathcal{K}$ is easily given in terms of the horizontal water velocity v_x,

$$
\delta\mathcal{K} = \frac{1}{2}\rho\,\delta x\,\delta y\,h v_x^2.
\tag{4.20}
$$

It is convenient to write both energy contributions in terms of the same wave property, so we now express the kinetic energy as a function of the wave height h instead of the horizontal water velocity v_x. We have already seen in equation (3.34) that the rate of change of velocity may be written as

$$
\frac{\partial v_x}{\partial t} = -g\frac{\partial h}{\partial x}
\tag{4.21}
$$

and, if we consider a travelling wave $h(u)$, where $u \equiv x - v_\mathrm{p}t$, then

$$
\frac{\partial h}{\partial x} = -\frac{1}{v_\mathrm{p}}\frac{\partial h}{\partial t}.
\tag{4.22}
$$

As we noted in Section 3.3.2, there are several different velocities in the motion of a water wave. v_p is the phase velocity of the wave motion, $v_x \equiv \mathrm{d}\xi/\mathrm{d}t$ is the horizontal velocity of the water within the wave and $v_z \equiv \mathrm{d}h/\mathrm{d}t$ is the vertical velocity of the water surface. Both v_x and v_z vary during the motion and increase in magnitude with the wave amplitude, whereas, for a given water depth and density and gravitational acceleration, v_p remains constant.

Consequently,

$$
\frac{\partial v_x}{\partial t} = \frac{g}{v_\mathrm{p}}\frac{\partial h}{\partial t},
\tag{4.23}
$$

so that, upon integration with respect to time, $v_x = (g/v_\mathrm{p})h + \text{constant}$. For a flat, motionless ocean, $v_x = 0$ and $h = h_0$, so we may write

$$
v_x = \frac{g}{v_\mathrm{p}}(h - h_0).
\tag{4.24}
$$

The kinetic energy of our column of horizontal dimensions δx and δy thus becomes

$$
\begin{aligned}
\delta\mathcal{K} &= \frac{1}{2}\rho\,\delta x\,\delta y\,h\left(\frac{g}{v_\mathrm{p}}\right)^2 (h - h_0)^2 \\
&= \frac{1}{2}\rho\,\delta x\,\delta y\,g(h - h_0)^2,
\end{aligned}
\tag{4.25}
$$

where we have substituted $v_p = \sqrt{gh_0}$. As with the guitar string, we thus find the potential and kinetic energies, given by equations (4.19) and (4.25), to be equal. The total energy of the wave per unit horizontal (seabed) area is hence given by

$$\mathcal{E} = \rho g(h - h_0)^2. \tag{4.26}$$

4.3.2 Power of ocean waves

We may now determine the power $\mathcal{P}$ of the ocean wave, which we calculate per unit length along the wavefront (perpendicular to the direction of propagation). As in Section 4.2.1, this is the product of the energy density of equation (4.26) and the wave speed $v_p = \sqrt{gh_0}$,

$$\begin{aligned} \mathcal{P} &= \rho g(h - h_0)^2 v_p \\ &= \rho g^{3/2} h^{1/2} (h - h_0)^2 \\ &\approx \rho g^{3/2} h_0^{1/2} (h - h_0)^2. \end{aligned} \tag{4.27}$$

For seawater with $\rho \sim 1030$ kg m^{-3} at a depth of 100 m typical of European coastal waters, this means that a gentle 1-m swell conveys a power of over 30 kW for each metre of wavefront.

If the wave proceeds through regions of smoothly varying ocean depth, the power remains approximately constant, for there is little frictional loss of energy to the ocean as the wave travels through it. As a result, the wave magnitude $\psi \equiv (h - h_0)$ must vary as the wave depth varies. For the power of equation (4.27) to remain constant,

$$\psi \equiv (h - h_0) \propto h_0^{-1/4}. \tag{4.28}$$

Thus, as the wave enters shallower water towards a shore, the wave speed falls, and the wave amplitude increases. If the motion becomes large enough for the water surface to slope steeply, or for the amplitude of the wave displacement to become comparable to the water depth, then many of our initial assumptions become invalid and, in practice, the wave *breaks*.

The word *tsunami* comes from the Japanese for '*harbour wave*', reflecting the occurrence of the phenomenon in shallow coastal regions.

This is also the origin of the '*tidal wave*' or *tsunami*, for a relatively benign wave out at sea can grow to a devastating magnitude as it approaches the shore.

4.3.3 The tsunami of Boxing Day 2004

Just after midnight on 26 December 2004, a magnitude-9.1 earthquake along a 1200-km fault line, running along the India–Burma subduction zone from Sumatra to the Nicobar and Andaman Islands in the Bay of Bengal, caused a tsunami that was the greatest natural disaster for many decades. It is estimated that 280 000 people perished.

Some 30 km^3 of water are reckoned to have been displaced by the earthquake over a period of 10 minutes, generating waves that were observed by the

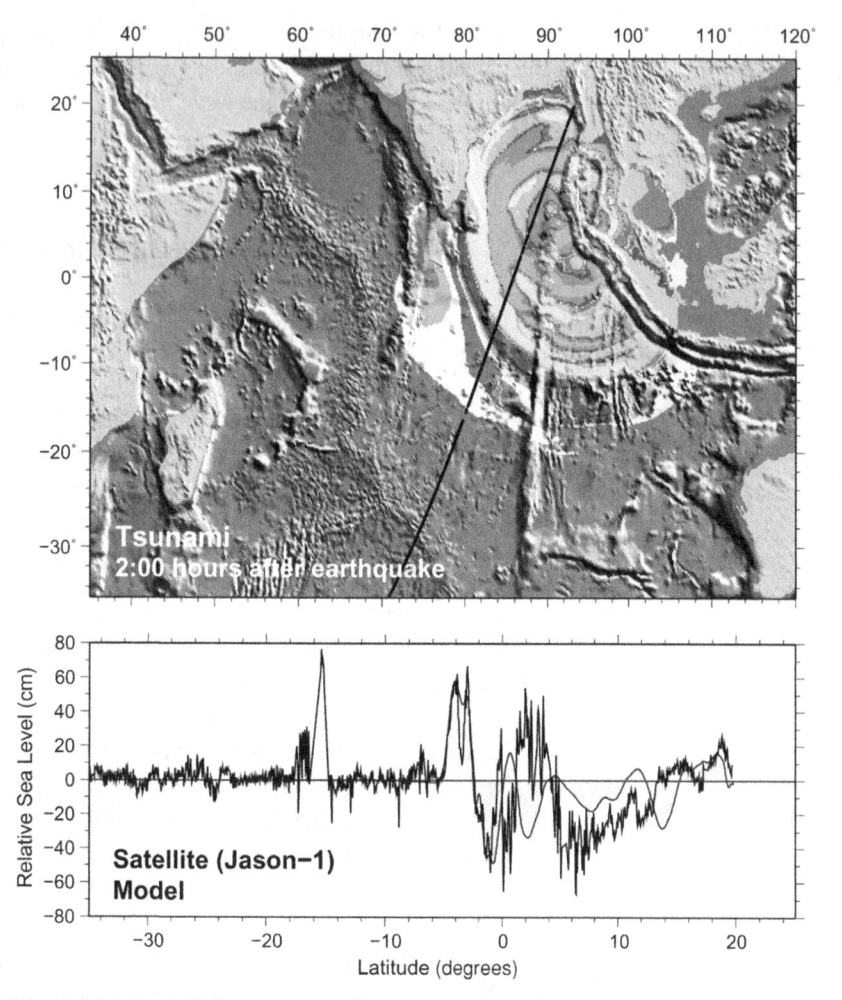

Wave heights recorded by instruments on the Jason-1 satellite, 2 hours after the earthquake that initiated the Sumatra tsunami on Boxing Day 2004; the figure compares measurements with predictions of a propagation model [79, 86].   Courtesy of NOAA

NOAA/PMEL satellite radar, as they crossed the 3.5-km-deep Indian Ocean, to be over half a metre high. As they approached the shores of many low-lying islands and peninsulas, the waves slowed and grew in amplitude. Figure 4.3 shows that individual mid-ocean wave heights of 0.65 m were recorded 2 hours after the earthquake itself.

Our calculation in Section 4.3.2 shows that a deep-sea wave height of ψ_1 in water of depth h_1 would be transformed by slowly shallowing water to a wave of height ψ_2 in water of depth h_2, where

$$\psi_2 = \psi_1 \left(\frac{h_1}{h_2} \right)^{1/4}.$$

(4.29)

From an initial height of 0.65 m where the ocean depth is 3.5 km, the wave will be 3 m in height by the time it reaches water 8 m deep. As the wave amplitude approaches the water depth, of course, our model requires some modification.

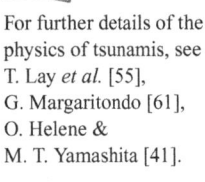

For further details of the physics of tsunamis, see T. Lay *et al.* [55], G. Margaritondo [61], O. Helene & M. T. Yamashita [41].

Such wave heights might not be unusual, but tsunamis have another crucial property: their wave durations are extremely long. Rather than passing in a second or two, and therefore containing only a small volume of water, tsunamis can have a duration of many minutes. Often, the sea initially recedes, exposing large areas of the shore that are usually submerged. The water then begins to rise, and remains high for tens of minutes, transporting huge quantities of water beyond coastal defences to engulf vast areas of low-lying land.

4.4 Normal modes, standing waves and orthogonality

We have seen that sinusoidal waves are just one class of solution to wave equations such as equation (2.18) for the idealized guitar string. The more general travelling-wave solution of equation (2.2) has happily allowed us to analyse the motion of a plucked guitar string; and in Section 4.2 we have seen that the energy and power associated with a wave motion can also be expressed in these more general terms. So why do we so often choose to identify the sinusoidal solutions?

We have already indicated several reasons for favouring such an analysis. First, many sources emit sinusoidal waves, because they consist of rotating bodies, charges in simple harmonic motion and suchlike; it is therefore helpful to describe the propagation of their particular motion. Secondly, we have seen in Section 4.1.3 that the velocity with which a wave propagates can vary between different sine waves depending upon their frequency, and that only when we are dealing with waves of a single frequency are we able to define the phase velocity v_p that is necessary in order for the general derivations of Section 4.2 to make sense. We shall consider this point in rather more depth in Chapters 5 and 13, but it turns out that there are many *linear, dispersive* systems for which sine wave solutions exist but for which the phase velocity varies with frequency. There are, however, two other compelling reasons for favouring sinusoidal wave solutions, and these concern *standing waves* and *orthogonality*.

4.4.1 Standing waves

The plucked guitar string of Section 2.2.3 is a classic example of a simple, linear, non-dispersive system; its solution is straightforward and makes intuitive sense. Yet the motion of the string after its release is in one respect relatively complex, for different sections of the string move at any time in quite different ways. Here, we shall seek a different class of solution, in which all parts of the string follow a common motion and differ only in their relative magnitudes – in other words, the motion $\psi(x, t)$ at any point is a common function of time, $T(t)$, multiplied by a function $X(x)$ that determines its magnitude and depends only upon position. We consider, therefore, solutions of the form

$$\psi(x, t) = X(x)T(t). \tag{4.30}$$

Inserting this form into the wave equation (2.18) for the guitar string, and noting that $X(x)$ is independent of t and $T(t)$ is independent of x, gives

$$MX(x)\frac{\mathrm{d}^2 T}{\mathrm{d}t^2} = WT(t)\frac{\mathrm{d}^2 X}{\mathrm{d}x^2}. \tag{4.31}$$

We now **separate the variables**, by collecting all functions of t on one side of the equation and all functions of x on the other:

Separation of variables

$$\frac{M}{W}\frac{1}{T(t)}\frac{\mathrm{d}^2 T}{\mathrm{d}t^2} = \frac{1}{X(x)}\frac{\mathrm{d}^2 X}{\mathrm{d}x^2}. \tag{4.32}$$

Now, the left-hand side does not depend upon x, and can therefore vary only with time; but the right-hand side does not depend upon t at all; so both sides must equal a constant that depends upon neither x nor t. We shall write this constant for later convenience as k^2 (where k may in general be complex):

$$\frac{M}{W}\frac{1}{T(t)}\frac{\mathrm{d}^2 T}{\mathrm{d}t^2} = -k^2 \tag{4.33a}$$

$$\frac{1}{X(x)}\frac{\mathrm{d}^2 X}{\mathrm{d}x^2} = -k^2. \tag{4.33b}$$

These are now simple second-order differential equations, rather than partial differential equations, and are solved in the usual way. If k and the functions $X(x)$ and $T(t)$ are real, then the solutions are sinusoidal,

$$X(x) = X_0 \sin(kx + \varphi), \tag{4.34a}$$

$$T(t) = T_0 \sin(\omega t + \vartheta), \tag{4.34b}$$

where

$$\omega = \sqrt{\frac{W}{M}}\, k \tag{4.35}$$

and φ and ϑ are constants. The overall solution is therefore given by

$$\psi(x, t) = \psi_0 \sin(kx + \varphi)\sin(\omega t + \vartheta), \tag{4.36}$$

where $\psi_0 = X_0 T_0$; note that it does not matter how the overall magnitude ψ_0 is divided between the functions of x and t.

Solutions of this type – whereby all points of the string follow a common motion that varies sinusoidally with time and has a magnitude that depends upon their position – are known as **standing waves**: they do not appear to travel, but instead all points rise and fall in unison. The motion still results, of course, from the propagation of an 'influence' between adjacent points, but it is not apparent as a propagation of the total amplitude – a phenomenon that causes much confusion elsewhere in the context of *superluminal* waves.

An important property of sinusoidal standing waves is that they can be formed from, and can in turn be converted into, sinusoidal travelling waves:

$$\sin(kx)\sin(\omega t) = \tfrac{1}{2}[\cos(kx - \omega t) - \cos(kx + \omega t)], \qquad (4.37a)$$

$$\cos(kx - \omega t) = \cos(kx)\cos(\omega t) + \sin(kx)\sin(\omega t). \qquad (4.37b)$$

Whether to represent a wave motion in terms of travelling or standing sinusoidal waves is therefore a somewhat arbitrary choice, and we may choose whichever form is the more convenient, for equations (4.37) allow us to convert either form into the other.

4.4.2 Standing-wave modes of a guitar string

Standing waves, as we shall see in more detail in Chapter 12, prove to be of particular value in describing the physical mechanisms of musical instruments such as the flutes and clarinets of the woodwinds, trumpets and trombones of the brass section, and stringed instruments including the violin, guitar and pianoforte. The reason is an important characteristic of sinusoidal standing-wave solutions in that, because of the position-dependent wave magnitude, there are positions at which the motion is always zero. For the general motion described by equation (4.36), the positions x_m of these **nodes** are defined by

$$kx_m + \varphi = m\pi, \qquad (4.38)$$

where m takes integer values. A guitar string executing such a motion, for example, could be held at these points without disturbing the wave, and when the string is set in motion it is therefore these modes that persist. If the guitar string is supported, say, at $x = 0$ and $x = l$, then the waves that are unaffected by the supports will be given by the condition $\varphi = 0$ together with one of a series of values for k, which we label k_n, where

$$k_n = n\frac{\pi}{l}. \qquad (4.39)$$

On re-writing $k_n \equiv 2\pi/\lambda_n$, we see that the distance l between the supports therefore corresponds to an integral number of half-wavelengths, $l = n\lambda_n/2$, as shown in Fig. 4.4.

When, as for our guitar string, there is a series of wave solutions, it is common to refer to them as the allowed **modes** of vibration, and they are labelled according to their *order* – in this case, the integer n. The simplest mode, which is that with the lowest order ($n = 1$ here), is known as the **fundamental** mode of vibration. Higher-order modes are known as **harmonics** or **overtones**, with the slightly confusing convention that the second mode ($n = 2$) is the second harmonic but the first overtone. For non-dispersive systems, the frequencies of the various modes are also integer multiples of the fundamental frequency,

Modes are often more fully referred to as the *normal modes* or *proper modes*, where 'normal' and 'proper' signify 'own' or 'individual' (Fr. *propre*) rather than indicating perpendicularity. In quantum mechanics, we refer to *eigenmodes* and *eigenfunctions*, using the German prefix *eigen-* which has the same meaning.

$$\omega_n = vk_n = n\frac{v_p\pi}{l}. \qquad (4.40)$$

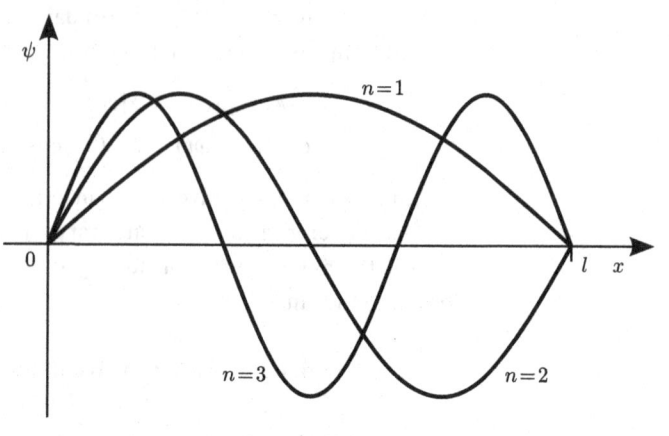

Fig. 4.4 The first three sinusoidal modes of a guitar string.

We could also have determined the modes of the guitar string by first considering the motion of an unbounded string, given by a superposition of counter-propagating travelling waves $\psi(x, t) = \psi_+(u_+) + \psi_-(u_-)$, where $u_\pm \equiv x \mp |v_p|t$ and hence $u_- = 2x - u_+$, with equal but opposite magnitudes to ensure a node at $x = 0$, so that,

$$\psi_+(u) = -\psi_-(-u). \tag{4.41}$$

For the finite string, we have mentioned in Section 2.2.3 that the further need for a node at $x = l$ imposes an additional condition (equation (2.32)),

$$\psi_+(u) = -\psi_-(2l - u), \tag{4.42}$$

so that, on combining equations (4.41) and (4.42), we have

$$\psi_+(u) = \psi_+(u - 2l). \tag{4.43}$$

For sinusoidal waves $\psi_\pm(u_\pm) \equiv \pm\psi_0 \cos(ku_\pm)$, this gives

$$\psi_0 \cos(ku_+) = \psi_0 \cos[k(u_+ - 2l)], \tag{4.44}$$

which is true provided that $2kl = 2n\pi$, i.e. for values $k = k_n$ given by equation (4.39) above.

The superposition of the counter-propagating waves may be written, as in equation (4.37b), as

$$\begin{aligned}
\psi(x, t) &= \psi_0\{\cos[k(x - |v_p|t)] - \cos[k(x + |v_p|t)]\} \\
&= \psi_0[\cos(kx)\cos(|v_p|t) + \sin(kx)\sin(|v_p|t) - \cos(kx)\cos(|v_p|t) \\
&\quad + \sin(kx)\sin(|v_p|t)] \\
&= 2\psi_0 \sin(kx)\sin(|v_p|t), \tag{4.45}
\end{aligned}$$

reproducing the standing-wave form of equation (4.36).

We shall examine the derivation of the normal modes of a system more fully when we consider boundary conditions in Chapter 12.

4.4.3 Orthogonality and normal modes

We mentioned in Section 2.2.2, and have implicitly used in Section 4.4.2, the property of *linearity*: that, if $\psi_1(x, t)$ and $\psi_2(x, t)$ are solutions to a linear wave equation, then so, for arbitrary constants a and b, is the *superposition*

$$\psi(x, t) = a\psi_1(x, t) + b\psi_2(x, t). \tag{4.46}$$

We shall address linearity and superpositions more fully, and in particular their value in the analysis of dispersive systems, in Chapter 13; here, however, we consider a simple superposition of arbitrary solutions as given above.

Suppose that we wish to calculate the energy contained within the motion of a guitar string that can be written as the superposition of two wavefunctions as in equation (4.46). Applying equation (4.15), the energy density will be

$$
\epsilon = W\left(\frac{\partial \psi}{\partial x}\right)^2 = W\left[a\frac{\partial \psi_1}{\partial x} + b\frac{\partial \psi_2}{\partial x}\right]^2
$$
$$
= W\left[a^2\left(\frac{\partial \psi_1}{\partial x}\right)^2 + b^2\left(\frac{\partial \psi_2}{\partial x}\right)^2 + 2ab\frac{\partial \psi_1}{\partial x}\frac{\partial \psi_2}{\partial x}\right]. \tag{4.47}
$$

We may integrate this over the string of length l to give the total wave energy

$$
\mathcal{E} = W\left[a^2\int_0^l \left(\frac{\partial \psi_1}{\partial x}\right)^2 \mathrm{d}x + b^2\int_0^l \left(\frac{\partial \psi_2}{\partial x}\right)^2 \mathrm{d}x + 2ab\int_0^l \frac{\partial \psi_1}{\partial x}\frac{\partial \psi_2}{\partial x}\,\mathrm{d}x\right]
$$
$$
= a^2\mathcal{E}_1 + b^2\mathcal{E}_2 + 2Wab\int_0^l \frac{\partial \psi_1}{\partial x}\frac{\partial \psi_2}{\partial x}\,\mathrm{d}x, \tag{4.48}
$$

where $\mathcal{E}_1$ and $\mathcal{E}_2$ are the energies for the cases of the individual components ψ_1 and ψ_2 alone. The total energy is therefore in general not simply the sum of these individual contributions.

Suppose, however, that the component functions ψ_1 and ψ_2 are both sinusoidal modes, with wavenumbers k_m and k_n, respectively, according to equation (4.39). The third integral in equation (4.48) hence becomes

$$
2ab\int_0^l \frac{\partial \psi_1}{\partial x}\frac{\partial \psi_2}{\partial x}\,\mathrm{d}x \rightarrow 2ab\int_0^l \sin(k_m x)\sin(k_n x)\mathrm{d}x
$$
$$
= ab\int_0^l \cos[(k_m - k_n)x] - \cos[(k_m + k_n)x]\mathrm{d}x, \tag{4.49}
$$

The integral of equation (4.49) may be written as

$$
\int_0^l \cos[(k_m - k_n)x]\mathrm{d}x = l\delta_{mn},
$$

where δ_{mn} is the **Kronecker δ-function**, defined by

$$\delta_{mn} = 0 \quad (m \neq n),$$
$$\delta_{mn} = 1 \quad (m = n).$$

which proves to be identically zero except in the trivial case $m = n$. For the case of superposed sine waves, then, the energy of the wave motion is indeed simply the sum of the contributions from the component sine waves alone,

$$\mathcal{E} = a^2\mathcal{E}_1 + b^2\mathcal{E}_2. \tag{4.50}$$

Functions whose integrated product is identically zero are said to be **orthogonal**, and form the obvious building blocks for more complex functions because their contributions to many properties may be evaluated in a simple fashion as for the energy of the vibrating guitar string above. Sine waves are not unique in having this property, but orthogonality is another of the crucial characteristics that make sinusoidal functions so useful in the analysis of wave motions.

Exercises

4.1   A wave of frequency 20 s^{-1} has a velocity of 80 m s^{-1}.
   1. Determine the wavelength of the wave.
   2. How far apart are two points whose displacements differ in phase by 30°?
   3. At a given point, what is the phase difference between two displacements occurring at times separated by 0.01 s?

4.2   A transverse wave travelling along a string is given by

$$y = 0.3 \sin[\pi (0.5x - 50t)], \tag{4.51}$$

where y and x are in centimetres and t is in seconds.
   1. Find the amplitude, wavelength, wavenumber, frequency, period and velocity of the wave.
   2. Find the maximum transverse speed of any particle in the string.

4.3   Sketch, on the same graph, the wave described by equation (4.51) at times $t = 0$ and $t = 5$ ms, in the range $x = 0$ to $x = 2$. Identify a point at which the particle velocity is zero at $t = 0$, and indicate how the tension of the string at that time results in a net force, which is apparent in its subsequent displacement.

4.4   Show by substitution that equation (2.18), governing shallow waves on a long, thin guitar string of mass ρ per unit length and under a tension W, has sinusoidal travelling-wave solutions

$$y(x, t) = y_0 \cos(kx - \omega t + \varphi), \tag{4.52}$$

and explain the significance of the parameters k, ω and φ.

4.5   By substituting the sinusoidal travelling waves of equation (4.52) into equation (2.18), determine the *dispersion relation* between k and ω for shallow waves on the guitar string. Hence determine how the wave speed depends upon ρ and W.

Complex wavefunctions

5.1 Complex harmonic waves

We have seen in Chapter 4 that the property of orthogonality, together with a particular relevance to waves from oscillating and rotating sources, makes sinusoidal waves a convenient set of solutions to many common wave equations. Indeed, the alternation of successive differentiations of a sinusoidal function between sine and cosine means that sinusoidal functions of the wave coordinates $x_i \equiv t, x, y, z, \ldots$ will in general be solutions to partial differential wave equations of the form

$$\sum_{n_i, n_j, \ldots} a_{n_i, n_j, \ldots} \frac{\partial^{n_i + n_j + \cdots} \psi}{\partial x_i^{n_i} \partial x_j^{n_j} \cdots} = 0, \tag{5.1}$$

provided that the orders of the derivatives (i.e. the terms $(n_i + n_j + \cdots)$) are either all odd or all even. Even in cases where the constant coefficients $a_{n_i + n_j + \cdots}$ of equation (5.1) are replaced by polynomial functions of the wave coordinates, sinusoidal functions may still provide the basis of solutions to the wave equation when multiplied by an appropriate polynomial factor.

Although many wave equations are indeed of the form given generically in equation (5.1), there are also many important exceptions, including the Schrödinger equation for a quantum particle, the thermal diffusion equation that we shall meet in Chapter 10, and equations describing dissipative systems like that considered below. Each of these contains both first and second derivatives of ψ, and cannot therefore be solved by a simple sinusoidal wave alone. It proves convenient in such cases to work in terms of **complex harmonic** waveforms, in which the wavefunction ψ again has a well-defined frequency and wavenumber but in which its amplitude may have *imaginary* terms. We shall see that these complex waveforms also to some extent remove the distinction between travelling and standing waves, in that complex travelling waves may be written as separable products of functions of time and functions of position.

Our approach is once again to use the property of linearity: that if $\psi_1(x, t)$ and $\psi_2(x, t)$ are solutions to a linear wave equation, then so is the superposition

$$\psi(x, t) = a\psi_1(x, t) + b\psi_2(x, t). \tag{5.2}$$

Here, the arbitrary constants a and b represent the amplitudes of the two components, but there is nothing in the wave equation that restricts them to being real numbers. We therefore temporarily suspend disbelief that physical quantities may be associated with complex amplitudes, and explore what happens when these constants take complex or imaginary values.

We consider a specific complex wave superposition, in which the two components are sine and cosine functions with the same frequency and the constants differ by a factor of $i = \sqrt{-1}$:

$$\psi(x, t) = a \cos(kx - \omega t) + i a \sin(kx - \omega t). \qquad (5.3)$$

This particular superposition can be written in a simpler fashion as a complex exponential function

$$\psi(x, t) = a \exp[i(kx - \omega t)]. \qquad (5.4)$$

On an Argand diagram, this is a vector of length a that rotates clockwise with angular frequency ω and anticlockwise as the position coordinate x is advanced. That this complex exponential function is a solution to the wave equation may be readily confirmed by direct substitution, recalling that, for any continuous variable u,

$$\frac{d}{du} e^{iau} \equiv i a e^{iau}. \qquad (5.5)$$

Indeed, the differentiation of complex exponentials is generally rather more compact and tidy, and less prone to error, than that of sinusoidal functions.

The complex exponential wave given in equation (5.4) is a travelling wave of the form $\psi(u)$, where $u \equiv k(x - |v_p|t)$, that we have considered previously: it propagates with a phase velocity $v_p = \omega/k$ and shows all of the mathematical properties characteristic of sinusoidal solutions to the wave equation. It is a solution to *any* wave equation of the form given in equation (5.1), irrespective of the orders of the derivatives. It may also be written as a separable product of functions of time and position in the form of equation (4.30),

$$\psi(x, t) = a \exp(ikx) \exp(-i\omega t). \qquad (5.6)$$

Lest at this point the suspension of disbelief should show signs of waning, we note that real, sinusoidal waves may readily be formed from superpositions or manipulations of complex exponentials, for example,

$$\cos(kx - \omega t) = \frac{1}{2}\{\exp[i(kx - \omega t)] + \exp[-i(kx - \omega t)]\}$$
$$= \frac{1}{2}\{\exp[i(kx - \omega t)] + [\exp[i(kx - \omega t)]]^*\}$$
$$= \text{Re}\{\exp[\pm i(kx - \omega t)]\}. \qquad (5.7)$$

Physical properties, including the energy associated with wave motions, will always correspond to real expressions, but we shall see that it is often easier to perform much of the wave analysis using complex exponential waveforms, and convert them to real quantities later.

5.2 Dispersion in dissipative systems

A particular example of the utility of the complex exponential notation is the solution of the wave equation for a **dissipative** system – one in which, like all classical systems when examined in sufficient detail, energy is lost through friction and the like. For our example, we return to the guitar string of Section 2.2 and, as before, derive the wave equation for the guitar string by applying Newton's third law to give a relation between the acceleration at a given position and the second spatial derivative of the wavefunction at that point. This time, however, we include a frictional term, which we take to be proportional to and opposing the velocity of the string:

$$M\frac{\partial^2 \psi}{\partial t^2} = W\frac{\partial^2 \psi}{\partial x^2} - \gamma\frac{\partial \psi}{\partial t}. \tag{5.8}$$

A drag force of γ times the string velocity is therefore assumed to exist per unit length of the string; this could be due to air resistance (we can often hear the vibrations and whistles of strings even without the aid of a soundboard), or due to flexural resistance resulting in heating of the string.

It is clear that a simple sine wave does not satisfy the wave equation (5.8) – nor should we expect it to, for the sine wave continues undiminished for ever, with no sign of the attenuation that we would expect for the damped system. We would instead expect the wave to show some sort of decay as it propagates, and this we shall indeed observe.

The wave equation above may be solved simply by inserting a complex exponential wave of the form given in equation (5.4) to give, after cancellations,

$$M(-\omega^2) = W(-k^2) - \gamma(-i\omega), \tag{5.9}$$

which, with a little rearrangement, yields

The relationship, such as equation (5.10), between the angular frequency ω and wavenumber k is generally referred to as the *dispersion relation*.

$$k = \pm\sqrt{\frac{M}{W}}\omega\sqrt{1 + i\frac{\gamma}{M\omega}}. \tag{5.10}$$

This reduces as expected to $k = \pm\sqrt{M/W}\,\omega$ if the frictional term γ may be neglected. It remains simply to work out the rather messy complex square root – a derivation that we now present in full.

We may write k in terms of its real and imaginary parts,

$$k = k_1 + ik_2 = k_0(p + iq), \tag{5.11}$$

where p and q are real and $k_0 \equiv \sqrt{M/W}\,\omega$, so that, from equation (5.10),

$$1 + i\frac{\gamma}{M\omega} = (p + iq)^2$$
$$= p^2 - q^2 + 2ipq. \tag{5.12}$$

On equating the real and imaginary parts,

$$p^2 - q^2 = 1, \tag{5.13a}$$

$$pq = \frac{\gamma}{2M\omega}. \tag{5.13b}$$

We rearrange equation (5.13b) to give p in terms of q, and substitute this into equation (5.13a) to yield a quadratic for p^2:

$$\left(p^2\right)^2 - p^2 - \left(\frac{\gamma}{2M\omega}\right)^2 = 0, \tag{5.14}$$

giving (for real p)

$$p^2 = \frac{1}{2}\left[1 + \sqrt{1 + \left(\frac{\gamma}{M\omega}\right)^2}\right]. \tag{5.15}$$

If $\gamma \ll M\omega$ – corresponding to weak damping – then $p \approx 1$ and

$$q = \frac{\gamma}{2M\omega}. \tag{5.16}$$

The damped-wave solution is therefore

$$\psi(x, t) = a \exp[i(k_0 x - \omega t)]\exp(-k_0 q x), \tag{5.17}$$

so that the complex harmonic wave $a \exp[i(k_0 x - \omega t)]$ decreases exponentially in magnitude as it propagates. The real decay constant $k_0 q \equiv \gamma/(2\sqrt{M})$ can appear in our solution only because we allowed the wavenumber k to be complex, in turn by allowing the wavefunction itself to be a complex exponential. A real wavefunction is easily formed by adding equation (5.17) to its complex conjugate to give

$$\psi(x, t) = a\{\exp[i(k_0 x - \omega t)] + \exp[-i(k_0 x - \omega t)]\}\exp(-k_0 q x)$$
$$= 2a \cos(k_0 x - \omega t)\exp(-k_0 q x). \tag{5.18}$$

Equation (5.18) describes the *spatially* decaying waves that diminish as they propagate away from a harmonic source, but there is an alternative motion corresponding to the *temporally* decaying motion of a string that is set oscillating along its length and allowed to die away. Rather than a real value for ω and complex k that we implied in our solution above, this requires a real value for k and complex ω: while equation (5.17) remains valid, such a substitution is rather messy and it is simpler to return to equation (5.9) and solve the quadratic equation for ω to give

$$\omega = \frac{-i\gamma \pm \sqrt{4MWk^2 - \gamma^2}}{2M}$$
$$\approx k\sqrt{\frac{W}{M}} - i\frac{\gamma}{2M}, \tag{5.19}$$

where the final step again assumes the damping to be weak so that $\gamma \ll k\sqrt{MW}$. Writing $\omega_0 \equiv k\sqrt{W/M}$, the temporally damped solution is then

$$\psi(x,t) = a\exp[\mathrm{i}(kx - \omega_0 t)]\exp[-(\gamma/(2M))t]. \qquad (5.20)$$

We could, of course, have solved our new wave equation (5.8) simply by substituting a generic form of equation (5.18) or equation (5.20). Alternatively, from first principles, we could have followed Section 4.4.1 and sought a separable solution $\psi(x,t) = X(x)T(t)$ in which the functions $X(x)$ and $T(t)$ were allowed to be complex: after separating the variables, we would arrive at a second-order differential equation

$$\frac{M}{W}\frac{1}{T(t)}\frac{\mathrm{d}^2 T}{\mathrm{d}t^2} + \frac{\gamma}{W}\frac{1}{T(t)}\frac{\mathrm{d}T}{\mathrm{d}t} = k, \qquad (5.21)$$

whose solutions are those we have found above. Separation of variables to look for the modes of a wave system thus reduces the partial differential wave equation to a simple differential equation of the sort that describes a damped oscillator, to which our wave system corresponds when we eliminate the spatial dependence by making it implicit.

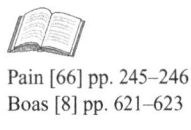

Pain [66] pp. 245–246
Boas [8] pp. 621–623

5.3 Phasors and geometric series

Feynman [22] Chapter 2

Complex numbers also appear in wave physics as part of the detailed model of how a wave motion propagates from point to point. The displacement due to a sinusoidal wave motion is represented at any point by an arrow or **phasor**, which we draw in two-dimensional space with its length proportional to the amplitude of the oscillation and its orientation indicating the phase of that oscillation relative to some datum. The influence of this displacement will be felt at another point only after a delay depending upon the speed and distance of propagation, and this phase lag is indicated by a rotation of the phasor. To determine the propagated field at a given point, all the contributing phasors are added together as vectors: that is, they are joined, head to tail, so as to form a chain, whose resultant is simply the phasor joining its ends. This method of determination embodies essentially the same physics as the Huygens method of wave analysis that we consider in the next chapter.

The graphical depiction of phasors closely resembles the graphical representation of complex numbers on an Argand diagram, and phasors may indeed be represented algebraically by complex exponentials whose modulus and argument correspond to the amplitude and phase of the wave displacement. The combined effect of a number of contributions is then calculated by evaluating the sum of a series of complex exponentials, which can in many cases be treated as a complex extension of a geometric progression.

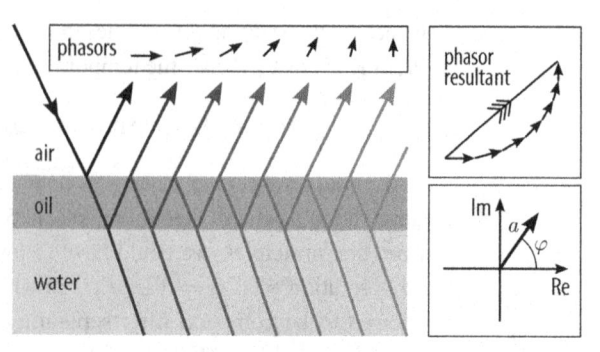

Fig. 5.1 Multiple contributions to the reflection of light by an oil film, and their representation as phasors.

Figure 5.1 shows the use of phasors in analysis of the light reflected by a thin film of oil of the sort found floating on many an urban puddle. Light striking the film is partially reflected, and partially transmitted to then be reflected again at the oil–water interface. This light is partially transmitted, to accompany the original air–oil reflection, and partially reflected back into the oil to begin a further journey from which it may again emerge, after one or more reflections, to contribute to the overall reflected wave. The reflection observed is hence composed of an infinite series of reflections that may be represented by an infinite sequence of phasors, each of which is smaller than the previous one by a constant factor accounting for the extra pair of reflections undergone, and rotated with respect to it by a constant angle representing the additional path travelled. In principle, a careful scale drawing would allow the phasor resultant, and hence the intensity of the observed reflection, to be found in a purely graphical fashion.

Mathematically, the contributions are of the form

$$a_0, \; \alpha a_0 e^{i\varphi}, \; \alpha^2 a_0 e^{2i\varphi}, \; \alpha^3 a_0 e^{3i\varphi}, \; \ldots \tag{5.22}$$

and the total light field reflected at the angle shown may therefore be written

$$
\begin{aligned}
E &= a_0 + \alpha a_0 e^{i\varphi} + \alpha^2 a_0 e^{2i\varphi} + \alpha^3 a_0 e^{3i\varphi} + \cdots \\
&= a_0 (1 + r + r^2 + r^3 + \cdots)
\end{aligned}
\tag{5.23}
$$

where $r \equiv \alpha \exp(i\varphi)$. The trick with geometric progressions such as this is to note that most of the terms also appear in

$$rE = a_0(r + r^2 + r^3 + \cdots) \tag{5.24}$$

and therefore that, on subtracting equation (5.24) from equation (5.23), we obtain

$$E(1 - r) = a_0, \tag{5.25}$$

so that

$$E = \frac{a_0}{1 - r} = \frac{a_0}{1 - \alpha \exp(i\varphi)}. \tag{5.26}$$

We shall use both the pictorial, phasor representation and the equivalent algebra of complex amplitudes to examine a number of examples of wave propagation in the following chapters.

Exercises

5.1 Show that the wave equation (2.18) can be solved by arbitrary, sinusoidal and complex-exponential travelling waves,

$$\psi(x, t) = \psi(x - v_\mathrm{p}t), \tag{5.27}$$

$$\psi(x, t) = \psi_0 \cos(kx - \omega t + \varphi), \tag{5.28}$$

$$\psi(x, t) = \psi_0 \exp[\mathrm{i}(kx - \omega t)]. \tag{5.29}$$

In each case, find the constraints upon (i.e. any relationships between) the parameters v_p, k and ω.

5.2 Explain how the different forms of wave solutions of equations (5.27)–(5.29) are related, and how these travelling waves can be used to construct *standing-wave* solutions.

5.3 Show that the wave equation

$$\mathrm{i}\, m\frac{\partial \psi}{\partial t} = -\frac{\partial^2 \psi}{\partial x^2} \tag{5.30}$$

(where $\mathrm{i}^2 = -1$ and m is a constant) has complex harmonic travelling-wave solutions of the form

$$\psi(x, t) = \psi_0 \exp[\mathrm{i}(kx - \omega t)] \tag{5.31}$$

and determine the *dispersion relation* between the parameters k and ω.

5.4 A travelling wave has two components of the form given in equation (5.31), equal in magnitude and with frequencies $\omega_0 \pm \delta\omega$ and wavenumbers $k_0 \pm \delta k$. Show that the wave may be written in the form

$$\psi(x, t) = \psi_1 \exp[\mathrm{i}(k_0 x - \omega_0 t)]\cos(\delta k\, x - \delta\omega\, t) \tag{5.32}$$

and thus takes the form of a complex exponential travelling wave that is modulated by a slowly varying, real periodic function.

5.5 Show that the velocities of the two functions in equation (5.32) – i.e. $\exp[\mathrm{i}(k_0 x - \omega_0 t)]$ and $\cos(\delta k\, x - \delta\omega\, t)$ – differ by a factor of two.

5.6 A section of a long rope of mass ρ per unit length is held subject to a tension W. The physical mechanisms governing the motion of the rope resemble those that describe waves on a guitar string but, because of the thickness of the rope, frictional terms are less easily neglected and lead to a force per unit length of $-\gamma\, \partial\psi/\partial t$ that is proportional to and opposes the velocity of the rope at each point.

By considering the net force acting on an element of the string (which may be considered approximately rigid), derive the wave equation governing its transverse motion,

$$\frac{\partial^2 \psi}{\partial t^2} = \frac{W}{\rho}\frac{\partial^2 \psi}{\partial x^2} - \frac{\gamma}{\rho}\frac{\partial \psi}{\partial t}. \tag{5.33}$$

Show that the wave equation (5.33) does not support sinusoidal travelling waves

$$\psi(x, t) = \psi_0 \cos(kx - \omega t + \varphi) \tag{5.34}$$

but can (fortunately) be solved by the complex exponential wave

$$\psi(x, t) = \psi_0 \exp[i(kx - \omega t)], \tag{5.35}$$

and find how k depends upon ω.

5.7 If the rope of Exercise 5.6 is moved up and down in a sinusoidal fashion at the point $x = 0$, with an angular frequency ω, so that $\psi(0, t) = a_0 \cos(\omega t)$, show that the resulting wave, for $x > 0$, is of the form

$$\psi(x, t) = a_0 \cos(k_0 x - \omega t)\exp(-\alpha x), \tag{5.36}$$

where $k_0^2 - \alpha^2 = \rho\omega^2/W$ and $2\alpha k_0 = \gamma\omega/W$.

5.8 The *tidal bore* that flows most notably along the River Severn is a wave pulse whose shape changes little as it propagates over 30 km upstream, and is possible only because the river waves show both *dispersion* and *nonlinearity*. The phenomenon, famously described after being observed on the Union Canal near Edinburgh by John Scott Russell in 1844, is nowadays categorized as a form of **soliton**; its optical equivalents are of great topical interest to communications engineers trying to achieve the fastest data rates down optical fibres.

J. S. Russell [76]

For systems such as the River Severn that show soliton behaviour, the important terms in the wave equation comprise (with appropriate scaling) the **Korteweg–de-Vries** equation,

$$\frac{\partial \psi}{\partial t} + 6\psi\frac{\partial \psi}{\partial x} + \frac{\partial^3 \psi}{\partial x^2} = 0. \tag{5.37}$$

1. Show that, if two waveforms $\psi_1(x, t)$ and $\psi_2(x, t)$ are individually solutions to equation (5.37), then an arbitrary superposition $\psi(x, t) = a\psi_1(x, t) + b\psi_2(x, t)$ will **not** in general be a solution – i.e. the Korteweg–de-Vries equation is *nonlinear*.

2. Show that the travelling-wave pulse (with propagation speed c)

$$\psi(x, t) = 2\alpha^2 \operatorname{sech}^2[\alpha(x - ct)] \tag{5.38}$$

   may, however, be a solution to the equation, but only for a particular value of the parameter α.

3. Find how the parameter α depends upon the propagation speed c of the wave pulse.

4. Sketch the waveform described by equation (5.38) at time $t = 0$.

6 Huygens wave propagation

6.1 Huygens' model of wave propagation

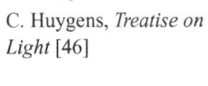

C. Huygens, *Treatise on Light* [46]

One of the most remarkable scientific manuscripts of the seventeenth century must be the *Traité de la lumière* published by Christiaan Huygens in 1690 (but largely complete by 1678). Without any knowledge of electromagnetism, the transverse nature of electromagnetic waves or any other such details that we now take for granted, and presuming the presence of an æther, Huygens was able to present a geometrical construction that completely described the propagation of light through space, its reflection and refraction, and its passage through birefringent media. Even atmospheric mirages were accounted for. The illustrations, which to our modern eyes have considerable charm, are models of clarity, and some have been redrawn as the figures of this chapter.

Huygens estimated the speed of light, in the reference unit of the day, to be some 110 million *Toises* per second; since the Toise was slightly under 2 m, this was remarkably close to today's defined value.

Huygens' principle, shown in Fig. 6.1, is exceptionally straightforward. A wavefront propagates away from an initial disturbance at A at the speed of light, and it does so equally in all directions, so that after a given time it has formed a spherical wavefront HBGI with a radius proportional to the time interval allowed. Crucially, each point b on that wavefront then acts as a source of successive wavefronts, which propagate in the same fashion. Where these secondary wavefronts coincide at DCEF, the new primary wavefront is formed.

The mechanism outlined by Huygens is essentially identical to that which we understand today: it corresponds conceptually to the description presented in Chapter 1 of the propagation of Mexican waves, and it lies at the heart of the theory of quantum electrodynamics (QED) developed by Richard Feynman to describe the more general propagation of quantum-mechanical interactions. There are some details – principally concerning the possibility of backward-travelling waves – that Huygens omitted to address, but the essence of his description serves so completely to account for classical wave phenomena that we shall use it here in their elucidation.

6.2 Propagation in free space

The Huygens construction for the propagation of a plane wave is shown in Fig. 6.2, which may be regarded simply as a section of the spherical wavefront of

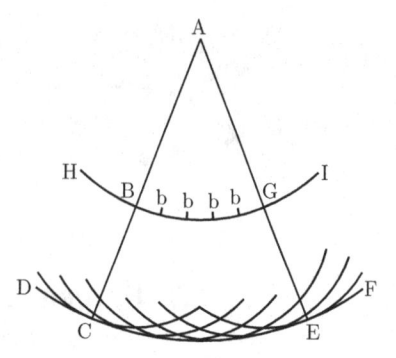

Huygens' depiction of the propagation of a spherical wavefront.

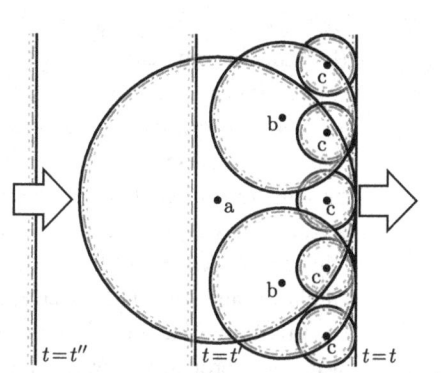

Huygens construction for the propagation of a plane wave. The incident wavefront, shown at successive times t'', t' and t, triggers secondary sources a, b and c as it passes; at time t, the secondary wavefronts reconstruct the propagating original.

Fig. 6.1 for which the section depicted is small in comparison with the distance from the source. As the wavefront advances, according to Huygens' model, it gives rise to secondary sources, each of which radiates a secondary wavefront. These sources appear continuously, and at all positions, as the wavefront passes, and the only places at which their radiated contributions coincide are along the advancing wavefront itself.

Huygens' principle is a consequence of the phenomenon of **interference**, whereby, when propagated wave disturbances arrive along different paths, their amplitudes are simply added as a superposition. Where the wavefronts radiated by secondary sources coincide, their wave displacements add and a new wavefront is created and reinforced; elsewhere, these out-of-step contributions either cancel out or, at best, produce a uniform background. Remarkably, Huygens developed his principle before such phenomena were widely understood.

Huygens recognized that the coincidence of many secondary wavefronts resulted in the formation of the new wavefront, but presented his model without any direct reference to the principle of wave interference. This turns out to be the

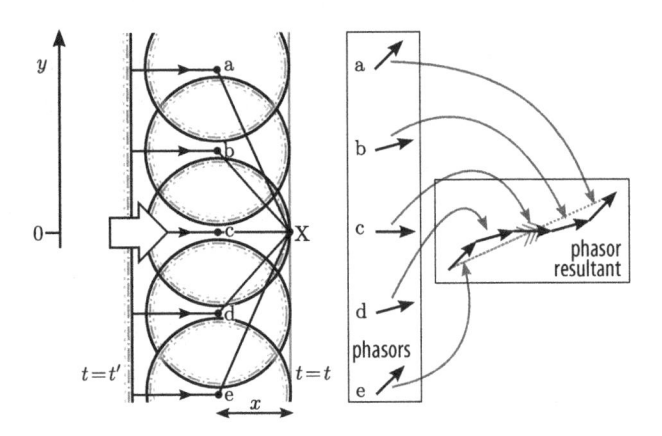

Fig. 6.3 Phasor contributions to a plane wavefront by secondary waves from sources a–e.

key to why there is no backward-travelling wave, for the various contributions can subtract as well as add, and on average they combine to cancel each other out completely. Fresnel and Kirchhoff added these details to refine Huygens' description into the rigorous formalism that we use today.

We can add a little qualitative detail to the Huygens treatment by evaluating the total contribution to the disturbance at a given point from a line of radiating secondary sources, five of which are shown in Fig. 6.3. The sources, a, b, c, d and e, are all displaced in the direction of wave propagation by a distance x from the point X where their combined contributions will be evaluated. Their perpendicular displacements are given by the coordinate y. The distance of each source from X is therefore given by

$$r(x, y) = \sqrt{x^2 + y^2}, \tag{6.1}$$

and, for each wavelength λ, the phasor corresponding to the contribution arriving from that source might therefore be written algebraically as

$$a(x, y) = \frac{\exp(\mathrm{i}kr)}{r}, \tag{6.2}$$

where $k \equiv 2\pi/\lambda$. The $1/r$ term here accounts for the weakening influence of the source the further it is from X: if a is the wave *amplitude*, then this represents an inverse-square dependence of the wave *intensity* upon r. If the most adjacent contribution, from c, corresponds to a phasor that is horizontal, the phasors for sources displaced from c will be smaller and rotated to account for the extra path length travelled. As $|y|$ increases, the phasors become ever smaller and are rotated by ever increasing angles. The total contribution from all sources in the line is found by integrating over y:

$$a(x) \equiv \int_{-\infty}^{\infty} a(x, y)\mathrm{d}y = \int_{-\infty}^{\infty} \frac{\exp(\mathrm{i}kr)}{r}\, \mathrm{d}y. \tag{6.3}$$

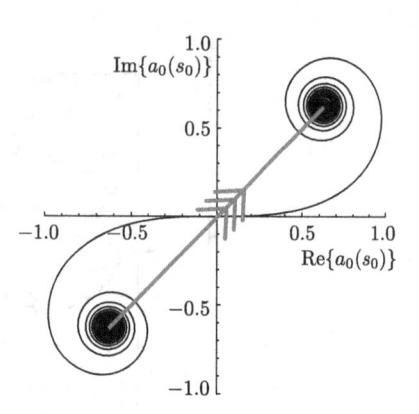

An Argand diagram showing the Cornu or Euler spiral $a_0(s_0)$, for $k = 1$, and its phasor resultant.

This may be evaluated by making the substitution $r \to x + s^2$ so that $y^2 \equiv s^4 + 2xs^2$ and the integral becomes

$$
\begin{aligned}
a(x) &= \int_{-\infty}^{\infty} \frac{\exp[ik(x+s^2)]}{x+s^2} \frac{2(x+s^2)}{\sqrt{2x+s^2}}\, ds \\
&= \sqrt{\frac{2}{x}} \exp(ikx) \int_{-\infty}^{\infty} \frac{\exp(iks^2)}{\sqrt{1+s^2/(2x)}}\, ds.
\end{aligned}
\tag{6.4}
$$

If x is greater than a few wavelengths, so that $kx \gg 1$, the variation of the integrand with s will be determined principally by the phase of its contribution; over the range of significant contributions, the denominator will vary little and may be approximated to $1/\sqrt{2x}$. The integral of equation (6.4) then becomes

$$
\int_{-\infty}^{\infty} \exp(iks^2)\, ds,
\tag{6.5}
$$

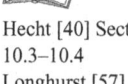

Hecht [40] Sections 10.1 and
10.3–10.4
Longhurst [57]
Sections 10.1–10.7
Pain [66] Appendix 2
Lipson *et al.* [56] Section 7.4

which is known as the **Fresnel integral**. This may be calculated for a range of three-dimensional geometries, but its detail is not required here. The important concept is that the wave disturbance at a given point may be regarded as the superposition of wave disturbances travelling by all possible routes.

The real and imaginary components of the Fresnel integral are shown graphically in the Argand diagram of Fig. 6.4, in which the curve, known as the **Cornu** (or **Euler**) **spiral** or **clothoid**, traces out the result of the finite integral

$$
a_0(s_0) \equiv \int_0^{s_0} \exp(iks^2)\, ds
\tag{6.6}
$$

as s_0 is varied from $-\infty$ to ∞. The Fresnel-integral part of equation (6.5) is therefore given by the vector joining the ends of the spiral and takes the value $\sqrt{\pi/k}\,\exp[i(\pi/4)]$. The approximation that we made in deriving equation (6.5) changes slightly the rate at which the spiral approaches its limit but, provided that $kx \gg 1$, makes little difference to the end points and hence has little effect upon the value of the Fresnel integral itself.

6.2.1 Subtleties in the Huygens description

The eagle-eyed will have noticed some troubling anomalies in the above précis of an often heavily mathematical description. First, the secondary sources radiate in all directions, and it is therefore often stated that the superposition of disturbances from a line of point sources, as in Fig. 6.3, will produce not only the forward-travelling wavefront indicated but also a backward-travelling wavefront formed by the opposite tangents. Secondly, because all of the contributions travel at least as far as the direct one via c, the phase of the resultant (that is, the angle of the phasor resultant) is retarded by $\pi/4$ with respect to the direct component, suggesting that a repeatedly scattered wave propagates more slowly than a direct component.

For the first point, we may make an equivalent observation regarding the guitar string: that, while the initial plucking action results in both forward- and backward-travelling wave components, each component subsequently travels happily in its own direction. An analysis of this specific phenomenon illuminates the more general conundrum. Our derivation, in Section 2.2, showed that the initial shape of the plucked guitar string could be formed from any of an infinite variety of combinations of forward- and backward-travelling waves, provided that the two initially added to give the observed displacement. The particular solution, with equal forward- and backward-travelling components, followed from the additional information that the string was released from rest. Had the same initial shape been a snapshot of a travelling wave, then only that travelling component would have propagated.

The initial velocities (rates of change of displacement) are therefore as important as the displacement itself. It proves to be a crucial feature of wave physics that processes depend upon a combination of the wave displacement and its rate of change. We shall see plenty of examples of this in Chapters 11 and 12, and it is a fundamental reason for the use of complex wavefunctions in quantum mechanics, where the displacement and its rate of change are combined into the real and imaginary parts of a single complex quantity containing everything there is to know about the quantum system.

Feynman [23] Vol. I,
pp. 28-1–28-4 and 30-10–30-12

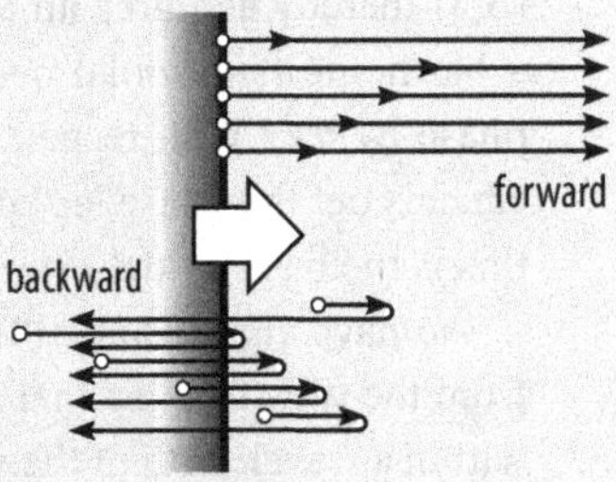

All forward-scattered contributions to a plane wavefront at a given time originate in the same part of the wave, whereas backward-travelling waves may originate anywhere.

Fig. 6.5

forward

backward

Huygens' general description omitted – and in many cases preceded – a detailed description of specific wave-propagation mechanisms, and is indeed open to different interpretations that merit further consideration. The argument outlined above applies to what we might consider a *snapshot* mechanism, in which a wave propagates until a specific instant, when we deftly replace the initial wavefront by secondary sources whose combined emissions then account for the subsequent evolution of the wave. In this case, we have seen that it is necessary to match both the displacement before and after the snapshot, and its rate of change. For an electromagnetic wave, these two terms correspond to the electric and magnetic fields, whose vector product defines the *Poynting vector* in the direction of energy propagation.

An alternative interpretation is that, as in Fig. 6.3, the propagating wavefront *constantly* triggers secondary sources *everywhere* along its path, and the wave reaching a given point is therefore the sum over an infinite number of routes with any number of secondary sources in each. (The snapshot interpretation describes the net effect of this sum for a given source point, so the two interpretations are equivalent.) In the forward direction, as shown in Fig. 6.5, these paths are all equally long, so the wavefronts arrive in step irrespective of their route; a backward-travelling wave, however, can take a path of any length to return to a given point, and the different contributions therefore arrive out of step (alternatively, the contributions arriving at any particular time come, as illustrated, from different parts of the waveform); if there are as many peaks as troughs, the overall combination will be zero.

The puzzling $\pi/4$ phase of the Fresnel integral, while partly the result of interpreting the Huygens model a little too literally, may also be resolved, although some initially surprising phenomena remain. It should first be noted, however, that our analysis has so far been restricted to a single transverse dimension, whereas a real wavefront will be two-dimensional. Equation (6.3) should therefore be replaced by a double integral, which will double the phase of the phasor resultant to $\pi/2$. A rigorous derivation from Maxwell's equations shows that the disturbance to be attributed to each secondary source depends not upon the wave displacement but upon its gradient, and equation (6.2) thereby acquires an additional factor of i; the rate of change follows a harmonic (sinusoidal or complex exponential) wave that is advanced in phase by $\pi/2$ with respect to the displacement, and this advance precisely cancels out the phase lag of the phasor resultant. With a few further caveats, this is the basis of the *Fresnel–Kirchhoff* theory of diffraction.

We have also assumed the observation point X to be many wavelengths from the wavefront so that $kx \gg 1$. If $kx \sim 1$ – i.e. the point X of Fig. 6.3 is within a wavelength of the wavefront – then the variation of the denominator dominates the integral in equation (6.4), the spiral collapses towards the real axis, and the phasor resultant approaches the phase of the wavefront. (The

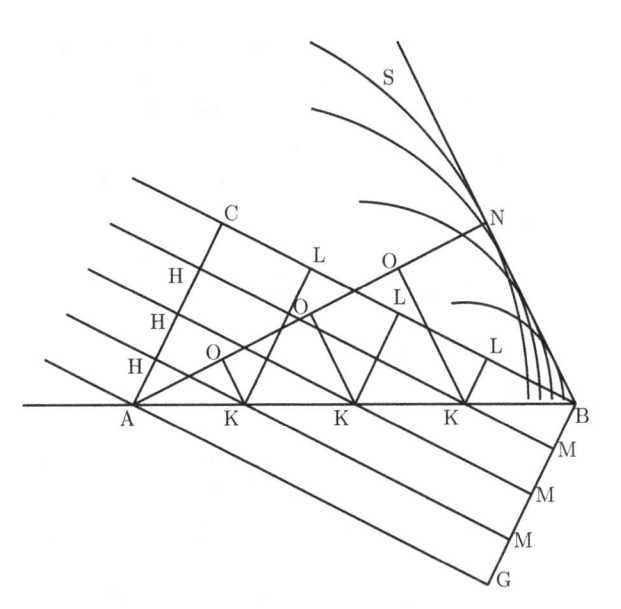

L. G. Gouy [34, 35]
Lipson *et al.* [56] Section 7.3

Fig. 6.6   Huygens' depiction of the reflection of a plane wavefront.

Fresnel–Kirchhoff theory undergoes an accompanying change in the phase factor to compensate.) This variation with x of the phase of the reconstructed wave proves to be related to the intriguing Gouy shift around the waist of a focussed beam.

6.3 Reflection at an interface

Huygens' description is helpful in describing the general mechanisms of wave propagation, as we have seen above; but it is most striking in its ability to describe the particular phenomena that occur when the progress of the wave is disrupted by a change of medium or by obstructions that are placed in its path. All of the phenomena of classical optics, and many of more modern subtlety, are accounted for with great clarity by the Huygens approach, and the improvements by Fresnel and Kirchhoff add more to its rigour and detail than to its fundamental concepts.

The process of **reflection** by a conducting surface is accounted for by a simple modification to the explanation of free-space propagation. Whereas a backward-scattered wave normally comprises elements originating at every point in the waveform, which combine to cancel each other out identically, the presence of a mirror surface hides some of these regions, destroying the cancellation. The true action of the mirror is to set in motion real charge oscillations whose re-radiated waves are sufficiently strong to cancel out the forward-travelling incident wave, and whose effect is to create the reflected wave. The reflected wave may therefore be calculated by considering the combined effect of the oscillating charges themselves, taking into account their relative phases that follow the different phases with which the wave arrives at different points in the mirror.

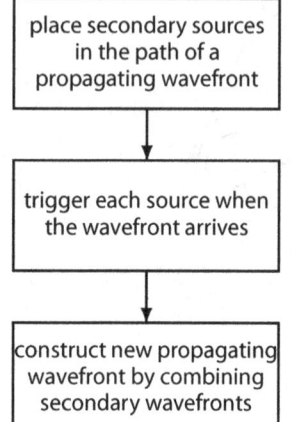

Huygens' illustration, redrawn as Fig. 6.6, depicts a plane wave AC that propagates in the direction CB and is incident upon a plane reflecting surface AB. Secondary sources K within the reflecting surface are considered to be triggered by the arrival of the incident wavefront. Arcs are drawn about each of the points K, such that when added to the corresponding path HK they each sum to the same length. These arcs, which therefore correspond to the same distance of travel from the wavefront, form the new wavefront BN. A little geometry shows that the incident and reconstructed wavefronts make the same angle to the mirror surface, and hence that the incident and reflected angles are the same. The reflected wavefront is the geometrical reflection in the mirror surface of that, BG, which would have occurred had the mirror been absent.

6.4 Refraction at an interface

The Huygens description of **refraction**, shown in Fig. 6.7, follows precisely the same construction as for reflection and indeed describes the partial reflection of the incident wave at the interface between two media. Like a mirror, the refracting interface AB disturbs the forward-travelling wave, so that the usual cancellation of the backward-travelling wavefronts is disrupted. The disruption is less complete than with a mirror, however, so the reflection is only partial. We shall see in Chapter 11 that the fraction of the wave that is reflected depends simply upon the ratio of the wave speeds v_1 and v_2 before and after the interface – in other words, upon the ratio of the refractive indices.

We are principally concerned, however, with the fraction of the wave that is transmitted into the second medium and, because the wave is presumed to travel more slowly in the region below the interface AB, we must be careful to draw arcs that define paths of the same time rather than merely distance. By considering in each case the triangles AKH and AKO, and noting that the lengths HK and AO must be in inverse proportion to the corresponding wave speeds, we find that

$$\frac{1}{v_1}\text{HK} = \frac{1}{v_2}\text{AO} \tag{6.7}$$

and hence, with a little trigonometry, that

$$\frac{1}{v_1}\text{AK}\sin(\angle\text{EAD}) = \frac{1}{v_2}\text{AK}\sin(\angle\text{FAN}). \tag{6.8}$$

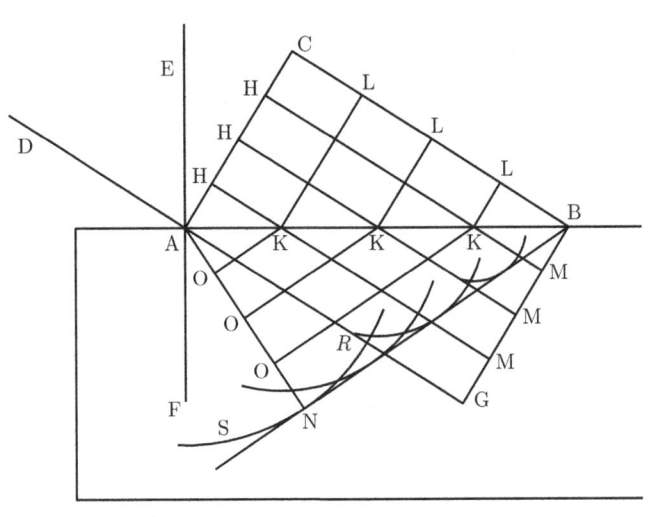

Fig. 6.7   Huygens' depiction of the refraction of a plane wavefront.

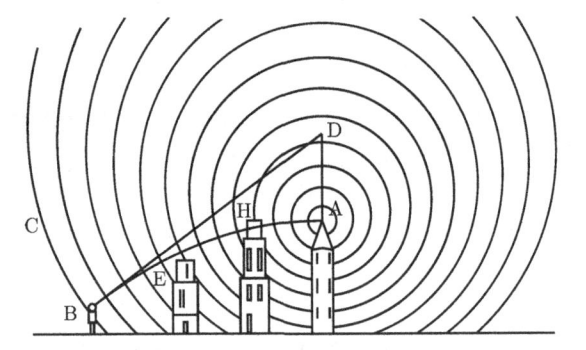

Fig. 6.8   Huygens' illustration of mirage formation by refraction in the atmosphere.

If ∠EAD is labelled ϑ_1 and ∠FAN ϑ_2, we therefore obtain

Snell's law

$$\frac{1}{v_1}\sin\vartheta_1 = \frac{1}{v_2}\sin\vartheta_2. \tag{6.9}$$

This is **Snell's law of refraction**. In optics, it is usual to write Snell's law in terms of the **refractive index** η of each medium, defined by

Refractive index

$$\eta \equiv \frac{c}{v_{\mathrm{p}}}, \tag{6.10}$$

where c is the speed of light in vacuum.

If the wave speed changes gently, rather than abruptly at a well-defined interface, refraction still occurs, although the reflected wave is eliminated. Huygens noted one example of this, with the apparent displacement of distant objects observed through the atmosphere due to variations in density and thus refractive index. Huygens' illustration of this process, redrawn in Fig. 6.8, shows wavefronts emanating from a single point at the top of a tower. The

Fig. 6.9   Ocean waves approaching a gently shelving shore.
Courtesy of National Park Service, Indiana Dunes National Lakeshore

path of a ray of light, defined to be always at right angles to the wavefronts, changes so as to reach the ground at a more acute angle, so that an observer at that position perceives the tower to be rather higher than it truly is and thus, in Huygens' illustration, visible above buildings that would normally block it from view. The same diagram describes the acoustic refraction that allows sound to travel further when radiative cooling overnight leaves a layer of cold air at ground level. The mountain-lee waves of Section 3.5.1 are similarly enhanced if the wind blows more strongly, thus elevating the net wave speed, at altitude.

A second example of progressive refraction concerns shallow-water ocean waves, whose propagation speed we have seen in Section 3.3.1 to be given by

$$v_{\mathrm{p}} = \sqrt{gh_0}, \tag{6.11}$$

where g is the acceleration due to gravity and h_0 the depth of water above the seabed. As waves approach the shore, their speed of propagation falls, and the nearer end of each wavefront travels more slowly than does the end that is further from the shore. The wave is commonly sufficiently refracted that, as in Fig. 6.9, the wavefront is almost parallel to the shore by the time it reaches the beach.

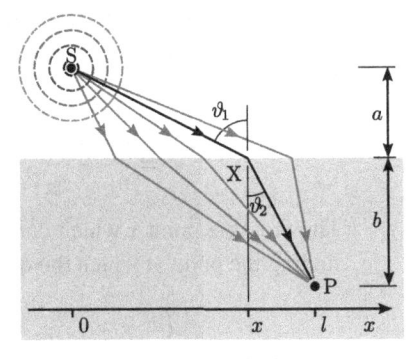

Fig. 6.10   Fermat's principle, applied to refraction at a plane interface.

6.5 Fermat's principle of least time

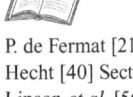

P. de Fermat [21]
Hecht [40] Section 4.5
Lipson *et al.* [56] Section 2.6.3
Feynman [23] Vol. I,
pp. 26-3–26-8

The refraction of waves at an interface illustrates another key aspect of wave propagation, which is entirely consistent with the Huygens construction and provides a neat and thought-provoking rule for propagation paths that are free of obstacles. Usually attributed to Fermat, it is the principle that

> in travelling between two points, a wave follows (with the normal to its wavefronts) the path that takes the least time.

Figure 6.10 indicates how we may analyse refraction by applying this principle. A point source at S emits waves, shown as dotted circular wavefronts, in a uniform medium (such as air) that extends down to a level interface a distance a below the source; beneath this is a different uniform medium (such as glass or water) of different wave speed. We wish to find the path that will be taken by waves that reach the point P in the second medium. By 'path' here we mean the **ray path**, which would be followed by a narrow beam of waves formed by placing a slit somewhere near the source; equivalently, this is the path taken by normals to the wavefronts.

Within an uninterrupted continuous medium, as we have concluded in Section 6.2, a wave will propagate in a straight line; but there are many possible paths from S to P that comprise straight line elements that meet at the interface between the two media. Five such possible paths are shown in Fig. 6.10, in which the vertical distances between points S and P and the interface are a and b, and the horizontal displacement x is measured from $x = 0$ beneath S to $x = l$ abeam P. If a given path crosses the interface at x, then the length of the path from S will be

$$l_S = \sqrt{a^2 + x^2} \tag{6.12}$$

and the length of the path from the crossing to P will be

$$l_P = \sqrt{b^2 + (l - x)^2}. \tag{6.13}$$

If the wave propagation speeds above and below the interface are v_1 and v_2, respectively, then it follows that the total time taken to travel from S to P via the crossing at x will be

$$\tau = \frac{l_S}{v_1} + \frac{l_P}{v_2} = \frac{\sqrt{a^2 + x^2}}{v_1} + \frac{\sqrt{b^2 + (l-x)^2}}{v_2}. \qquad (6.14)$$

The crossing point x which corresponds to the shortest time will be found by finding the point at which the derivative $d\tau/dx$ is zero, where

$$\frac{d\tau}{dx} = \frac{\frac{1}{2}\left(a^2 + x^2\right)^{-1/2} 2x}{v_1} + \frac{\frac{1}{2}\left(b^2 + (l-x)^2\right)^{-1/2}(-2(l-x))}{v_2}$$

$$= \frac{1}{v_1}\frac{x}{\sqrt{a^2 + x^2}} - \frac{1}{v_2}\frac{l-x}{\sqrt{b^2 + (l-x)^2}}. \qquad (6.15)$$

The initially discouraging expressions here correspond conveniently to the sines of the angles of incidence and refraction $\vartheta_1 = x/\sqrt{a^2 + x^2}$ and $\vartheta_2 = (l-x)/\sqrt{b^2 + (l-x)^2}$. We hence obtain, once again, Snell's law,

$$\frac{1}{v_1}\sin\vartheta_1 = \frac{1}{v_2}\sin\vartheta_2. \qquad (6.16)$$

The operation of Fermat's principle may be understood by considering the phasor summation corresponding to all of the possible routes (varying x) by which the wave may reach P, which will look similar to Fig. 6.4. At values of x for which the total time varies rapidly with x, successive phasors will be greatly rotated and will tend to cancel each other out. Only those phasors around the middle of the figure contribute significantly to the resultant, and the condition for least variation is the same as for the minimum time: $d\tau/dx = 0$.

Fermat's principle applies equally to the simple path from S to the interface, and the result is the straight line that we have assumed for this calculation. For such free-space propagation, the symmetry of the problem makes the solution rather more obvious.

Fermat's principle for the propagation of waves is in many ways a specific manifestation of the *principle of least action* that defines trajectories in classical dynamics. Here, as throughout our examination of wave physics, we see a strong correspondence between the rules governing the behaviour of waves and those governing that of particles.

Exercises

6.1   With reference to a carefully labelled sketch, explain how the Huygens construction can be used to account quantitatively for the refraction of a plane wave at the interface between two media.

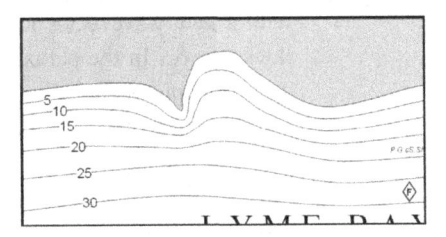

Fig. 6.11 Contours showing water depths near the southern coast.

6.2 A Mexican wave travels through the home supporters at 12 m s^{-1} until it reaches the less athletic fans of the away team (against which a third goal has just been scored), amongst whom the speed is 10 m s^{-1}. If the wavefront makes an angle of $20°$ to the dividing line as it approaches the visiting fans, and assuming that Mexican waves obey the usual rules of wave physics, what is the direction of motion after the dividing line and through what angle is its motion deflected?

6.3 Assuming that the speed v of shallow-water waves is given (see Section 3.3.1) by $v^2 = gh_0$, where g is the acceleration due to gravity and h_0 is the water depth, explain why ocean waves are almost always nearly parallel to the shoreline when they break.

6.4 The propagation speed v of long-wavelength water waves is given in terms of the depth of water h_0 and the acceleration due to gravity g by $v = \sqrt{gh_0}$. The roughly parallel wavefronts of ocean waves approach a shoreline whose gently shelving seabed is indicated by the contour lines (showing depths in metres) in Fig. 6.11. Explain how the deep-ocean waves will be affected as they approach the shore, and explain where would be a good spot to anchor your yacht for a tranquil Sunday lunch.

6.5 Graded-index ('GRIN') lenses – which nowadays are common elements in optical-fibre-based communication systems – comprise plane-faced cylinders of glass whose refractive index varies with the radial distance r from the axis of the cylinder.

By considering a thin slice of such a lens, assuming the wave propagation to be nearly axial (the *paraxial approximation*) and addressing a small region in which the refractive index $\eta(r)$ varies linearly with the transverse coordinate r, show that the refractive properties of the graded-index medium may be expressed as the *eikonal equation*

More generally, the eikonal equation may be written

$$\frac{\mathrm{d}}{\mathrm{d}z}(\eta \mathbf{k}) = -k \, \nabla \eta.$$

$$\frac{\mathrm{d}}{\mathrm{d}z}(\eta k \sin \vartheta) = -k \frac{\mathrm{d}\eta}{\mathrm{d}z}, \tag{6.17}$$

where $\mathbf{k}$ is the wavevector of magnitude $k = |\mathbf{k}| = 2\pi/\lambda$, ϑ the angle between the ray and the lens axis and z the coordinate along the axis.

It may help to consider two light paths (rays) that are parallel, with wavevector $\mathbf{k}$, within the medium, as they enter the lens slice at different radial positions r_1 and $r_2 = r_1 + \delta r$, and which are still parallel, but with a new wavevector $\mathbf{k}_r = \mathbf{k}_i + \delta\mathbf{k}$, just before they emerge after a thickness δz. In the paraxial approximation, $\cos\vartheta \approx 1$ and $\sin\vartheta \approx \vartheta$. The principle is that all paths from the initial source to the final image take the same time.

6.6   A horizontal ray enters a spherical raindrop of radius r and refractive index η at a point X, a distance x above its centre. The ray undergoes refraction as it enters the raindrop, and proceeds at an angle ϑ_2 to the normal until it reaches the edge of the raindrop, where it is reflected and crosses the raindrop a second time. At the next encounter with the surface, the ray is transmitted, undergoing refraction once again, and continues through the air to form part of the primary rainbow observed by a fortunate viewer.

1. Show that the angle of incidence ϑ_1, between the incident ray and the normal to the surface at X, will be given by $\sin\vartheta_1 = x/r$.

2. Show that the internal rays make the same angle ϑ_2 to the normal each time they strike or are reflected from the surface.

3. Hence show that the total angle φ through which the ray is deviated by the raindrop will be $\varphi = 2\vartheta_1 - 4\vartheta_2 + 180°$.

4. Show, by differentiating Snell's law, that

$$\frac{d\vartheta_2}{d\vartheta_1} = \frac{\cos\vartheta_1}{\eta\cos\vartheta_2}. \tag{6.18}$$

5. Hence show that, for the ray that is deviated through the greatest angle φ (as x and hence ϑ_1 are varied),

$$2\cos\vartheta_1 = \eta\cos\vartheta_2. \tag{6.19}$$

6. Combine this result with Snell's law to give the angle of incidence for the rainbow ray,

$$\vartheta_1 = \cos^{-1}\sqrt{\frac{\eta^2 - 1}{3}}. \tag{6.20}$$

7. Hence find the total deviation angle φ for raindrops with red and green light, given that $\eta \approx 1.332$ and 1.337, respectively.

Geometrical optics

7.1 Ray optics

We have seen in Chapter 6 that, in an uninterrupted continuous medium, a wave propagates away from its source along a straight line; and that, when it encounters a change in the properties of the medium, it may be reflected or refracted. These principles are the foundations of the subject of **geometrical optics** and, although they apply generally to any wave system, the language in such cases usually borrows heavily from the science and technology of optics for which, in so many cases, they were first developed. In the following pages, we therefore apply Snell's law to determine the behaviour of lenses and imaging systems, which are both of huge significance in their own right and will also form the basis for some of the experimental arrangements to which we shall refer in future chapters.

A grounding assumption of geometrical optics is that we may represent the propagation of a wave by a **ray**, which is drawn normal to the wavefront and generally indicates the direction of energy flow. Experimentally, a ray may be considered to be a narrow beam that is formed when the wave passes through an aperture whose dimensions are small in comparison with the dimensions of the optical components but large in comparison with the wavelength of the wave. The latter requirement ensures that the aperture may not be considered a serious 'interruption' to the continuous medium, and therefore that there are sufficient distinct secondary sources for Huygens' approach to be sensible; more precisely, it ensures that the phasor sum of Fig. 6.4 will be little changed when truncated by the aperture. Provided that the ray approximation is valid, we may forgo a detailed Huygens or phasor analysis and regard Snell's law as an adequate summary that describes everything that we may wish to know about the paths taken by waves through complex systems.

How the wave energy is divided between the various reflections and transmitted paths is a separate question that we shall address in Chapter 11, and the *diffraction* that occurs when the paths are partially obstructed is the subject of Chapters 8 and 9.

We begin by applying Snell's law to refraction at a curved surface, and thus deriving the *lensmaker's formula* for refraction by a curved surface.

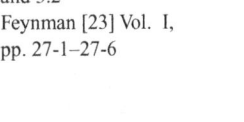

Hecht [40] Section 5.2
Lipson *et al.* [56] Sections 3.1 and 3.2
Feynman [23] Vol. I, pp. 27-1–27-6

In anisotropic materials such as crystals used for their nonlinear optical properties, the direction of energy flow need not be exactly normal to the wavefront.

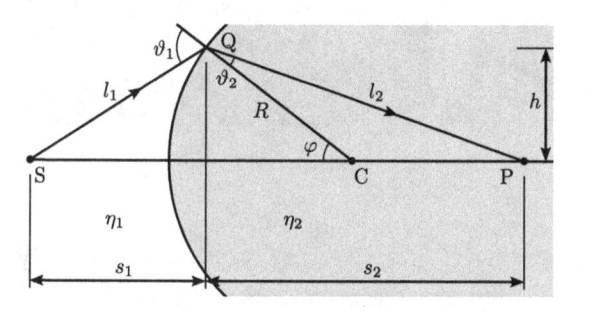

Refraction of light rays at a spherical surface.

7.2 Refraction at a spherical surface

We consider, as illustrated in Fig. 7.1, the propagation of a ray that, having originated at a source S, undergoes refraction at a point Q on the spherical interface between two media of refractive indices η_1 and η_2, the incident and refracted angles being related by equation (6.9). The centre of the sphere, which is of radius R, lies at C, and we follow the ray until it crosses the axis SC containing the source, to end at P. Our aim is to find the position s_2 of P in terms of s_1 and the lens properties R, η_1 and η_2. Snell's law tells us that the angles of incidence and refraction, ϑ_1 and ϑ_2, are related by

$$\eta_1 \sin \vartheta_1 = \eta_2 \sin \vartheta_2, \tag{7.1}$$

and a little geometry gives us

$$h/s_1 = \tan(\vartheta_1 - \varphi), \tag{7.2a}$$

$$h/s_2 = \tan(\varphi - \vartheta_2), \tag{7.2b}$$

$$g/R = \tan \varphi, \tag{7.2c}$$

where φ is the angle between the axis and the normal at the point of incidence Q. These four equations are sufficient to eliminate h, φ, ϑ_1 and ϑ_2, although the result and its derivation are somewhat messy. We therefore make two approximations, both of which are completely reasonable for most situations: the **paraxial approximation**, namely that the angle made by any ray to the axis is small; and what we shall later know as the **thin-lens approximation**, namely that the distance h of all rays from the axis is much smaller than the lens radius of curvature R. Our equations are therefore simplified to

$$\eta_1 \vartheta_1 = \eta_2 \vartheta_2, \tag{7.3a}$$

$$h/s_1 = \vartheta_1 - \varphi, \tag{7.3b}$$

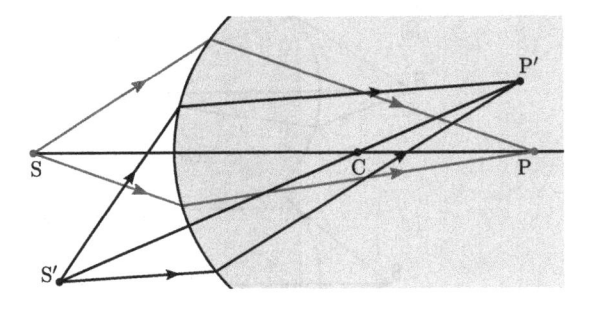

Fig. 7.2   Image formation by refraction at a spherical surface.

$$h/s_2 = \varphi - \vartheta_2, \tag{7.3c}$$

$$h/R = \varphi, \tag{7.3d}$$

from which it is straightforward to show that

$$\frac{\eta_1}{s_1} + \frac{\eta_2}{s_2} = \frac{\eta_2 - \eta_1}{R}. \tag{7.4}$$

The distances s_1 and s_2 here are measured from the points where the rays cross the axis to the point on the axis level with the point of refraction. In practice we shall prefer to measure to the point where the interface crosses the axis, but in the regime to which our approximations apply the difference will be negligible.

The important feature to note in the above result is that there is no dependence upon h: that is, any ray that leaves S and is refracted by the curved interface will pass through P, irrespective of its initial inclination to the axis. P is therefore known as the **focus** of the rays from S and, in the context of light waves, all the light leaving S, provided that it is heading towards the area of the curved interface, will be focussed onto the point P. The geometry is reversible: any light originating at P would be focussed onto the point S; the two points are **images** of each other.

If we move the source S above or below the axis to a point S′, as shown in Fig. 7.2, then the effect is the same as rotating the system about C so as to incline the original axis. The image point P is then also rotated, to P′. This applies for any displacement of the source, provided that our approximations remain valid, and we thus see that for all points S there is a unique image point P, which cannot be reached by any other source point S′ that is the same distance $(s_1 + R)$ from the centre C. This one-to-one mapping from points S to points P is the process of **image formation** upon which vision, photography and microscopy all rely.

It should be noted that, although the thin-lens and paraxial approximations are valid in many practical situations, it is not difficult to find systems in which

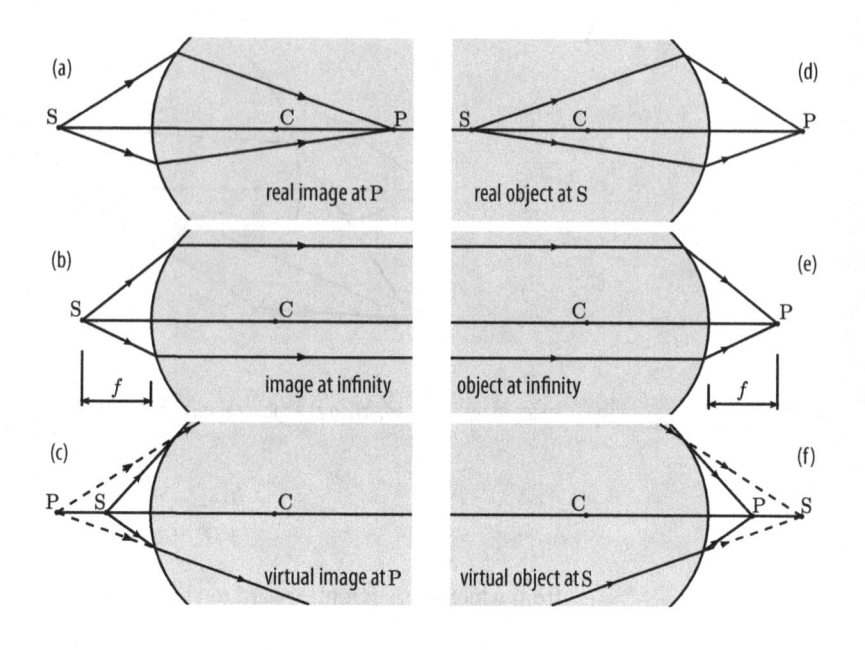

Fig. 7.3   Real and virtual images and objects. (a) With a sufficiently distant object S, refraction at the curved interface causes the rays to converge and form a *real image* at P. As the object S is moved closer, image P retreats until, in (b), the refracted rays are parallel and the image is at infinity; the distance of the object from the lens is the *focal length*, f. For still closer objects (c), the refracted rays diverge and *appear* to come from a *virtual image* at P. (d)–(f) When the ray direction is reversed, the apparent position of the source S determines whether it is considered to be a *real* or *virtual* object.

the approximations are poor, and the image point can then vary with ray angle. This, and a range of other imperfections in the imaging properties, are known as **aberrations**, and their nature and minimization are exhaustively covered in many classic texts on geometrical optics.

7.2.1 Real and virtual images and sign conventions

The geometry depicted in Fig. 7.2 results in the formation of what is known as a **real image**, whereby the rays from the source S *converge* upon the point P at which, if the surrounding medium permits, a detector could be placed to record it. With an array of such detectors, this geometry would represent a photographic camera or a prototype of the human eye.

We are not always concerned with such arrangements, however: for example, the lenses of a pair of spectacles serve to move the *apparent* positions of an object such as the source S, and are therefore used to produce *divergent* rays that *appear* to emanate from a position in front (in our diagrams, to the left) of the curved surface. Figures 7.3(a)–(c) illustrate the transition between the two configurations as the source S is moved ever closer to the refracting surface. Equation (7.4) shows that, as the distance s_1 is reduced, s_2 must increase in compensation; and, if s_1 falls below a critical value, equation (7.4) may be solved only if s_2 is negative. This does not mean that the rays travel backwards but that, as in Fig. 7.3(c), they diverge from rather than converge upon the point P, which is known as the **virtual image**.

It is apparent from equation (7.4) that a negative value for s_2, and hence a virtual image, results for any positive value of s_1 if R is negative – that is, the centre of curvature C is to the left of the interface – or, if R is positive, for $\eta_2 < \eta_1$. In other cases, it depends upon the value of s_1. There is nothing about the derivation of equation (7.4) that precludes any of these possibilities, and the equation is valid for all cases.

Just as it is possible to have both real and virtual *images*, we may have real and virtual sources, or *objects*. Figures 7.3(d)–(f) show these cases. In (d) we have essentially the same case as (a): rays diverging away from a **real object** S strike the refracting surface and, in this case, form a real image. Significantly, while the source S could be a physical source, it could just as well be the real image of a preceding imaging arrangement such as (a), for the rays diverge away from both in exactly the same way. Figure 7.3(f) shows what happens in such a case if the second refracting surface is placed to the left of the real image of the previous system: the refracting surface in this case sees a **virtual object** at S, from which it forms a real image at P.

To distinguish easily between real and virtual images or objects, we adopt a **sign convention**, whereby distances $s_{1,2}$ are positive if the source or image is real, and negative if they are virtual. Similarly, R is considered positive if, as in Figs. 7.3(a)–(c), the centre of curvature is beyond the interface, but negative if, as in Figs. 7.3(d)–(f), it lies before it. Measured in the direction of propagation, positive distances are therefore from the object to the interface, from the interface to the image and from the interface to the centre of curvature. Some thought is necessary in the application of these conventions to specific cases, but with care and practice they allow imaging problems to be solved with relative ease.

Figures 7.3(b) and (e) show two intermediate cases, in which the rays emerging from or approaching the refracting surface are parallel and the image or object distances are therefore infinite. These prove to be important cases, for with astronomical objects, and, indeed, distant objects on Earth, the object distance s_1 is effectively infinite; conversely, since we usually find it comfortable to look at such distant objects, optical instruments such as microscopes are often adjusted so that the viewed image appears to be infinitely distant. These are also mathematically convenient cases, for equation (7.4) then yields $s_{1,2} = \eta_{1,2}R/(\eta_2 - \eta_1)$; this is known as the **focal length** f in the medium of refractive index $\eta_{1,2}$.

In many applications, it is inconvenient for the source and its image to be located in different media, and we therefore combine two interfaces, with

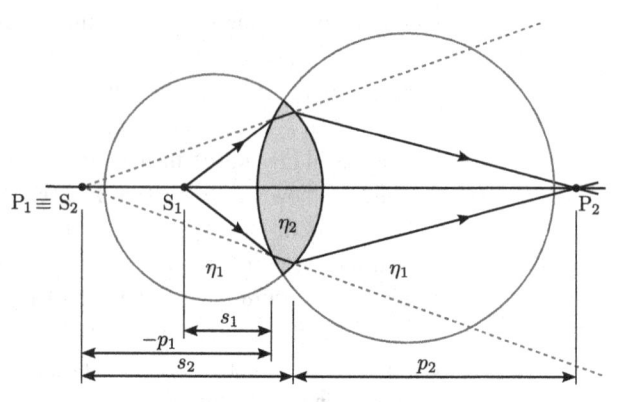

Fig. 7.4 The thin lens formed by the intersection of two spherical surfaces.

different curvatures, to form a **lens**. We shall see in the next section that the definition of the focal length is easily extended to such cases of multiple refractive surfaces, and that they also provide a specific illustration of the use and importance of the virtual image.

7.3 The thin lens

The most common technological example of a lens is the shaped piece of glass or other transparent material whose curved faces refract and focus light to form the real or virtual images of photographic cameras, microscopes, telescopes, data-projectors and so on. Lenses are not limited to visible light, however: polyethylene lenses are used to focus infrared radiation for thermal imaging in motion detectors; polymer lenses are used in ultrasound medical imaging, and have been used under water for high-resolution sonar; and the sand-bars of shoaling water can act as lenses for ocean waves. While we shall in the following section refer to optical lenses, the same principles may be applied to imaging of other wave types.

We consider a region of space, such as air or vacuum, of refractive index η_1, within which is a lens of refractive index η_2, bounded by two spherical surfaces with different curvatures, as shown in Fig. 7.4. Light from a source S_1 strikes the first interface, with radius of curvature R_1, where it is refracted to form a virtual image at P_1. This acts as a real object S_2, from which the refracted rays appear to diverge until they meet the second interface, of radius of curvature R_2, where they are further refracted to converge on the image point P_2.

We apply equation (7.4) to refraction at each surface, giving

$$\frac{\eta_1}{s_1} + \frac{\eta_2}{p_1} = \frac{\eta_2 - \eta_1}{R_1}, \tag{7.5a}$$

$$\frac{\eta_2}{s_2} + \frac{\eta_1}{p_2} = \frac{\eta_1 - \eta_2}{R_2}. \tag{7.5b}$$

In the *thin-lens approximation*, we may assume that the distance travelled by the ray between the two surfaces is negligible in comparison with the other ray distances and hence, taking into account our sign convention, that $s_2 \approx -p_1$. We thus obtain the **lensmaker's formula**

Lensmaker's formula

$$\frac{1}{s_1} + \frac{1}{p_2} = \frac{\eta_2 - \eta_1}{\eta_1} \left(\frac{1}{R_1} - \frac{1}{R_2} \right). \tag{7.6}$$

The reciprocals of the source and image distances s_1 and p_2 therefore always sum to a value determined entirely by the fixed properties and dimensions of the lens. By considering the case when the distance s_1 to the source is infinite, or at least sufficiently great that its reciprocal in equation (7.6) may be neglected, we find that this value defines the *focal length*,

$$f = \left[\frac{\eta_2 - \eta_1}{\eta_1} \left(\frac{1}{R_1} - \frac{1}{R_2} \right) \right]^{-1}, \tag{7.7}$$

in terms of which equation (7.6) yields the **Gaussian lens formula**

Gaussian lens formula

$$\frac{1}{s} + \frac{1}{p} = \frac{1}{f}, \tag{7.8}$$

where we have dropped the indices to give source and image distances for the lens as a whole. For a **converging lens** such as that illustrated in Fig. 7.4, whose overall effect is to deflect rays towards the lens axis, the focal length is positive. In the case of a **diverging lens**, when $(\eta_2 - \eta_2)(1/R_1 - 1/R_2) < 0$, rays will be deflected away from the axis, and the focal length is negative.

Just as we applied equation (7.4) twice to obtain the lensmaker's equation (7.6) for a thin lens, so equation (7.6) or equation (7.8) may be applied successively to determine the effect of a series of lenses forming a telescope, microscope or compound camera lens. If we look through a series of lenses, it transpires, we shall always see some sort of image of whatever lies at the other end; the effect of each lens is simply to change the size and position of that image.

7.3.1 The spherical mirror

Our analysis in Section 7.2 of refraction at a spherical interface may, with a little imagination, also describe the action of a spherical mirror. The reversal of the propagation direction upon reflection, without any change in the refractive index of the medium, may be represented in equation (7.4) by writing $\eta_2 = -\eta_1$ (so that Snell's law gives $\vartheta_2 = -\vartheta_1$) and changing the sign of s_2, giving

$$\frac{1}{s_1} + \frac{1}{s_2} = -\frac{2}{R} = \frac{1}{f}. \tag{7.9}$$

The focal length of a spherical mirror is therefore half the radius of curvature. In equation (7.9), we have maintained the notation that a positive image distance indicates a real image, and that the radius of curvature is positive if the centre of curvature is beyond the mirror surface, and negative if it lies before. For a concave mirror, which focusses parallel light to a real image, R will be negative. It is possible to find, or invent, alternative sign conventions; provided that they are self-consistent, they are equally valid.

Hecht [40] Section 5.4.3

7.3.2 Position–angle transducers

Figures 7.3(b) and (e), and the Gaussian lens equation (7.8), demonstrate an important property of converging lenses and concave mirrors: that parallel rays defined by their angle to the lens or mirror axis ϑ will be focussed to a point whose position $y = f\vartheta$. The plane, a focal length f from the lens, in which the focus is formed is known as the **focal plane**.

It is generally only in the paraxial approximation that the image surface is truly a plane; deviations from the plane are the abberation known as Petzval field curvature.

In a sense that we shall later find rather useful, the lens or mirror therefore maps the angle of incidence ϑ of a ray to a position y within the focal plane – and, conversely, converts a ray position y in the focal plane to a propagation angle ϑ. We shall see in Chapter 9 that this allows *Fraunhofer diffraction* to be observed without the inconvenience of a large physical distance and, in Chapter 14, that position and angle may be regarded as *conjugate variables*.

7.4 Fermat's principle in imaging

Our deductions about geometrical optics followed from the application of Snell's law, which in turn we deduced from Fermat's principle. The path of any ray through an optical instrument is that which takes the least time; the time taken hence shows a turning point with respect to small variations in the path.

We have seen that, by definition, all rays leaving a source S to pass through an optical system will arrive at the corresponding image point P, regardless of the direction in which they leave the source (provided that the ray remains within the optical system). We have used Fermat's principle to determine the route of the ray once it has left the source, but find that it cannot determine the initial direction taken by a ray that reaches the image, for all directions are equally valid. This is not true for an arbitrary end point X $\neq$ P, for which a unique ray path exists and all alternatives require a greater time of travel. The implication, which may be proved explicitly, is that all the ray paths from a source S, regardless of their initial direction, take the same time to reach the image point P.

That the time taken to travel from a source to its image is independent of the path taken is an important general property of imaging systems. It is immediately obvious for the case of a source placed at the centre of curvature of a concave mirror ($s_1 = s_2 = -R$ in equation (7.9)); for a thin lens, it occurs because rays that are inclined to the axis, although longer outside the lens, are shorter within the lens where the refractive index is high.

What we have done in our brief analysis of geometrical optics, then, has been equivalent to considering all of the possible paths by which a disturbance may propagate between two points, and we have thereby found the point at which all the components from a given source add together in phase. This, quite fundamentally, is what we mean by the *image*.

Exercises

7.1   A symmetrical *biconvex* lens (bulging out on both sides) is ground on both sides to form sections of spheres with the same radii of curvature R. Measurements of the refraction of orange–yellow light using a block of the same glass show the sines of the angles of incidence and refraction to be in the ratio 17 to 11. If the lens is found to have a focal length $f = 2.12$ m, calculate the radius of curvature R.

(These data are taken from Newton's description [65] of his investigations of the interference rings now named after him, which we shall examine further in Exercise 8.4.)

7.2   A pair of spectacles contains identical lenses whose outer and inner surfaces have radii of curvature of 10 cm and 7.5 cm, respectively. Both centres of curvature lie on the wearer's side of the spectacles. The lenses are made of high-index polymer with a refractive index of 1.6.

1. Sketch a diagram summarizing the information above, and deduce whether the lenses are *converging* or *diverging*, and therefore whether the wearer is *long-sighted* (hyperopic) or *short-sighted* (myopic).

2. Use the lensmaker's formula of equation (7.6) to determine the focal length f of each lens, and hence determine the range of distances over which the wearer would be able to see sharply without spectacles.

7.3   The eye may be taken to comprise a lens of variable focal length f_e that lies a fixed distance d_e in front of the retina and an adjustable distance d behind the spectacle lens of Exercise 7.2.

1. Sketch this arrangement and show that, if initially parallel rays are brought to a focus on the retina, the focal length f_e will be given by

$$f_e = \frac{d_e(d - f)}{d_e + d - f}. \tag{7.10}$$

2. Show that the spectacle lens alone will cause a distant object that subtends an angle ϑ to form an image of size $h = -\vartheta f$, and hence

that the final image on the retina will be of size

$$h_e = \vartheta \frac{d_e f}{d - f},\qquad(7.11)$$

where a negative value of h_e indicates that the image will be inverted.

3. Show that if the spectacle lens is close to the eye then $h_e \approx -\vartheta d_e$ and estimate the image size on the retina for the 79-mm-high digit of a car registration plate, at a distance of 20 m (the legal eyesight requirement for UK driving), if the diameter of the eyeball is 25 mm.

4. Show that the image will decrease in size if the spectacles are moved away from the eyes. Calculate the distance at which the image will be half of its usual size (i.e. the size when the lenses are close to the eyes), and hence suggest a method of finding the focal length of the spectacle lenses if the only technical instrument available is a ruler or tape measure.

5. How does the use of a *magnifying glass* differ from the arrangement considered above?

7.4 The compound microscope comprises two lenses, known as the *objective* and *eyepiece*, which are separated by a tube or *barrel* of fixed length D that is rather greater than the sum of the focal lengths of the lenses. The objective forms a real image of an object placed a short distance in front of it, and the eyepiece then collimates the rays emerging from this real image so that the image seen by the eye, on the other side of the eyepiece, appears to be infinitely distant. The focal lengths of the objective and eyepiece may be taken to be f_1 and f_2, respectively.

1. Show that, if the eventual image is to be at infinity, then the distance p from the objective to the real image within the instrument will be $p = D - f_2$.

2. Hence show that the *working distance, s,* of the object from the objective will be given by

$$s = \frac{L + f_1}{L} f_1 = f_1 + \frac{f_1^2}{L},\qquad(7.12)$$

where the *tube length, $L \equiv D - (f_1 + f_2)$.*

3. Show that the real image formed within the instrument is therefore magnified, with respect to the object, by a factor $M_1 = -L/f_1$, and hence that an object of length x will be presented to the eye as a distant image subtending an angle ϑ, where

$$\vartheta = -\frac{L}{f_1 f_2} x.\qquad(7.13)$$

7.5 It is common to compare the apparent size of the object when viewed through the microscope with that if it were viewed directly (i.e. without the microscope) from a distance of $L_0 = 10$ inches (254 mm), which typically corresponds to the nearest point at which the eye may comfortably focus.

With reference to your answers to Exercise 7.4, show that the microscope magnifies the object by a factor of $M = -M_1 M_2$, where $M_1 = L/f_1$ is referred to as the *magnification of the objective* and $M_2 = L_0/f_2$ is referred to as the *magnification of the eyepiece*. (These figures are commonly marked on the lens housings.)

7.6   For the standard tube length L of 160 mm, determine the focal lengths of a '×25' objective (i.e. $M_1 = 25$) and a '×10' eyepiece (i.e. $M_2 = 10$).

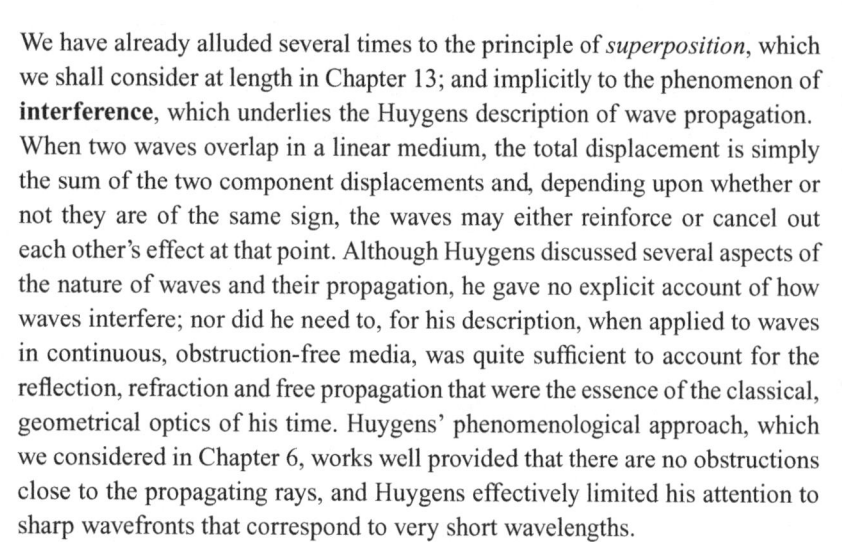

8   Interference

8.1   Wave propagation around obstructions

Pain [66] Chapter 12
French [29] pp. 280–284
Main [60] pp. 320–324
Hecht [40] Chapter 9
Lipson *et al.* [56] Chapter 9
Feynman [23] Vol. I,
pp. 28-6 and 29-5–29-7

We have already alluded several times to the principle of *superposition*, which we shall consider at length in Chapter 13; and implicitly to the phenomenon of **interference**, which underlies the Huygens description of wave propagation. When two waves overlap in a linear medium, the total displacement is simply the sum of the two component displacements and, depending upon whether or not they are of the same sign, the waves may either reinforce or cancel out each other's effect at that point. Although Huygens discussed several aspects of the nature of waves and their propagation, he gave no explicit account of how waves interfere; nor did he need to, for his description, when applied to waves in continuous, obstruction-free media, was quite sufficient to account for the reflection, refraction and free propagation that were the essence of the classical, geometrical optics of his time. Huygens' phenomenological approach, which we considered in Chapter 6, works well provided that there are no obstructions close to the propagating rays, and Huygens effectively limited his attention to sharp wavefronts that correspond to very short wavelengths.

In this chapter, we consider what happens when there are such obstructions close to the propagating rays, and we shall see that Huygens' description continues to assist us, provided that we treat interference properly. The essence of his approach is that we can determine the field at any point by adding the contributions corresponding to the different ways in which the disturbance can reach that point. Specifically, we may take any surface surrounding the point and add the displacements or fields at that surface, taking into account the time taken for their effects to propagate onwards to the point of interest. Depending upon our approach, as discussed in Section 6.2.1, we may have to account for an additional phase and even a geometrical factor when the routes involve scattering through sharp angles, but for most of the situations that we shall consider these subtleties are of no consequence: the phase is a common factor to which our measurements are insensitive, and we shall usually be concerned with only shallow angles (the *paraxial approximation* once again). It is merely necessary to be aware of these constraints, and remember that a more detailed treatment (such as that due to Kirchhoff) may be needed should we ever wish to determine, for example, the field very close to an obstructing object.

G. R. Kirchhoff [53]

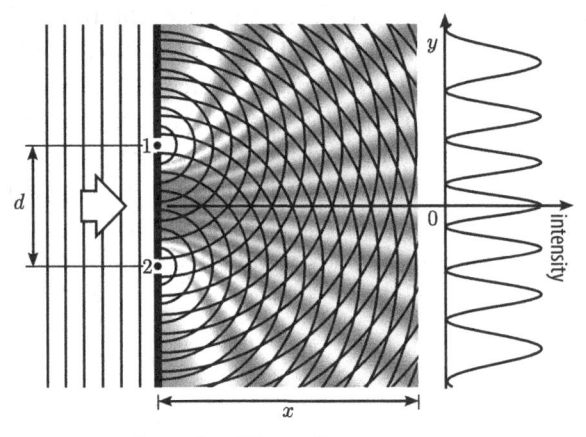

Fig. 8.1
Double-slit experiment, equivalent to that of Thomas Young.

We shall make one simplification in the following chapter, by limiting our-selves at any time to waves of a particular frequency. This allows us to use the phasor method of analysis but, as we shall see later, proves not to be a restriction at all for, provided that the system is linear, any arbitrary wave can be broken down into sinusoidal components.

Huygens could perhaps have addressed the phenomenon of interference himself, for he refers in his treatise to Hooke's magnificent *Micrographia*, published in 1665, in which there is a fine description of the coloured fringes observed in a thin glass wedge; Boyle had described similar observations the previous year, and Grimaldi reported observations of diffraction at about the same time. However, despite the painstaking analysis by Newton of what are now known as *Newton's rings*, permitting what we may in retrospect regard as a determination of the wavelength of orange–yellow light as $2/89\,000$ inch (570 nm), a complete description had to await Young's description of '*the interference of two portions of light*', which we consider next.

R. Boyle [9]
R. Hooke [43]
F. M. Grimaldi [37]
I. Newton [65]

8.2 Young's double-slit experiment

T. Young [94]

One of the most significant experiments in modern physics was recounted by Thomas Young in his lecture to the Royal Society in 1803. Wavefronts of sunlight, whose uninterrupted propagation had been described so successfully by Huygens, were disrupted by an opaque mask that, apart from a small hole divided by a thin piece of card, blocked all but two narrow and closely spaced beams of light. Young's arrangement is equivalent to, and usually described inaccurately as having been, a pair of narrow slits, as shown in Fig. 8.1.

Young observed that the propagated wave was no longer uniform, but instead showed a periodic variation in intensity. The reason, following Huygens' model, is that the mask around the slits blocks all but two of the secondary sources that we should otherwise place along the intercepted wavefront. The propagated disturbance hence corresponds to the phasor sum of the secondary waves from just the two sources labelled 1 and 2 in Fig. 8.1. According to where the wave is observed, the two components may arrive in or out of phase with each other, and their effects alternate between addition and cancellation.

If the centres of the two slits lie at $y = \pm d/2$, and the wavelength of the disturbance is λ, then, following equation (6.2), we may write the amplitudes of the contributions arriving at a point X, with coordinates (x, y), as

$$\psi_{1,2}(x, y) = \frac{\psi_0}{r_{1,2}} \exp(ikr_{1,2}), \tag{8.1}$$

where $k = 2\pi/\lambda$, the amplitude of the disturbance is represented by ψ_0, and the distances to X from the slits are

$$r_{1,2} = \sqrt{x^2 + \left(y \mp \frac{d}{2}\right)^2}. \tag{8.2}$$

The total disturbance will therefore be

$$\psi(x, y) = \psi_1(x, y) + \psi_2(x, y). \tag{8.3}$$

If d is sufficiently small, then we may use the binomial expansion to write r as

$$r_{1,2} = r_0 \sqrt{1 + \frac{\mp yd + d^2/4}{r_0^2}} \approx r_0 \left(1 \mp \frac{d}{2r_0^2}y\right), \tag{8.4}$$

where $r_0^2 \equiv x^2 + y^2$; and we may also approximate $\psi_0/r_{1,2}$ to ψ_0/r_0. The amplitude of the total disturbance will therefore be

$$\psi(x, y) \approx \frac{\psi_0}{r_0} \exp(ikr_0) \left[\exp\left(ik\frac{d}{2r_0}y\right) + \exp\left(-ik\frac{d}{2r_0}y\right) \right]$$
$$= 2\frac{\psi_0}{r_0} \exp(ikr_0)\cos\left(k\frac{d}{2r_0}y\right) \tag{8.5}$$

and, following Section 4.2, the associated intensity may hence be written as

$$I(x, y) \approx I_0 \cos^2\left(k\frac{d}{2r_0}y\right). \tag{8.6}$$

Alternatively, if we define the point X by the angle it subtends with the x axis, $\vartheta \equiv \sin^{-1}(y/r_0)$, then

$$I(\vartheta) \approx I_0 \cos^2(kd \sin \vartheta/2). \tag{8.7}$$

This result describes exactly the intensity of light observed by Young after the double slit, and differs from the smooth distribution that would have been expected had light been composed of a stream of particles. Young's experiment thereby confirmed the wave-like nature of light, whereby contributions arriving via different paths (the two slits) could subtract from each other as well as add. These processes are known as **destructive** and **constructive interference**.

8.2.1 Achieving true stereo sound reproduction

When we listen to stereo recordings through headphones, each ear hears exactly what it should: the left ear receives only the sound from the left headphone, and the right only that from the right. If the recording has been mixed correctly, then the listener will hear an accurate representation of the spatial distribution of the original sources, which the brain infers from the relative intensities and timings of the sounds as they arrive at the ears.

If, as is perhaps more common, we listen to recordings through a pair of loudspeakers, then the effect is imperfect, for each ear hears sounds originating at both loudspeakers. It may be, of course, that the left loudspeaker is better heard by the left ear, but in general there will be considerable *cross-talk*. The result is that, although a degree of stereo imaging persists, the effect is far less pronounced than when using headphones; in general, sources that with headphones appeared to be spread all around the listener now seem all to be located between the two loudspeakers.

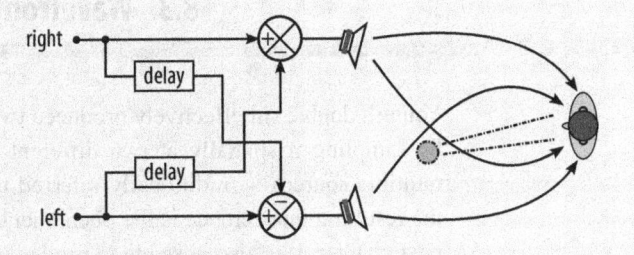

Fig. 8.2

Cross-talk compensation in stereo audio systems.

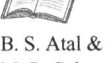
B. S. Atal &
M. R. Schroeder [2]

Interference allows this defect, at least in principle, to be corrected, for we can use the principle of Young's double-slit fringes to achieve a cancellation of the sound from the left channel that reaches the right ear, and vice versa. We achieve this, as illustrated in Fig. 8.2, by feeding the left channel to both loudspeakers – just as in Young's experiment the same wave arrived at both slits – but we introduce a phase shift, whose effect is to shift the fringe pattern to the left or right, so that a node in the interference pattern coincides with the position of the right ear. The left channel is thus heard only by the left ear. The converse arrangement similarly limits the right channel to the right ear.

Sound, of course, comprises a range of frequencies, each of which requires the appropriate phase difference for this effect to work. Fortunately, the principle can be understood equally simply for arbitrary signals: our phase shift is actually a delay, sufficient to cancel out that with which simultaneous sounds would arrive at the right ear, together with an inversion (multiplication by −1). Sound from the left channel therefore reaches the right ear via both loudspeakers at the same time, but with opposite signs, cancelling out identically. The right channel is then added, processed in the same way with a delay and inversion before it is fed to the left-hand loudspeaker. The result is that the left ear is positioned at a node for the right channel, and the right ear at a node for the left channel.

There are, of course, some disadvantages to this trick. First, there is a slight delay between the two versions of the left channel heard by the left ear: not sufficient to be apparent as an echo, but enough to disturb the purists. More critical is that the head must be positioned precisely with respect to the loudspeakers. For these reasons, most implementations attenuate the correction, achieving a smaller but still significant enhancement over a wider range of conditions. This is the mechanism behind the 'wideness' or 'surround' setting on portable 'ghetto-blasters', and was the starting point behind more recent developments in 'surround-sound' technology.

8.3 Wavefront dividers

Young's double slit effectively produced two sources from the same wavefront, by sampling it spatially at two different points. This method of producing multiple sources is traditionally referred to as *division of the wavefront*, and the resulting interference is the phenomenon of *diffraction* – the subject of the next chapter. It is also possible to produce two waves from a single wavefront by *division of amplitude* by using devices that reflect a fraction of the wave while transmitting the remainder. The distinction between these two processes is neither absolute nor important, but rather indicates a pragmatic classification of traditional interference geometries. We shall make little further reference to these distinctions but, since they are referred to in many textbooks, the following brief explanation is offered.

Interference by division of the wavefront is regarded as a method for comparing different parts of the wave at a given time: the sampling points are reached simultaneously by the wavefront, but are displaced in a direction *transverse* to the wave propagation. For stable fringe patterns to be observed, the two samples must have a fixed phase relationship, or *coherence*. Instruments that use division of the wavefront are therefore sensitive to the **transverse** (or **spatial**) **coherence** of the wave: the extent to which laterally separated points maintain a fixed relative phase, which may be shown to depend upon the shape and size of the wave source. It follows that interference between sources generated by division of the wavefront may be used to measure the size of a source – as in the *Michelson stellar interferometer* or, arguably, any imaging system including telescopes and microscopes.

Interference by division of amplitude, on the other hand, is generally used to compare parts of the wave that are separated in their *times of arrival* at the *same* point in space – that is, they are displaced in the *longitudinal* direction, but not in the transverse. Instruments that employ this process are therefore sensitive to the **longitudinal** (or **temporal**) **coherence** of the wave – the extent to which longitudinally separated points maintain a fixed relative phase. The longitudinal coherence of a wave may be shown to depend upon the wave *spectrum*. Division of amplitude is therefore used to determine the spectral properties of a source, in optical instruments such as the *Michelson interferometer*, *Fabry–Perot étalon* and *interference filter*.

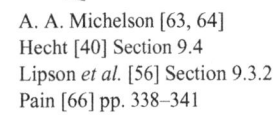

Hecht [40] Sections 9.2 and 9.3
Pain [66] pp. 333–340

Although the distinction between these two categories of beamsplitter is therefore of practical significance, examination of the detailed mechanisms may sometimes render the separation rather less clear: *diffraction gratings*, for example, may fall into either category, and almost all interference is to some extent affected both by the spectrum and by the size of the source. Fundamentally, however, this is of no great significance, for in all cases we are simply sampling and comparing parts of the wave motion at different points in space.

8.4 The Michelson interferometer

To illustrate interference through the division of amplitude, and its use in investigating the spectral properties of a wave source, we consider the **Michelson interferometer** – an optical instrument, shown in Fig. 8.3, that is commonly used to investigate narrow-band emissions in the visible and infrared regions of the spectrum. The heart of the interferometer is a *beamsplitter*, typically comprising an optically flat plate of glass on which has been deposited a layer of metal that is sufficiently thin to be partially transmitting. Incident light strikes the beamsplitter at an angle of $45°$; part of the light is transmitted, while the remainder is reflected sideways at $90°$.

A. A. Michelson [63, 64]
Hecht [40] Section 9.4
Lipson *et al.* [56] Section 9.3.2
Pain [66] pp. 338–341

The transmitted and reflected beams subsequently strike the mirrors A and B, which reflect them back upon themselves towards the beamsplitter where, as before, they may be transmitted or reflected. The result is that incident light may travel through the interferometer via either mirror A or mirror B. The wave therefore has two routes to reach P from S, and the total wave amplitude at P is given by the phasor sum of the components emerging from each of these routes. If the times taken to follow the two routes were equal, then the corresponding wave components would always arrive at P with the same phase and we should find nothing remarkable; the Michelson interferometer therefore works by introducing a difference between the two path lengths.

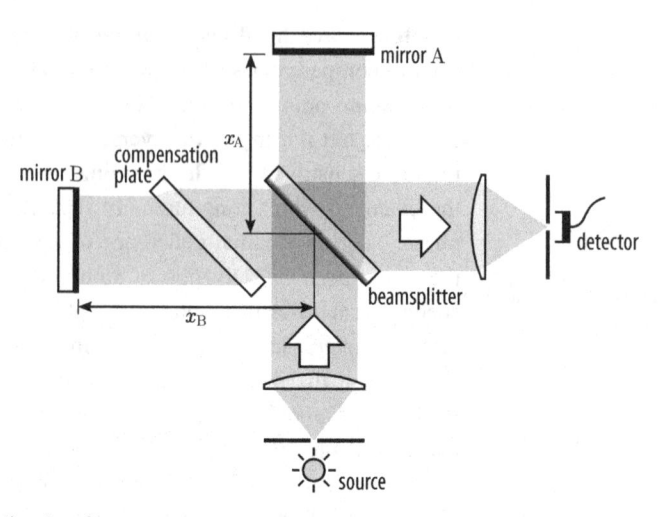

Fig. 8.3   The Michelson interferometer.

If the incident light is monochromatic – that is, of a single wavelength λ – then the electric field of the travelling light wave may be written as

$$E(x, t) = E_0 \cos(kx - \omega t), \tag{8.8}$$

where $k \equiv 2\pi/\lambda$, $\omega = ck$ and x is the distance travelled since leaving the source. If the beamsplitter reflects a fraction r of the incident amplitude and transmits a fraction t, then the electric field components reaching the detector via the two mirrors will be

$$E_{A,B}(t) = rtE_0 \cos(k(x_0 + 2x_{A,B}) - \omega t), \tag{8.9}$$

where x_0 is the common path length excluding the excursions from the beamsplitter to the two mirrors, x_A and x_B are, respectively, the distances of mirrors A and B from the centre of the beamsplitter face, and we neglect for the time being any glass etc. through which the beams may pass. The superposed electric field strength at the detector may hence be written as

$$
\begin{aligned}
E(t) &= E_A(t) + E_B(t) \\
&= rtE_0 \left[\cos\left(k(x_0 + 2x_A) - \omega t\right) + \cos\left(k(x_0 + 2x_B) - \omega t\right) \right] \\
&= 2rtE_0 \cos\left(k(x_0 + x_A + x_B) - \omega t\right) \cos\left(k(x_A - x_B)\right).
\end{aligned}
\tag{8.10}
$$

The intensity of the light falling upon the detector will therefore be

$$I = 4RT\, I_0 \cos^2\left(k(x_A - x_B)\right), \tag{8.11}$$

where I_0 is the initial intensity, $R \equiv r^2$ is the intensity reflectivity of the beamsplitter and $T \equiv t^2$ is the intensity transmissivity. Conservation of energy requires that $R + T \leq 1$; the maximum value of $4RT$ is therefore unity.

If the intensity is recorded as one of the mirrors is slowly moved towards or away from the beamsplitter, a series of sinusoidal fringes will therefore indicate that the source is monochromatic, with a wavenumber related to the fringe periodicity. If the source emits a range of wavelengths, then we shall record the superposition of the corresponding sinusoidal patterns: when the path difference $2(x_A - x_B)$ is zero, these patterns will coincide and large intensity changes will be observed, while at large path differences they will fall out of step and the intensity will tend to an average value. A detailed recording of the transmitted intensity as a function of path difference proves to contain complete information about the spectrum of the source, with a resolution that depends upon how far the mirror is displaced.

Some technical subtleties merit consideration. First, it should be noted that there is, in principle, no mechanism by which light may be absorbed by the instrument; all the incident light is therefore either transmitted to P or reflected to S. The beams reflected by A and B may be either transmitted or reflected upon their second encounter with the beamsplitter, so there are not only two routes by which the light may reach P, but also two by which it can return to S. In both places we therefore see a wave that results from interference between two paths round the instrument, and the crucial point is that the phase differences themselves differ by $180°$ – a fundamental property of any form of beamsplitter. When the mirror displacement is such as to produce a maximum in the intensity at P, there will therefore be a minimum in the intensity reflected to S, and vice versa. A corollary of this is that the maximum intensity that can emerge from either arm of the instrument is the same as the incident intensity, even though all paths through the instrument involve passing twice through the partially transmitting beamsplitter.

It might be thought that the beamsplitter must transmit and reflect equally for the instrument to operate correctly, but equation (8.11) shows that the signal has the same form in all cases. Both the path via A and that via B involve one transmission and one reflection at the beamsplitter, so any imbalance between R and T simply reduces the overall magnitude of the signal at P. The paths back to S are dissimilar, however, so an imbalance prevents complete cancellation at S: a non-equal beamsplitter causes a reduction in the modulated signal reaching both exits, because an unmodulated part of the wave is always reflected back to the source.

A further imbalance between the two paths is due to the difficulty of making extremely thin beamsplitters, which instead must be coated onto one surface of a glass plate. The path via mirror A involves three passes through this glass substrate, whereas the path via mirror B involves just one. The refractive index of the glass results in an apparent offset of the mirror displacement; and *dispersion* can cause the apparent mirror position for zero path difference to vary with wavelength. The first of these effects is of little practical consequence, but the second can cause significant complications, particularly with spectrally broad sources. It is therefore common to insert a compensation plate, with the same dimensions and orientation as the beamsplitter substrate, into the path B, so that both paths involve the same number of passes through a substrate.

Rays can also pass through the instrument inclined at small angles ϑ to the beam axis; equation (8.11) hence more generally becomes

$$I = I_0 \cos^2\left(k(x_B - x_A)\cos\vartheta\right) \tag{8.12}$$

and the intensity is seen to depend upon the inclination angle. If the output beam is projected by a lens onto a screen at its focus, then concentric circular fringes will be observed.

It is little noted that there is a further reason for the compensation plate, which for spectrally narrow sources would otherwise appear barely necessary. This is that the tilted plate introduces *astigmatism*, which is different for the two arms, and which has the rather alarming effect of rendering the usually circular fringes elliptical or hyperbolic.

Exercises

8.1 Two identical loudspeakers are placed 10 m apart, and play a single frequency of 440 Hz at the same power. An observer, facing and equidistant from the loudspeakers, is exactly 100 m away from a point midway between them. If he moves to either his left or his right he hears the sound become progressively quieter until, when he has moved 3.75 m from his initial position, the sound intensity is a minimum. Determine the wavelength of the sound waves, and hence the speed of sound in air under the prevailing conditions.

8.2 The amplitude of the light transmitted by a Michelson interferometer may be determined by summing the amplitudes resulting from the two routes through the instrument. If the partially reflecting beamsplitter divides incident light equally between the two paths, the difference in path length between the two routes for normal incidence is s, and the rays of wavelength λ make an angle ϑ to the mirror normals, derive expressions for the relative amplitudes of these two contributions.

Hence derive the overall fractions of the amplitude and intensity that are transmitted by the interferometer, as functions of λ, s and ϑ.

8.3 Two long transmitting slits of width a are separated by a non-transmitting region of width $(d - a)$. Show that, when the slits are illuminated by a parallel beam of monochromatic light at normal incidence, the intensity distribution of the light leaving the slits at an angle ϑ is given by

$$I(\vartheta) = I_0\left(\frac{\sin\alpha}{\alpha}\cos\beta\right)^2, \tag{8.13}$$

where $\alpha = (ka/2)\sin\vartheta$, $\beta = (kd/2)\sin\vartheta$ and $k = 2\pi/\lambda$. Sketch the intensity distributions for the special cases $d = 3a$, $d = 2a$ and $d = a$.

8.4   The symmetrical biconvex lens of Exercise 7.1 is placed so that one of
      its convex faces rests upon a flat piece of glass, and is illuminated from
      above with bright yellow light. When the lens is viewed from above, a
      concentric series of finely spaced bright and dark rings is observed.
      1. Explain the origin of the observed rings, and the reason why the centre
         of the fringe pattern is dark rather than bright.
      2. Show that, if the wavelength of the orange–yellow light is λ, the radius
         r_n of the nth dark fringe will be approximately given by

         $$r_n \approx \sqrt{nR\lambda}. \tag{8.14}$$

      3. The radius of the fifth dark ring is measured to be 2.57 mm. Using
         your result for the radius of curvature R from Exercise 7.1, deduce the
         wavelength of the orange–yellow light, and estimate the precision of
         your result.
      (These data are taken from Newton's description [65] of his investigations
      of the interference rings now named after him.)

Fraunhofer diffraction

9.1 More wave propagation around obstructions

Pain [66] pp. 366–407
French [29] pp. 284–298
Main [60] pp. 320–339
Hecht [40] Chapter 10
Lipson *et al.* [56] Chapter 8
Feynman [23] Vol. I,
Chapter 30

When a wave has a choice of paths by which to reach a given destination, we have seen in Chapter 8 that the components add or subtract according to their relative phases upon arrival. When the number of contributing paths is small, we generally refer to their *interference*; when the number is large, it is more common to describe this as *diffraction*.

As Feynman remarks in his *Lectures on Physics*, however, '*no one has ever been able to define the difference between interference and diffraction satisfactorily*'. Inasmuch as the interference fringes observed by Young were the **diffraction pattern** of the double slit, this chapter is therefore a continuation of Chapter 8. On the other hand, few opticians would regard the Michelson interferometer as an example of diffraction. Some of the important categories of diffraction relate to the interference that accompanies division of the wavefront, so Feynman's observation to some extent reflects the difficulty that we may have in distinguishing between *division of amplitude* and *division of wavefront*. There are nonetheless important classes – notably *Bragg diffraction* – in which the longitudinal dimension plays a crucial rôle.

Fortunately, perhaps, the **Fraunhofer diffraction** with which we shall generally be concerned here, which is one of the most important types of diffraction in general, occurs under particular conditions in which the confusion between division of amplitude and division of wavelength is often almost eliminated, and in which we are usually concerned with a very large number of alternative paths from the wave source to its destination. Fraunhofer diffraction, according to the most formal definition, is that which *is observed in the image plane of the source*, and we have already noted that all optical paths from the source to its image are equal in length. Fraunhofer diffraction, therefore, is almost entirely concerned with the transverse disruption, or division, of the wavefront. It *tends* to be associated with plane waves, small angles and distant images, although these are not essential. We shall examine further the definition of Fraunhofer diffraction in Section 9.6.

Diffraction involving division of the wavefront but falling beyond the Fraunhofer regime is known as **Fresnel diffraction**, and typically includes that observed close to the diffracting objects, near to the source, and so on. These are the conditions in which the distinction between diffraction and just plain

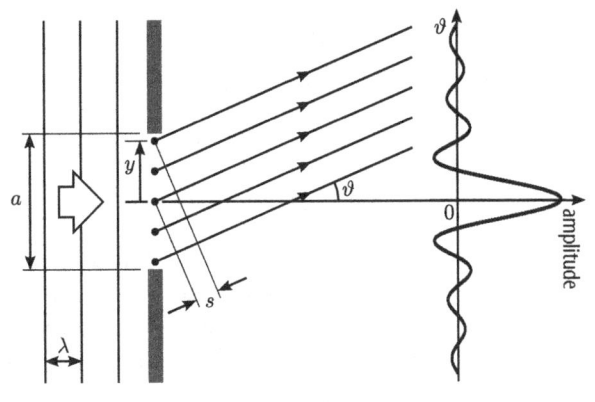

Fig. 9.1 Diffraction by a single slit in an otherwise opaque mask.

interference is less clear. **Bragg diffraction**, which we shall consider briefly in Section 15.5, further challenges this distinction. The problem is only one of nomenclature, however: the physics is simply that waves arriving via different routes will exhibit the effects of interference.

The algebra of Fresnel diffraction, however, is generally rather messy, and we shall reserve Bragg diffraction until our examination of waves in three dimensions in Chapter 15. In this chapter we shall therefore restrict our attention to diffraction of the Fraunhofer type.

9.2 Diffraction by a single slit

In our treatment in Section 8.2 of the interference pattern in light from a double slit, we have already examined one of the classic diffraction arrangements. For our second example, we shall consider diffraction by a single slit alone, but while each slit of Young's pair could be considered to be infinitely narrow, we must now allow the slit some width. We shall therefore model the light passing through our single, broad slit as if it came separately from a stack of narrow slits that together account for the same overall aperture.

We adopt a geometry, characteristic of Fraunhofer diffraction, in which a flat mask is illuminated with plane waves and observed after the diffracted waves have travelled a distance that is large in comparison with the diffracting object. Paths from the object to the observer will therefore be approximately parallel, and we shall describe the diffraction pattern through its dependence upon the angle ϑ through which the wave is diffracted, as in equation (8.7).

As with Young's double slit, our analysis uses Huygens' method, and we simply sum the contributions from secondary sources that are imagined to lie in the plane of the mask, as shown in Fig. 9.1. The slit, of width a, is illuminated at right angles with plane waves of wavelength λ, and the diffracted light is

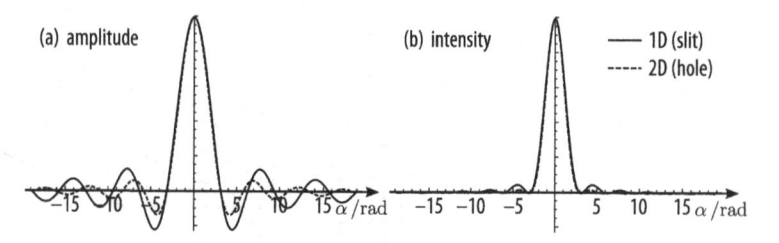

Fig. 9.2 Normalized diffraction-pattern amplitudes and intensities for apertures in one (slit) and two (circular hole) dimensions. The one-dimensional pattern is a sinc function $\sin\alpha/\alpha$ with $\alpha = (kd/2)\sin\vartheta$; the two-dimensional pattern $2J_1(\alpha)/\alpha$ is shown for an aperture of diameter $1.22d$ to give a central maximum of the same width.

observed at an angle ϑ. Taking as our reference the path via the centre of the slit, the path via a source with position y is longer by a distance $s(y) = -y\sin\vartheta$. The phasor for such a path is therefore represented by the complex quantity

$$\psi(y) = \psi_0 \exp(-iky\sin\vartheta), \tag{9.1}$$

where ψ_0 represents the observed amplitude, which is assumed to be the same for all secondary sources within the slit. The diffracted amplitude is now found by adding the contributions from all such sources,

$$
\begin{aligned}
\psi(\vartheta) &= \int_{-a/2}^{a/2} \psi_0 \exp(-iky\sin\vartheta)\,dy \\
&= \left[\frac{\psi_0 \exp(-iky\sin\vartheta)}{-iky\sin\vartheta} \right]_{-a/2}^{a/2} \\
&= \frac{2\psi_0}{k\sin\vartheta} \sin\left(\frac{ka\sin\vartheta}{2} \right) \\
&= \psi_0 a \frac{\sin\alpha(\vartheta)}{\alpha(\vartheta)},
\end{aligned}
\tag{9.2}
$$

where we have written $\alpha(\vartheta) \equiv (ka/2)\sin\vartheta$. The function $\sin\alpha/\alpha$ is known as the **sinc** function, and is plotted in Fig. 9.2(a). This, then, with appropriate scaling, is the amplitude of the diffraction pattern from the single slit; the intensity varies simply in proportion to its square, as shown in Fig. 9.2(b).

Although we have restricted our analysis to a mask that, along the direction of the slit, has translational symmetry and therefore requires only a one-dimensional integral, our treatment may be readily extended at slightly increased mathematical complexity to more arbitrary shapes. The two-dimensional equivalent of the one-dimensional slit is a circular aperture, and we find that the diffraction pattern in this case is an **Airy pattern**, shown in Figs. 9.2 and 9.3(b) and given by

$$I(r) = I_0 \left[\frac{2J_1(\alpha)}{\alpha} \right]^2, \tag{9.3}$$

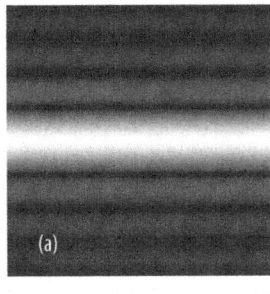
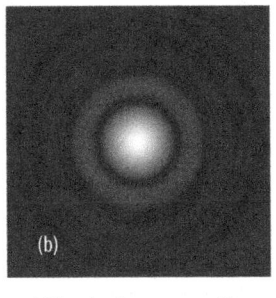

(a)   (b)

Fig. 9.3 Simulated diffraction patterns from (a) a horizontal slit and (b) a circular aperture. The weaker regions have been enhanced to render them visible.

where $\alpha \equiv (ka/2)\sin\vartheta$ and $J_1(\alpha)$ is a *Bessel function of the first kind*. This qualitatively resembles a sinc function of the radial coordinate, although the nodes of the Airy pattern lie at slightly different positions.

9.2.1 Phasors and the Cornu spiral

We have seen in Section 6.2 that the contributions from a propagating plane wavefront to the disturbance at a point a given distance ahead of it may be written mathematically as the *Fresnel integral* and represented graphically as the phasor summation known as the *Cornu* (or *Euler*) *spiral* or *clothoid*, which is reproduced in Fig. 9.4. Provided that the evaluation point is more than a few wavelengths ahead of the wavefront, this spiral proves always to take the same shape and is simply scaled according to the wavelength and distances involved. The varying curvature of the spiral occurs because the paths from the wavefront's secondary sources to the observation point are at different angles; this is therefore an example of *Fresnel diffraction* and does not satisfy the *Fraunhofer* criteria.

When the line of secondary sources is curtailed by the presence of a slit as in Fig. 9.1, the integral and Cornu spiral are also truncated. Figure 9.4 shows an example, in which the dark section of the spiral corresponds to the contributions from unobstructed sources. Instead of joining the ends of the full spiral, the phasor representing the overall disturbance now connects the ends of the truncated part. As the slit is narrowed, or the observation point is placed further away from the wavefront, the truncated spiral becomes ever flatter until, in what turns out to be the regime of *Fraunhofer diffraction*, the paths from the secondary sources to the observation point are essentially parallel and equal in length, and the phasor components stretch out in a straight line.

This is true, however, only if the observation point is directly behind the centre of the slit. If it is placed to one side, then, although the paths from the secondary sources remain parallel, they are no longer perpendicular to the wavefront; instead, they all make the same angle ϑ to the propagation direction and, as we have already illustrated in Fig. 9.1, the relative distance s to the

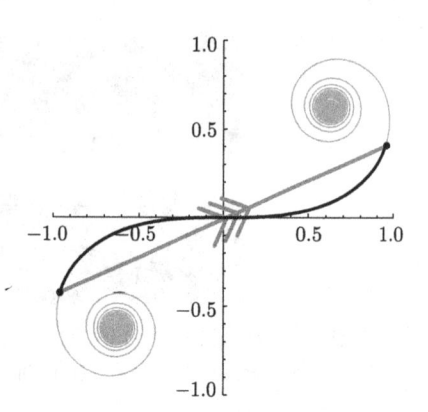

Fig. 9.4 Phasors from sources across an uninterrupted wavefront form a Cornu (or Euler) spiral. An aperture enclosing only the central sources contributes only the (dark) central section of the spiral, whose resultant is shown.

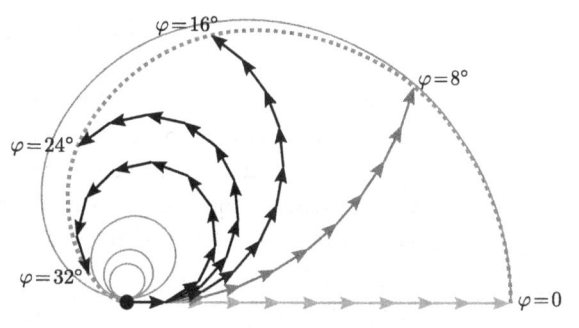

Fig. 9.5 Phasors corresponding to the diffracted amplitude for a single slit, shown for several phase angles φ. The dotted curve shows the locus traced out by the resultant; the smooth curve is the locus for infinitesimal phasors.

observer therefore depends upon the position of the secondary source. Instead of parallel phasors that form a straight line, each phasor will be rotated relative to its neighbour, and the resultant will be the section of a circle that becomes more tightly curved as the angle of observation ϑ is increased.

Figure 9.5 shows the result with just $N = 10$ secondary sources within our slit, for various values of the phase increment $\varphi \equiv k\,\Delta s$ between adjacent phasors. The arcs are all the same length, but their resultants shrink with increasing curvature until (at $\varphi = 360°/N$ with N sources) the arc forms a complete circle and the resultant is zero. This corresponds to the first zero of the sinc function of Fig. 9.2. If the arc is tightened further, part of it will continue beyond the full turn of the circle, and the resultant will grow and then diminish until the arc makes two complete turns of a rather smaller circle, corresponding to the second zero of the sinc function. The locus of the phasor resultant is shown dotted in Fig. 9.5: as φ is increased, the distance from the origin to the point on the locus is proportional to the overall amplitude of the diffracted

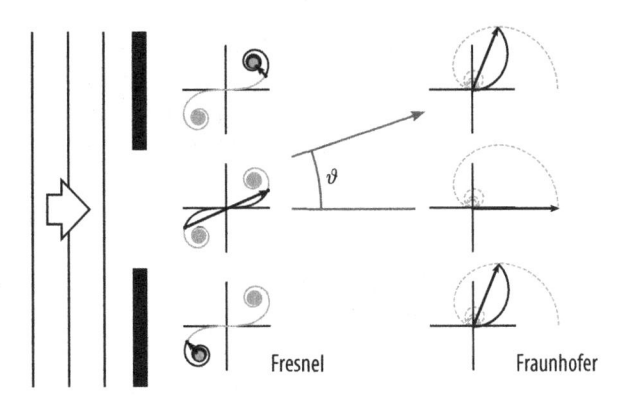

Fresnel   Fraunhofer

Fig. 9.6   Phasor constructions on and off axis in the near-field (Fresnel) and far-field (Fraunhofer) cases. Note that, if referred in phase to a path through the slit centre, the Fraunhofer resultants will be rotated to the horizontal.

wave. The solid curve, known as a *cochleoid*, shows the result when the number N of secondary sources is increased to reduce their spacing towards zero.

The phase angle φ depends, of course, upon the spacing a/N between our N imaginary sources, but may be written in terms of the viewing angle ϑ since $s \equiv -y \sin \vartheta$, giving $N|\varphi| \equiv ka \sin \vartheta \equiv 2\alpha$; the locus parameter hence corresponds directly to the argument of the sinc function of equation (9.2).

Figure 9.6 summarizes the phasor pictures on and away from the diffraction axis for the near-field (Fresnel) and far-field (Fraunhofer) cases. It is interesting to consider the phase of the phasor resultant with respect to the individual contribution from a secondary source at the centre of the aperture, for this changes by $45°$ (for our one-dimensional aperture; $90°$ for two dimensions) between the near and far fields. This is again the Gouy shift that we met in Section 6.2.

An important feature of the far-field case is that, as we increase the width of the slit and thereby approach an uninterrupted wavefront, the line of parallel phasors grows in length, and even the slightest curvature causes it to curl around and give a much smaller resultant. In the limit of an unobstructed beam, the amplitude and intensity of the diffracted wave are therefore zero except for $\vartheta = 0$, just as we would expect for a propagating plane wave. This has an important consequence in *Babinet's principle*, as we shall now see.

9.3 Babinet's principle

We have seen in Section 6.1 that Huygens' approach satisfactorily describes the propagation of an unobstructed plane wave through a uniform medium. Wavefronts that, for example, are initially aligned with the plane $x = 0$ will remain parallel to this plane and the wave will propagate along the x axis.

If we determine the wave amplitude as a function of the viewing angle ϑ relative to the x axis (e.g. by projecting the wave onto a detector array using a lens as an angle–position transducer as in Section 7.3.2), we shall find that the sum of all the secondary waves is zero for all observation angles except $\vartheta = 0$, and the wave amplitude will show a sharp peak at $\vartheta = 0$ with zero elsewhere – a shape that we shall in Chapter 14 refer to as a *Dirac δ-function*.

This is an important result, for it may be combined with our ideas of diffraction to yield something rather less obvious. We have seen in Section 9.2 that the diffraction pattern when the plane wavefront is obstructed is given by the wave contributions from the unobstructed Huygens sources; if we now add the contributions from the *obstructed* sources, then we reconstruct the *unobstructed wavefront* and the sum of all the secondary waves will be zero for all observation angles except $\vartheta = 0$. The wave diffracted from just the *obstructed* sources will, except along the x axis, therefore be equal but opposite in amplitude to that from the *unobstructed* sources, and will give (again, except along the x axis) an identical diffraction pattern. Apart from a narrow peak at $\vartheta = 0$, the diffraction pattern from a circular aperture, illustrated in Figs. 9.2 and 9.3(b), will therefore be identical to that around a circular mask of the same diameter; if our aperture is a hole in a plane mirror, for example, the diffraction pattern for the reflected light will be identical to that for the light that is transmitted.

Babinet's principle, that the Fraunhofer diffraction patterns of complementary masks are (except at $\vartheta = 0$) identical, proves to be of great utility in the calculation and interpretation of diffraction patterns. It also offers helpful insight into the phenomenon of **Poisson's bright spot**, the observation of which was another key indication of the wave nature of light. Poisson, reviewing Fresnel's work for the French Académie des Sciences, realized that an implication of his theory of diffraction would be the presence of a bright spot in the centre of the shadow of a small disc, corresponding to the Airy diffraction pattern of a circular aperture of the same size. The Académie committee was sceptical, generally favouring the corpuscular theory of light, which would have predicted a simple geometrical shadow, but had the scientific integrity to seek experimental investigation. This was carried out under Arago, the bright spot was indeed observed, and Fresnel was subsequently awarded the *Grand Prix* of the Académie.

Hecht [40] Section 10.3
Lipson *et al.* [56] Section 8.4.3

9.4 The diffraction grating

Huygens' theory of wave propagation, and its refinements by Kirchhoff, Fresnel and Fraunhofer, have already allowed us to explain experimentally observed diffraction patterns. As is usual in the development of a theory, the first efforts are to confirm the principles and investigate its mechanisms; then, once confidence and understanding have been achieved, it becomes a tool for the design of useful applications. One of the most significant designs to apply the theory of wave interference is the diffraction grating: a regular series of narrow slits or rulings that may be regarded as an extension of Young's double slit, but whose practical utility is far greater. Diffraction gratings are available both in transmissive and in reflective forms, the latter being equivalent to the combination of a transmissive grating with a plane mirror. We consider here the transmissive form; the only difference in principle is the direction of illumination.

The diffraction grating is shown schematically in Fig. 9.7. An array of slits, each of width w and with centres separated by d, fill an aperture of width a. Light of wavelength λ is incident at right angles on the array, and diffracted light is observed a large distance away at an angle ϑ to the direction of incidence. Proceeding exactly as in Section 9.2, we may write the diffracted amplitude as

$$\psi(\vartheta) = \int_{-a/2}^{a/2} \psi_0 t(y) \exp(-iky \sin \vartheta) \, dy, \tag{9.4}$$

where $t(y)$ is the amplitude transmission (here always either 0 or 1) of the grating at a distance y from the axis and ψ_0 is proportional to the incident amplitude. Explicitly, this hence becomes

$$\psi(\vartheta) = \int_{y_1-w/2}^{y_1+w/2} \psi_0 t(y) \exp(-iky \sin \vartheta) \, dy$$
$$+ \int_{y_2-w/2}^{y_2+w/2} \psi_0 t(y) \exp(-iky \sin \vartheta) \, dy$$
$$+ \int_{y_3-w/2}^{y_3+w/2} \psi_0 t(y) \exp(-iky \sin \vartheta) \, dy + \cdots, \tag{9.5}$$

where $y_1, y_2, y_3, \ldots$ are the coordinates of the centres of the slits, taken in turn across the whole aperture. This expression may be simplified by realizing that each integral is identical but for a phase factor $\exp(-iky_n \sin \vartheta)$:

$$a(\vartheta) = \exp(-iky_1 \sin \vartheta) \int_{-w/2}^{w/2} \psi_0 t(y) \exp(-iky \sin \vartheta) \, dy$$
$$+ \exp(-iky_2 \sin \vartheta) \int_{-w/2}^{w/2} \psi_0 t(y) \exp(-iky \sin \vartheta) \, dy$$
$$+ \exp(-iky_3 \sin \vartheta) \int_{-w/2}^{w/2} \psi_0 t(y) \exp(-iky \sin \vartheta) \, dy + \cdots. \tag{9.6}$$

This may be written as the product of an integral and a geometric series

$$a(\vartheta) = \left(\sum_{n=-N/2}^{N/2} \exp(-iknd \sin \vartheta) \right) \psi_0 \int_{-w/2}^{w/2} t(y) \exp(-iky \sin \vartheta) \, dy, \tag{9.7}$$

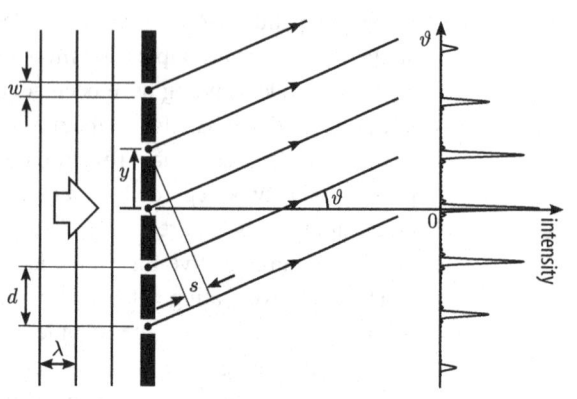

Fig. 9.7 The transmissive diffraction grating and its far-field (Fraunhofer) diffraction pattern.

where we have substituted

$$y_n \equiv nd, \tag{9.8}$$

with $n = -N/2 \to N/2$, and $Nd = a$. The geometric progression is evaluated in the usual fashion, giving

$$
\begin{aligned}
\sum_{n=-N/2}^{N/2} & \exp(-iknd \sin \vartheta) \\
&= \frac{\exp\left(ik\left(\frac{N}{2}+1\right)d \sin \vartheta\right) - \exp\left(-ik\frac{N}{2}d \sin \vartheta\right)}{\exp(ikd \sin \vartheta) - 1} \\
&= \frac{\exp\left(ik\frac{N+1}{2}d \sin \vartheta\right) - \exp\left(-ik\frac{N+1}{2}d \sin \vartheta\right)}{\exp\left(\frac{1}{2}ikd \sin \vartheta\right) - \exp\left(-\frac{1}{2}ikd \sin \vartheta\right)} \\
&= \frac{\sin\left(\frac{N+1}{2}kd \sin \vartheta\right)}{\sin\left(\frac{1}{2}kd \sin \vartheta\right)}.
\end{aligned}
\tag{9.9}
$$

If we count the number of slits from $-N/2$ to $N/2$, we find that there are in fact $N' = N + 1$, since there is one with the number $n = 0$. Equation (9.9) may therefore be written more helpfully as

$$\sum_{n=-N/2}^{N/2} \exp(-iknd \sin \vartheta) = \frac{\sin\left(\frac{N'}{2}kd \sin \vartheta\right)}{\sin\left(\frac{1}{2}kd \sin \vartheta\right)} \tag{9.10}$$

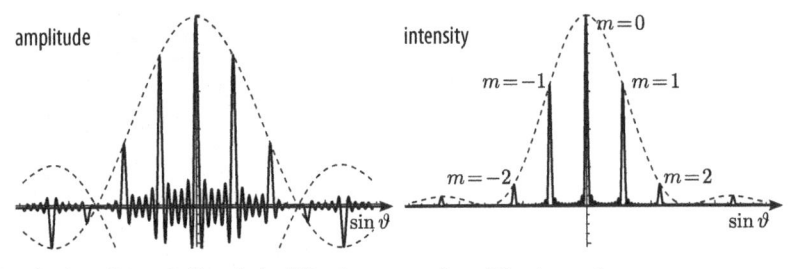

Fig. 9.8 Amplitude and intensity Fraunhofer diffraction patterns for a diffraction grating.

and, using the result for the integral calculated in Section 9.2, the diffracted amplitude is hence

$$\psi(\vartheta) = \frac{\sin\left(\dfrac{N'}{2}kd\sin\vartheta\right)}{\sin\left(\tfrac{1}{2}kd\sin\vartheta\right)}\,\psi_0 d\,\frac{\sin\left(\tfrac{1}{2}kw\sin\vartheta\right)}{\tfrac{1}{2}kw\sin\vartheta}. \tag{9.11}$$

The diffracted amplitude is therefore that for a single slit of width w, multiplied by the function of equation (9.10), which turns out to be a series of narrow peaks occurring every time that the denominator approaches zero – i.e. when

$$\frac{kd\sin\vartheta}{2} = m\pi. \tag{9.12}$$

If we write k in terms of the wavelength λ, we hence obtain

$$d\sin\vartheta = m\lambda \tag{9.13}$$

and the integer m serves as a label for each **diffraction order**.

We shall see in Chapter 14 that the shape of the peak corresponding to each order is simply the diffraction pattern for a single slit of width a – the overall width of the diffraction grating. The diffracted amplitude and intensity distributions for a typical diffraction grating are shown in Fig. 9.8.

The utility of the diffraction grating rests in its ability to direct different wavelengths to different angular positions, as we may see by rearranging equation (9.13) to give

$$\sin\vartheta = \frac{m\lambda}{d}. \tag{9.14}$$

A beam of light containing several wavelengths is therefore *dispersed* so that an appropriately placed photographic film or detector will record its spectrum. Usually, a lens is used to focus the parallel rays that leave the grating at a given angle onto a particular point on the film or detector, as described in Section 7.3.2. In this arrangement, the diffraction is recorded at the position where an image of the incident beam of light (for example, from a distant point source) would have been focussed. We therefore record the diffracted wave in the *image plane of the source* – which we mentioned earlier to be the criterion for the observation of Fraunhofer diffraction, as we address further in Section 9.6.

UHF television ($\sim$650 MHz, top), VHF radio ($\sim$100 MHz, middle) and DAB ($\sim$220 MHz, bottom) aerials.

9.4.1 The television aerial

An everyday example of the diffraction grating is the directional aerial with which we receive television or radio signals, several examples of which may be seen in Fig. 9.9. Each aerial is a linear array of elements, designed to diffract back along its own axis so as to enhance the signal at each element, and the illumination direction is therefore at a right angle to that considered above. The elements themselves are of the correct length to build up a standing wave along them (determined by the boundary condition, of the sort we consider in Chapter 12, that no current can flow beyond their ends), and therefore resonate at the required frequency, further enhancing the sensitivity. The exact lengths and spacings of the elements are often adjusted to allow the aerial to operate over a range of frequencies.

9.4.2 Phased-array radar

A classic example of the application of the Huygens principle of wave propagation is the giant *phased-array radar* that has replaced the famous 'golf-balls' that housed earlier mechanically rotated radar dishes near Fylingdales moor in North Yorkshire. The three faces of this 32-m-high pyramid, similar to that in Fig. 9.10, each contain 2560 radar transducers that partly occupy an array of 3584 positions filling a roughly circular area some 26 m in diameter. The beam produced is steered, not by moving this vast antenna, but by electronically introducing an appropriate delay between the transducers. Each array can thereby be steered through 120° of azimuth and 85° of elevation.

It is instructive to check these figures. Each transducer position accounts for some 0.14 m² of the antenna face, giving a typical separation between

Fig. 9.10 Phased-array radar in Eldorado (TX) for early warning of ballistic missiles. Each of the 1792 transducers radiates 0.25–16-ms pulses at 340 W in the range 420–450 MHz.

© 1999 Courtesy of Raytheon Company

hexagonally packed transducers of 35 cm. Frequencies of 420–450 MHz correspond to wavelengths of 0.69 ± 0.02 m. If the transducers are driven in phase, then equation (9.14) gives, at the centre frequency,

$$\sin \vartheta = \frac{m \times 0.69}{0.35}. \tag{9.15}$$

Only $m = 0$ gives a valid solution, so the antenna will produce a single beam.

If a diffraction grating is illuminated at an angle φ to the normal, equation (9.13) is modified to become

$$d(\sin \vartheta_m - \sin \varphi) = m\lambda, \tag{9.16}$$

where the primary, zeroth-order beam ($m = 0$) emerges at $\vartheta_0 = \varphi$. Inserting an appropriate phase shift between transducers of the radar array has the same effect as inclining the illumination in this manner, so we can use this expression to establish how far the main beam may be deflected before higher diffraction orders are produced. The critical angle is found by setting

$m = -1$, $\vartheta_{-1} = -90°$, giving

$$\sin\varphi = \frac{0.69}{0.35} + \sin(-90°) = 0.97 \qquad (9.17)$$

and hence, depending upon the exact frequency, the angle $\vartheta_0 = \varphi$ of the primary ($m = 0$) beam must be increased beyond about 70° for a side-beam ($m = -1$) to appear. A full field of view, divided among the three faces of the radar, will only require beams to be produced at angles up to $\varphi = 60°$ (a little higher if the beam is also inclined from the normal vertically), and any beams produced at more than 30° will be visible to the array on another face and may thus in principle be distinguished. In practice, the directionality of the transducers, the packing arrangement within the array and so on also play important rôles.

We may estimate the angular resolution of the radar by assuming the main beam to have a width determined by its single-slit (or, strictly, single-circular-aperture) diffraction pattern. For a slit width of 26 m and a wavelength of 0.69 m, equation (9.2) predicts a first zero in the diffraction pattern when

$$\alpha = \frac{kd}{2}\sin\vartheta = \pi, \qquad (9.18)$$

giving $\sin\vartheta = \lambda/d$ and hence $\vartheta = 1.5°$ for the half-width of the beam at $\varphi = 0$. At the specified range of 5000 km, this corresponds to an area of the order of 10^{10} m^2; even a large missile would therefore intercept and reflect only a few mW of the 800 kW radiated. Although we have here assumed a rectangular aperture, a more careful calculation for a circular aperture would yield roughly similar results. It is common to 'thin out' the array towards the edges to give the primary beam a smoother shape through the process known as *apodization*.

9.4.3 Phased-array sonar

Just as a phased array of transducers allows radar systems to be scanned without mechanical movement, and thus enables the high resolution and sensitivity of very large transducer arrays, so the same technique can be applied for the same reasons to **sonar**, the acoustic equivalent. Strong attenuation at high frequencies, and the difficulty of making small, high-power transducers, generally limit operating frequencies to below 1 MHz, with sub-kHz frequencies being used for the best sensitivity for long-range detection. Electronically phased arrays are again used to steer the transmitted beam and detection direction.

High-resolution, short-range survey sonars, such as that shown in Fig. 9.11, which operate into the MHz range, are typically of the order of a metre in length, and may be mounted within a single small vessel. For very-low-frequency systems, however, *towed arrays* are trailed behind a ship or submarine below

Fig. 9.11

NOAA Ordnance Reef survey team retrieving a side-scan sonar used to find discarded munitions on the Hawaiian seabed.   Courtesy of NOAA

the sea surface; the transducer arrays of military surveillance sonars can be over a kilometre in length.

Phased arrays of still greater scale extend the resolution of radio astronomy by linking telescopes around the world to give an effective aperture measuring many thousands of kilometres through the technique of *very-long-baseline interferometry (VLBI)*. It is nowadays common to omit a direct signal link between the telescopes, and instead record each signal separately against a precisely synchronized timebase, and computationally combine the signals from different telescopes later.

9.5 Wavefront reconstruction and holography

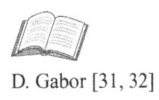

D. Gabor [31, 32]

An instructive perspective on the diffraction grating is that it transmits the incident illumination only in those regions where the phase is also correct for a diffracted order. This insight, recognized by Dennis Gabor many years before the invention of the laser, led him to propose that *any* propagating wave could be reconstructed using a given source by applying a mask to transmit only those parts of the incident wavefront that match the phase of the original wave. If the recorded wave were light that had been scattered by an object, then the reconstructed wave would form a virtual image of the object, even though the object itself had since been removed. Further, the virtual image would change with viewing angle, exactly as for the original three-dimensional object. This is the principle behind **holography**.

Production of the required mask (or *hologram*), which at first sight would appear a significant challenge, can be achieved simply by recording the interference pattern between the original propagating wave and the wavefront that will be used for reconstruction. With reference to Fig. 9.7, it is apparent that, were the diffracted beam already to exist, then it would interfere with the plane-wave illumination to produce fringes that coincide exactly with the slits of the grating. A photographic plate placed at the position of the grating would record these fringes, and could subsequently be developed to convert them into transmissive regions. In accordance with Babinet's principle, a holographic negative will have the same effect.

It may be intuitively helpful, although mathematically challenging, to regard any scattered light as having approximately plane wavefronts over a sufficiently small region, within which interference with the plane *reference* wave to be used for reconstruction will produce regular fringes exactly as described above. When recorded and illuminated with the reference beam, this section of the hologram will function just like a diffraction grating, producing a diffracted ray that retraces the path of the original, hence reconstructing the image.

Holograms that function as described so far, by modulating the transmitted amplitude of the reference beam, are known as *amplitude holograms* and, provided that the apparatus can be held stable to within fractions of a wavelength during the recording process, are relatively straightforward to produce. A variation on this principle is the *phase hologram*, whereby the entire reference wave is transmitted but its phase is modulated, by the varying thickness of the hologram, so that it at all points has the phase of the original object wave. A photographically produced amplitude hologram can be converted to a phase hologram by appropriate bleaching during the development process.

Because amplitude holograms reproduce only certain points of the original wave, it is clear that some information regarding the original wave must have been lost. This turns out to be apparent in the production of a second, *real* image that, in the diffraction grating analogy, corresponds to the negative diffraction order. The phase hologram, by reconstructing the entire wavefront, does not suffer from this (although the real image can, of course, be useful in its own right).

It will be apparent from the above explanation that holograms may only be recorded and replayed using monochromatic light: with a range of wavefronts, multiple fringe patterns would overlap and destroy the recording process, while different reconstructing wavelengths will produce images of varying sizes. Information theory again explains this through the reduction of three-dimensional information into a two-dimensional hologram. The solution is therefore to record the hologram within a volume of photographic emulsion: such *volume holograms* may indeed be viewed in white light.

9.5.1 Phased arrays at visible wavelengths: computer-created holograms

Centimetre-sized liquid-crystal displays, when illuminated with plane waves from a laser, have the effect of imposing a controllable phase shift at each of the million or so pixel positions. As with phased-array radar, these act as Huygens secondary sources and, with suitable programming, can produce diffraction patterns of considerable complexity. This is an example of *computer-generated holography*.

(a)   (b)

Images produced by diffraction from a spatial light modulator. (a) The crest of Berlin's Humboldt Universität is reproduced in two sizes according to the two illumination wavelengths; the spot between the two first-order images is the zeroth-order beam. (b) Portrait of Augustin-Jean Fresnel, reproduced using a phase-only spatial light modulator.   ©️ Courtesy of Holoeye Photonics AG

Examples of diffraction from a computer-controlled pixel array, demonstrating the wavelength dependence of the diffraction, are shown in Fig. 9.12. Spatial light modulators such as this are used both for holographic image formation and for the production of dynamically adjustable light beams for the remote, contact-free manipulation of biological cells and the like, using a technique known as *optical tweezers* that is based upon a force that is mediated entirely optically.

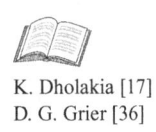

K. Dholakia [17]
D. G. Grier [36]

9.6 Definition of Fraunhofer diffraction

We have seen that the calculation of diffraction patterns is relatively straightforward when a flat mask is illuminated with plane wavefronts and the diffraction is observed as a function of angle. This is an example of the *Fraunhofer diffraction* condition, which merits a little further examination.

It is instructive to consider how our classic diffraction arrangement could be achieved in a finite space and using a point source. As shown in Fig. 9.13, we require the addition of two lenses: the first, placed a focal length away from the source, produces a parallel beam of plane wavefronts from the spherical waves emanating from the source, while the second, placed a focal length from a screen (or photographic film or photodetector) images parallel rays of diffracted light

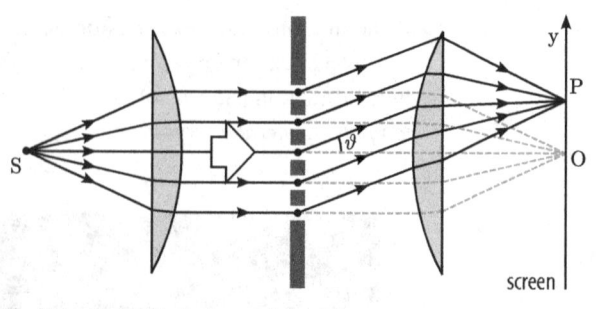

Fig. 9.13 Fraunhofer diffraction, observed in the image plane of the source.

onto individual points. Both lenses therefore act as angle–position transducers as in Section 7.3.2.

Now, if we were to omit the diffracting mask from this arrangement, we should simply have a pair of lenses placed so as to form on the screen an image of objects in the transverse plane containing S, and indeed the point source at S would form a point image on the screen at O. This is what we mean when we say that the screen is *in the image plane of the source*.

The presence of the diffracting mask causes some light to appear at points other than O on the screen, and thus form the Fraunhofer diffraction pattern. Since diffraction causes light that would otherwise have been focussed on a single point to be spread elsewhere, the diffraction pattern is referred to by lens designers as the *point-spread function* (to which, for real lenses, aberrations may also contribute). Every point on the image is affected in this way: diffraction is therefore apparent in the image through a hotel net curtain in Fig. 9.14, although only with narrow, bright reflections from the joints of the roof are the diffracted orders distinguishable.

It is clear that moving the diffracting mask to the left and right between the lenses will have no effect upon the diffraction pattern's intensity distribution; and a little thought will show that, provided it remains within the beam, neither will movement up and down. It turns out that, remarkably, the pattern is the same wherever the mask is placed – even if it is outside the pair of lenses – provided that the paraxial approximation remains valid and the mask remains within the beam.

The mathematical statement of the Fraunhofer condition for this geometry is that

the length of the path from the source S to a point P is a linear function of the mask coordinate y.

The practical interpretation of this is that

Fraunhofer diffraction is observed in the image plane of the point source.

Composite photograph of the roof of the Cairns Casino, Australia, viewed directly and through a net curtain, showing diffraction of sunlight reflected from joints in the glazing.

While we are commonly interested only in the *intensity* of the diffraction pattern observed at the screen, there are times when we may also be concerned about its *phase*. For arbitrary lens separations, this incurs a quadratic dependence upon ϑ; if the lenses are both also positioned a focal length away from the mask (in what, for lenses of equal focal lengths, is known as a **4f arrangement**), then the quadratic dependence is eliminated, and only quartic and higher terms may remain.

9.7   The resolution of an imaging system

Figure 9.13 offers an immediate insight into the resolution with which a lens may form an image, for the image of each point of an object will be spread by diffraction from the effective aperture corresponding to the finite size of the lens. Practical imaging systems must optimize the combination of diffraction, which is improved by increasing the lens diameter, and aberration, which is better with smaller lenses. It is interesting that the diffraction limit to the resolution of the eye is approximately the limit imposed by the sampling resolution, which is determined by the density of rods and cones on the retina.

Exercises

9.1   The first-order beam from a diffraction grating occurs when paths from the source to the image through adjacent rulings differ in length by a single wavelength. Show by geometry that the first-order diffraction angle ϑ for a diffraction grating of line spacing d when normally illuminated with light of a wavelength λ is given by

$$\vartheta = \sin^{-1}\left(\frac{\lambda}{d}\right) \tag{9.19}$$

and determine this angle when light from a sodium lamp with a wavelength of 589 nm is diffracted by the fabric of an umbrella woven from threads with a centre–centre spacing of 0.04 mm.

9.2 Describe qualitatively what you might see when viewing the sodium lamp of Exercise 9.1 through the fabric of the umbrella. If the sodium lamp is 20 cm across and the distance from the umbrella is 100 m, calculate by how many multiples of the lamp width the diffraction order, viewed through the umbrella, will appear to be displaced.

9.3 A naval towed-array sonar comprises a line of 100 transducers, equally spaced every 3 m, that is towed behind a ship so that it lies in a straight line just below the surface of the water. An adjustable phase delay can be introduced electronically for each transducer, allowing the sonar beam to be steered without physically moving the array. The speed of sound in salt water may be taken to be about 1500 m s^{-1}.

1. If the transducers are driven in phase at a constant frequency f, estimate the angular width of the (zeroth-order) sonar beam.

2. A phase delay $\delta\varphi$ is now introduced between successive transducers. Determine how the angle ϑ through which the beam is steered depends upon $\delta\varphi$.

3. Find the maximum frequency that may be used if only one diffraction order is ever to be present as the beam is scanned from $\vartheta = -90°$ to $\vartheta = 90°$. (It may help here to consider the dependence upon the function $u \equiv \sin\vartheta$ instead of ϑ itself.)

4. Hence determine the smallest angular width that can be obtained unambiguously with such a system.

5. How many such beams can be resolved within the 180° angular range? (It may again be helpful to consider the dependence of the diffraction pattern upon the parameter $u = \sin\vartheta$ rather than ϑ alone.)

10 Longitudinal waves

10.1 Further examples of wave propagation

Pain [66] pp. 151–161
French [29] pp. 170–178
Main [60] pp. 174–180
Feynman [23] Vol. I,
pp. 47-3–47-8

Our attention so far has focussed largely upon guitar strings, light, water ripples and even Mexican waves, and it would be easy to infer that all wave motions are *transverse* – that is, the disturbance undergoing the wave motion is a physical displacement in a direction perpendicular to that of the wave propagation. There are indeed many examples of transverse waves but, as already discussed in Section 1.4, other wave motions occur and, from the point of view of wave physics, there is little that is special about transverse motion. In this chapter, we shall therefore address an important class of waves that are apparent in the *longitudinal* motion of a medium; we shall subsequently encounter in Chapters 16 and 17 other waves that do not correspond to physical displacements at all.

10.2 Sound waves in an elastic medium

For our first example of longitudinal waves, we consider **sound**: a one-dimensional motion within an elastic medium such as the column of air within a wind instrument. As in our examination of transverse waves on strings in Section 2.2, we begin by establishing the mechanisms governing the movement of small elements of the medium, and derive from them a wave equation governing wave propagation in that medium.

Sound waves are propagating compressions and rarefactions of an elastic medium. The equilibrium situation, shown in the upper part of Fig. 10.1, is that the medium – air, water or metal, for example – uniformly fills the indicated space, such as a tube of uniform cross-section, from O to R. We consider an element that lies between x and $x + \delta x$, bounded by points P and Q. This may be displaced during the wave motion so that at a given time, as shown in the lower part of Fig. 10.1, it is bounded by the points P$'$ and Q$'$, whose positions are $x + \xi$ and $(x + \delta x) + (\xi + \delta \xi)$. The wave motion is therefore associated with both a displacement of the element and a change in its thickness.

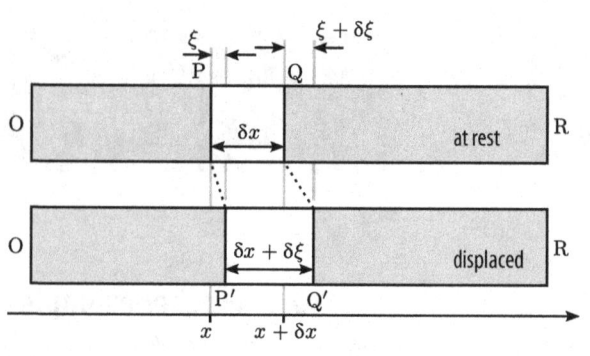

Fig. 10.1   Longitudinal displacement within an elastic medium.

Any extension of an elastic medium is accompanied by a tension, in accordance with *Hooke's law*. If the tube has a cross-sectional area A, then we may write the tension W as

$$W(x) = EA \frac{\delta \xi}{\delta x}, \tag{10.1}$$

where E is *Young's modulus* (or the *modulus of elasticity*) for the medium. If the density of the medium is ρ, then the mass of the element will be

$$\delta m = \rho A \, \delta x. \tag{10.2}$$

The net force on the element will be

$$W(x + \delta x) - W(x) = EA \left[\frac{\delta \xi}{\delta x}(x + \delta x) - \frac{\delta \xi}{\delta x}(x) \right]. \tag{10.3}$$

Newton's second law equates this to the product of the mass of the element and its acceleration,

$$\delta m \frac{\partial^2 \xi}{\partial t^2} = W(x + \delta x) - W(x), \tag{10.4}$$

and hence

$$\rho A \, \delta x \frac{\partial^2 \xi}{\partial t^2} = EA \left[\frac{\delta \xi}{\delta x}(x + \delta x) - \frac{\delta \xi}{\delta x}(x) \right]. \tag{10.5}$$

We now divide by the free length δx of the element, and take the limit as this is reduced to zero:

$$\rho A \frac{\partial^2 \xi}{\partial t^2} = EA \lim_{\delta x \to 0} \frac{\frac{\delta \xi}{\delta x}(x + \delta x) - \frac{\delta \xi}{\delta x}(x)}{\delta x}$$

$$= EA \frac{\partial^2 \xi}{\partial x^2}, \tag{10.6}$$

so that the wave equation for longitudinal sound waves is

$$\rho \frac{\partial^2 \xi}{\partial t^2} = E \frac{\partial^2 \xi}{\partial x^2}. \tag{10.7}$$

Equation governing longitudinal waves in an elastic medium

Validity: small amplitudes, low frequencies, negligible damping

This wave equation is of exactly the same form as that we derived for the oscillations of a guitar string; its method of solution is therefore identical.

This derivation is applicable to the transmission of vibrations along a bar, the passage of sound through the atmosphere and the pulse of the heartbeat around the body (with modifications, for elasticity in the blood-vessel walls introduces a transverse component to the wave) and even as an approximation to variations in traffic speed on a motorway. In each case, the speed of wave propagation v_p will be given by

$$v_p = \sqrt{\frac{E}{\rho}}. \tag{10.8}$$

For metals, typical densities and values of Young's modulus are 7000 kg m^{-3} and 10^{11} Pa; the speed of sound is therefore about 4 km s^{-1}. We have deliberately considered here only one-dimensional effects; in real cases, the material's *Poisson's ratio*, which accounts for changes in volume as a medium is stretched, may have to be taken into account.

10.2.1 The speed of sound in air

Young's modulus is rarely given for gases, but our knowledge of the *ideal-gas law* (equation (3.69)) renders its derivation straightforward. For an element of volume $\delta V = A \, \delta x$, at ambient pressure P_0 and temperature Θ_0, we have

$$P_0 \, \delta V = \delta n \, R \Theta_0, \tag{10.9}$$

where $R = 8.31$ J mol^{-1} K^{-1} is the *ideal-gas constant* and δn is the number of moles of gas within the volume element. When we expand the volume to a new volume

$$\delta V' = \delta V \left(1 + \frac{\delta \xi}{\delta x} \right), \tag{10.10}$$

the new pressure, measured at a new temperature Θ', will be $P_0 + \delta P$, where

$$(P_0 + \delta P) \delta V' = \delta n \, R \Theta'. \tag{10.11}$$

The deviation from ambient pressure δP corresponds simply to the negative of the tension per unit area, W/A, in equation (10.1).

We shall initially suppose that the expansion or compression process occurs at constant temperature – i.e. $\Theta' = \Theta_0$, giving

$$W(x) = -A\,\delta P$$

$$= -A\left(\frac{P_0\,\delta V}{R\Theta_0}\frac{R\Theta_0}{\delta V'} - P_0\right)$$

$$= -AP_0\left[\left(1 + \frac{\delta\xi}{\delta x}\right)^{-1} - 1\right]. \tag{10.12}$$

For shallow waves, for which $\delta\xi/\delta x \ll 1$, this becomes

$$W(x) \approx AP_0\frac{\delta\xi}{\delta x}, \tag{10.13}$$

and hence, by comparison with equation (10.1), we could write $E \equiv P_0$. The speed of sound in air of density $\rho = 1.29$ kg m^{-3} at a typical sea-level atmospheric pressure of 10^5 Pa would therefore be ~278 m s^{-1} – an order of magnitude lower than the speed in solid metals.

Generally, however, the compressions and rarefactions accompanying sound waves at audible frequencies are too rapid for heat to be exchanged with the surroundings, and the changes are therefore *adiabatic*: temperature changes will accompany the variations in pressure, and the relationship between pressure and volume will be, as in equation (3.71),

$$PV^\gamma = \text{constant}, \tag{10.14}$$

where γ is the ratio of the specific heats at constant pressure and constant volume. Using equation (10.14) in place of equation (10.9), and recalculating the pressure $P_0 + \delta P$ pertaining at a volume δV defined by equation (10.10), we now obtain

$$(P_0 + \delta P)\delta V'^\gamma = P_0\,\delta V^\gamma \tag{10.15}$$

and therefore

$$W(x) = -A\,\delta P$$

$$= -A\left[P_0\left(\frac{\delta V}{\delta V'}\right)^\gamma - P_0\right]$$

$$= -AP_0\left[\left(1 + \frac{\delta\xi}{\delta x}\right)^{-\gamma} - 1\right], \tag{10.16}$$

so that, again making the approximation of small displacements, we obtain

$$W(x) \approx AP_0\gamma\frac{\delta\xi}{\delta x}. \tag{10.17}$$

Air is principally composed of diatomic nitrogen, so we expect $\gamma \approx (1 + 5/2)/(5/2) = 1.4$. Our prediction for the speed of sound in air hence becomes

$$v_{\mathrm{p}} \approx \sqrt{\frac{\gamma P_0}{\gamma}}$$

$$= \sqrt{\frac{1.4 \times 10^5}{1.29}} = 329 \text{ m s}^{-1}. \tag{10.18}$$

The experimentally determined speed of sound at sea level is about 330 m s^{-1}. When true measured values of atmospheric pressure, density and γ are taken, the predicted and measured speeds of sound are indeed consistent.

10.2.2 Pressure or displacement?

We have already seen in Section 2.2 that, for transverse waves, both the displacement and its rate of change are important quantities. For the time being, we shall duck the question of whether it is the rate of change with time, $\partial \psi / \partial t$, or that with position, $\partial \psi / \partial x$, which is the more important, for in non-dispersive systems that support waves of the form $\psi (x - v_{\mathrm{p}} t)$ the two are related simply by the constant speed of propagation, as we saw in Section 2.2.2. The crucial point is that, to determine the future behaviour of a wave motion, we need to know not merely the displacement at a given time, but also some other function such as the temporal or spatial derivative.

With longitudinal waves, equation (10.1), or equation (10.13) or equation (10.17) together with the ideal-gas equation (10.9), shows that the spatial derivative of the displacement is, for a given medium, proportional to the tension or pressure within the medium. Often – particularly with sound waves – it will be simpler to measure the pressure rather than the displacement and, with a few alterations, our derivation could instead have produced a wave equation for the pressure. It is not really possible to say which is the more fundamental property.

10.3 Thermal waves

An example of the propagation of a disturbance that is neither a transverse nor a longitudinal coordinate is the flow of **heat** in a thermally conducting medium. Heat flows into a region whenever its surroundings are hotter, and the finite thermal conductivity causes a lag between the temperature gradient and its full effect. We shall see, though, that the absence of *inertia*, which in mechanical systems causes the medium to continue moving when the force is removed, results in a different form of wave equation and correspondingly

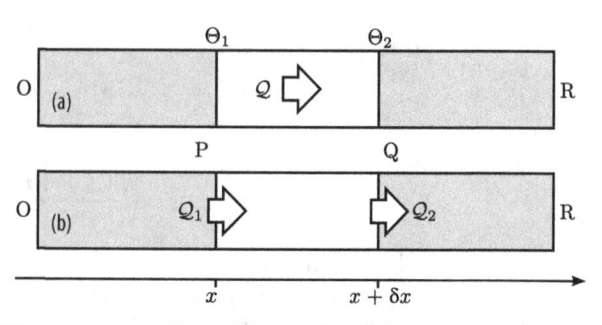

Heat flow within a conducting medium: (a) the heat flow $\mathcal{Q}$ depends upon the temperature difference $\Theta_1 - \Theta_2$ across the element; (b) if there is a net flow of heat $\mathcal{Q}_1 - \mathcal{Q}_2$ into the element, its temperature will rise.

different solutions. Thermal waves, we shall see, are fundamentally damped and dispersive.

Figure 10.2(a) shows a bar, which we take to have a uniform cross-sectional area A, thermal conductivity κ, density ρ and specific heat capacity C, and which we assume to be lagged to eliminate heat losses to the sides. We consider a slice of length δx. If the temperatures at x and $x + \delta x$ are Θ_1 and Θ_2, then the heat $\mathcal{Q}$ flowing per second through the slice will be given by

$$\mathcal{Q} = \frac{\kappa A}{\delta x}\left(\Theta_1 - \Theta_2\right) \tag{10.19}$$

and hence, taking the limit as the slice thickness δx is reduced to zero, we may write the heat flow at any point as

$$\mathcal{Q}(x) = -\kappa A \frac{\partial \Theta}{\partial x}. \tag{10.20}$$

The net heat flowing into the slice, as shown in Fig. 10.2(b), will be $\mathcal{Q}_1 - \mathcal{Q}_2$, and this will result in a change in the temperature of the slice, according to

$$\frac{\partial \Theta}{\partial t} = \frac{\mathcal{Q}(x) - \mathcal{Q}(x + \delta x)}{C\rho A\,\delta x}, \tag{10.21}$$

so that, again taking the limit as the slice thickness δx tends to zero,

$$\frac{\partial \Theta}{\partial t} = -\frac{1}{C\rho A}\frac{\partial \mathcal{Q}}{\partial x}. \tag{10.22}$$

Inserting equation (10.20) into equation (10.22) gives the wave equation, often known as the **diffusion equation**,

Diffusion equation governing thermal waves in a conducting medium

$$\frac{\partial \Theta}{\partial t} = \frac{\kappa}{C\rho}\frac{\partial^2 \Theta}{\partial x^2}. \tag{10.23}$$

This clearly resembles the wave equations that we have met before, but involves the first temporal derivative rather than the second. We must therefore solve this equation anew, using the same approach but expecting a different result.

Suppose that, as in Section 4.4.1, our wave may be written in the form

$$\Theta(x, t) = X(x)T(t). \tag{10.24}$$

The diffusion equation (10.23) may then be written

$$\frac{1}{T}\frac{\partial T}{\partial t} = \frac{\kappa}{C\rho}\frac{1}{X}\frac{\partial^2 X}{\partial x^2} \tag{10.25}$$

and hence gives two separate equations coupled by a yet-to-be-determined separation constant, which we shall write as $-i\omega$:

$$\frac{1}{T}\frac{\partial T}{\partial t} = -i\omega, \tag{10.26a}$$

$$\frac{1}{X}\frac{\partial^2 X}{\partial x^2} = -i\omega\frac{C\rho}{\kappa}. \tag{10.26b}$$

The first of these has complex exponential solutions $T = T_0 e^{-i\omega t}$ (which is why we wrote the separation constant as we did); and inserting the trial form $X = X_0 e^{ikx}$ for the spatial part gives

$$-k^2 = -\frac{i\omega C\rho}{\kappa} \tag{10.27}$$

and hence

$$k = \pm\sqrt{\frac{\omega C\rho}{2\kappa}}(1 + i). \tag{10.28}$$

The overall wave solutions are therefore, with $\Theta_0 = X_0 T_0$,

$$\Theta(x, t) = \Theta_0 \exp[i(\pm k_0 x - \omega t)]\exp(\mp k_0 x)$$
$$= |\Theta_0|\exp[i(\pm k_0 x - \omega t + \varphi)]\exp(\mp k_0 x), \tag{10.29}$$

where $\varphi \equiv \mathrm{Arg}(\Theta_0)$ and

$$k_0 \equiv \sqrt{\omega C\rho/(2\kappa)}. \tag{10.30}$$

Such solutions describe damped, travelling oscillations, and real-valued solutions correspond to the superpositions

$$\Theta(x, t) = \frac{|\Theta_0|}{2}\{\exp[i(\pm k_0 x - \omega t + \varphi)]\exp(\mp k_0 x)$$
$$+ \exp[i(\mp k_0 x - \omega t - \varphi)]\exp(\pm k_0 x)\}$$
$$= |\Theta_0|\cos(\pm k_0 x - \omega t + \varphi)\exp(\mp k_0 x). \tag{10.31}$$

Both complex components have the same decaying amplitude because changing the sign of ω in equation (10.28) changes $(1 + i)$ to $i(1 + i) \equiv -(1 - i)$.

Periodic variations in temperature therefore propagate through a medium as a wave motion, subject to damping with a $(1/e)$ decay length $1/k_0 = \lambda/(2\pi)$ – a factor of 2π smaller than the wavelength λ.

The phase velocity of the thermal waves,

$$v_{\rm p} \equiv \frac{\omega}{k_0} = \sqrt{\frac{2\kappa\omega}{C\rho}}, \tag{10.32}$$

varies with the wave frequency ω, and thermal wave propagation is therefore fundamentally *dispersive*. As we would intuitively expect, the phase velocity increases with the thermal conductivity and decreases with the heat capacity of the medium. A little thought will show that sinusoidal temperature variations at any point lag the heat flow by $45°$.

Exercises

10.1 Given that the modulus of elasticity E for water is about 2.2×10^9 Pa, and its density is 10^3 kg m^{-3}, estimate the speed of sound in water.

10.2 The energy density of a sound wave is, in common with other mechanical wave motions, composed in part by the kinetic energy $\frac{1}{2}\rho(\partial\xi/\partial t)^2$ and in part by the potential energy $\frac{1}{2}E(\partial\xi/\partial x)^2$, where ρ is the density of the medium and E is its elasticity, and ξ is the displacement at position x. By considering a sinusoidal sound wave of definite frequency, show that these two contributions are equal.

10.3 Given that the acoustic intensity is equal to the product of energy density and wave speed, find the amplitudes of displacement for sound waves in air that correspond to the limits of human hearing, 10^{-12} W m^{-2} and 1 W m^{-2}, at a frequency of 1 kHz. The density of air may be taken to be 1.29 kg m^{-3}, and the modulus of elasticity $E \approx 1.4 \times 10^5$ Pa.

10.4 Advances in aerodynamics and engine performance during World War II allowed aircraft in the mid 1940s to investigate high-speed dives in which their speed through the air approached that of sound. Early attempts revealed that *trans-sonic* flight was associated with problems of aerodynamic control and structural strength, resulting in the popular myth of a *sound barrier* that persisted until supersonic flight was relatively routine in the mid 1950s. Early difficulties were solved through a better understanding of high-speed aerodynamics, awareness of aerodynamically induced structural oscillations (*flutter*, whereby the flexure of an aerofoil affects the airflow that causes its deflection), and the resulting design of strong, streamlined aircraft with control systems appropriate to high-speed flight. The problems were collectively associated with *compressibility*, and the highly dangerous test flights to obtain the experimental data that led to their solution were known as the 'compressibility trials'.

1. By considering travelling waves whose displacement is of the form $\xi(x, t) = f(u)$, where $u \equiv x - v_0 t$, show that when such a wave

propagates through the atmosphere its spatial and temporal derivatives are related at each point by

$$\frac{\partial \xi}{\partial t} = -v_0 \frac{\partial \xi}{\partial x}. \tag{10.33}$$

2. We have seen in equation (10.17) how, for adiabatic compression of an ideal gas, the difference in pressure δP from the ambient value P_0 is related to $\partial \xi / \partial x$. Use this to determine how the pressure variation resulting from the motion of the air, caused in this case by the passage of an aircraft, may be written in terms of the speed $\partial \xi / \partial t$ with which the air is moved.

3. Thus show that, for low airspeed, there is little variation in pressure from ambient. (The air is said to be approximately *incompressible*.)

4. Show that, conversely, air starts to show *compressibility* – a notable pressure variation associated with its displacement – when the speed of the displacement approaches the speed of sound in the medium, and that our assumption in Section 10.2.1 that $\partial \xi / \partial x \ll 1$ will then no longer be valid.

11   Continuity conditions

11.1   Wave propagation in changing media

Pain [66] pp. 117–119 and
163–165
French [29] pp. 253–259
Main [60] pp. 299–307

For much of our analysis so far, we have been concerned with the propagation of waves through continuous, uniform media. We have then dealt with diffraction at obstacles by adopting the Huygens approach, and we have seen that this can also describe the reflection and refraction that occur at interfaces, for which it both provides an excellent conceptual description of the processes and allows us to calculate the directions of the reflected and refracted waves.

With appropriate mathematical dexterity, we could also in principle use the Huygens approach, in the form of the phasor or Fresnel-integral methods, to calculate the reflected and refracted amplitudes, but for most cases a more direct way to determine this behaviour quantitatively involves returning, at least in part, to the physical mechanisms behind the wave propagation. As we shall see in the following pages, the physics of a given situation determines *continuity conditions* that relate the otherwise freely propagating waves when they coincide at an interface. Although the conditions reflect the detailed rules and mechanisms of the specific system, we shall see that for the most part they do so in the same fashion, depending upon the relative *impedances* of the different media.

11.2   The frayed guitar string

We begin with a one-dimensional example by returning to the transverse travelling waves on a taut guitar string that we examined in Section 2.2, but we now consider what happens when the travelling wave encounters a change in the string properties. The heavier strings on a guitar are often constructed by winding a spiral wire around a lighter core to achieve a high mass per unit length while maintaining flexibility. With wear, the outer wire can become damaged and fray, leaving some places with just the inner part of the string and changing the sound of the instrument.

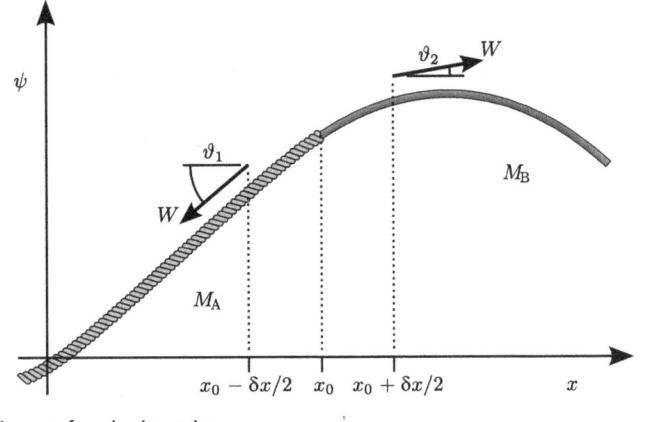

Fig. 11.1 Forces acting on a frayed guitar string.

Such an occurrence is shown schematically in Fig. 11.1. The intact section of string in the region $x < x_0$ is taken to have a mass M_A per unit length, while in the frayed section $x > x_0$ the corresponding value is M_B. As in Section 2.2 the string is taken to be under a tension W, and consideration of the forces on elements of width δx therefore reproduces the wave equation (2.18),

$$\frac{\partial^2 \psi}{\partial t^2} = \frac{W}{M} \frac{\partial^2 \psi}{\partial x^2}, \tag{11.1}$$

where M is the mass per unit length of the chosen element. For regions in which M is constant – that is, the intact region A and the frayed region B – we have already solved such wave equations in Section 2.2, so we know how the separate sections of the string will behave; we shall label these waveforms ψ_A and ψ_B to indicate their ranges of validity.

We now determine how the motions of the two sections depend upon each other at the interface, where M changes from M_A to M_B. We consider the element around $x = x_0$ which extends, as in Fig. 11.1, from $x - \delta x/2$ to $x + \delta x/2$. Following equation (2.16) and writing the mass of the element as

$$\delta m = \delta x \frac{M_A + M_B}{2}, \tag{11.2}$$

we find that the force on the element will be

$$W \left[\left(\frac{\partial \psi_B}{\partial x} \right)_{x_0 + \delta x/2} - \left(\frac{\partial \psi_A}{\partial x} \right)_{x_0 - \delta x/2} \right] \tag{11.3}$$

and hence that the acceleration of the element will be

$$\frac{\partial^2 \psi}{\partial t^2} = \frac{2W}{M_A + M_B} \frac{\left(\dfrac{\partial \psi_B}{\partial x} \right)_{x_0 + \delta x/2} - \left(\dfrac{\partial \psi_A}{\partial x} \right)_{x_0 - \delta x/2}}{\delta x}. \tag{11.4}$$

By taking the limit $\delta x \to 0$, we may now establish the **continuity conditions** that relate the wave motions ψ_A and ψ_B at the interface.

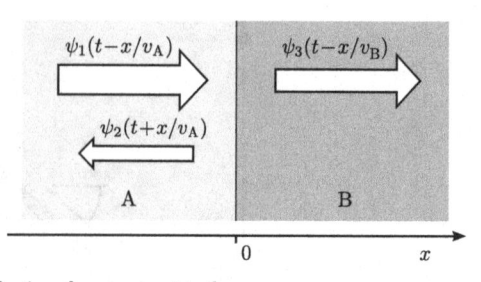

Fig. 11.2   Transmission and reflection of waves at an interface.

The first rule is rather obvious: the two parts of the string are attached to each other, and therefore the displacement ψ must be a continuous function:

Continuity condition 1:
continuity of displacement

$$\psi_A(x_0, t) = \psi_B(x_0, t). \tag{11.5}$$

The string displacement cannot suddenly jump from one value to another, and the derivative $\partial\psi/\partial x$ must be finite everywhere.

The second rule is only slightly more subtle: we require the acceleration of the string at all points to be finite. Equation (11.4) indicates that, as we reduce δx to zero, the gradient $\partial\psi/\partial x$ must be the same on both sides of the interface – in other words, the gradient itself must be continuous,

Continuity condition 2:
continuity of derivative of displacement

$$\frac{\partial\psi_A}{\partial x}(x_0, t) = \frac{\partial\psi_B}{\partial x}(x_0, t), \tag{11.6}$$

and the second derivative $\partial^2\psi/\partial x^2$ must be finite.

We have thus determined the continuity conditions for waves on a frayed string: that both the displacement and its spatial derivative must be continuous across the interface. These conditions are sufficient to determine the reflected and transmitted amplitudes when a wave is incident upon the frayed region.

We shall refer to the conditions applying to a wave motion at an interface – a knot, or a change of density, refractive index, etc. – as *continuity conditions*; in contrast, the conditions that apply when external constraints are imposed will be known as *boundary conditions*, and are examined further in Chapter 12. This distinction is often not made elsewhere, for the physical origins are essentially the same, but the conditions themselves can be quite different.

11.2.1 Determination of reflected and transmitted amplitudes

Just as the wave equation embodies the specific physics needed to determine the characteristics of wave propagation through a given medium, so the continuity conditions contain the information we require in order to determine how waves will be transmitted or reflected at interfaces. We continue for the time being to consider one-dimensional propagation, but our treatment is otherwise general, and the frayed string just a useful example.

We consider, as shown in Fig. 11.2, the case in which a travelling wave $\psi_1(x, t)$ is incident from the left upon an interface, which for convenience we position at $x = 0$. The regions A and B again correspond to the two sections of the string, with masses M_A and M_B per unit length and wave speeds v_A and v_B given in each region by $v^2 = W/M$. It proves convenient to write the incident wave in the form $\psi_1(u_1)$, where $u_1 \equiv t - x/v_A$, and we similarly write arbitrary reflected and transmitted waves $\psi_2(u_2)$ and $\psi_3(u_3)$, where $u_2 \equiv t + x/v_A$ and $u_3 \equiv t - x/v_B$. The total displacement in region A is given by the sum of the incident and reflected waves

Although it does not greatly matter whether we write our waves as $\psi_1(k_A x - \omega t)$, $\psi_1(t - x/v_A)$ or $\psi_1(x - v_A t)$, it is helpful to avoid the third form, for, at the interface ($x = 0$), the three waves will then vary with time in the same manner. Had we chosen the third form, the functions would have had to take into account the different rates of change of $v_A t$ and $v_B t$ with time, which is messy.

$$\psi_A(x, t) = \psi_1(t - x/v_A) + \psi_2(t + x/v_A), \tag{11.7}$$

while that in region B is simply the transmitted wave $\psi_3(x, t)$. Our continuity conditions, equations (11.5) and (11.6), hence give

$$\psi_1(t) + \psi_2(t) = \psi_3(t), \tag{11.8a}$$

$$\frac{1}{v_A}\left[\frac{d\psi_1}{dt}(t) - \frac{d\psi_2}{dt}(t)\right] = \frac{1}{v_B}\frac{d\psi_3}{dt}(t), \tag{11.8b}$$

where we have written $\partial\psi_1/\partial x \equiv (d\psi_1/du_1)(\partial u_1/\partial x) \equiv -(d\psi_1/du_1)/v_A$, etc., as in Section 2.2.2, and set $x = 0$ so that $u_1 = u_2 = u_3 = t$.

If ψ_2 and ψ_3 are assumed to be simply constant multiples of the incident wave ψ_1, then equations (11.8a) and (11.8b) may be regarded as a pair of equations with two unknowns at any time (the reflected and transmitted amplitudes), which are easily solved as such to give

$$\psi_2 = \frac{v_B - v_A}{v_B + v_A}\psi_1, \tag{11.9a}$$

$$\psi_3 = \frac{2v_B}{v_B + v_A}\psi_1 \tag{11.9b}$$

and hence, explicitly,

$$\psi_2(t + x/v_A) = \frac{v_B - v_A}{v_B + v_A}\psi_1(t - x/v_A),$$

$$\psi_3(t - x/v_B) = \frac{2v_B}{v_B + v_A}\psi_1(t - x/v_A). \tag{11.10}$$

If $v_A = v_B$ then, as we would expect, there is no reflection and the entire incident wave is transmitted.

Assuming ψ_2 and ψ_3 to be constant multiples of the incident wave ψ_1 does not really involve any sleight-of-hand. Rigorously, we should differentiate equation (11.8a), hence giving a pair of equations with two unknowns that can be solved to give $d\psi_2/dt$ and $d\psi_3/dt$ in terms of $d\psi_1/dt$. These may then be integrated to give ψ_2 and ψ_3 in terms of ψ_1 and a constant of integration, which application of equation (11.8a) will then eliminate.

We may summarize equations (11.10) through the *amplitude* reflection and transmission coefficients r and t for the frayed guitar string,

$$r \equiv \frac{\psi_2(0,t)}{\psi_1(0,t)} = \frac{v_B - v_A}{v_B + v_A}, \tag{11.11a}$$

$$t \equiv \frac{\psi_3(0,t)}{\psi_1(0,t)} = \frac{2v_B}{v_B + v_A}. \tag{11.11b}$$

The *intensity* reflection and transmission coefficients will be proportional to the *squares* of these terms, but include scaling factors that reflect the different dependences of the energy density upon the wave displacement ψ, and the different velocities of propagation, in the two regions, as in Section 4.2.1.

11.2.2 Reflection and transmission of sine waves

We may clarify the above treatment of general travelling waves by considering the example of sinusoidal disturbances. The procedure is identical, but we now assume a specific form $\psi_1(x,t)) = \psi_1 \cos(t - x/v_A)$. Equations (11.8) become

$$\psi_1 \cos(t - x/v_A) + \psi_2 \cos(t + x/v_A) = \psi_3 \cos(t - x/v_B), \tag{11.12a}$$

$$\frac{1}{v_A}[\psi_1 \sin(t - x/v_A) - \psi_2 \sin(t + x/v_A)] = \frac{1}{v_B}\psi_3 \sin(t - x/v_B), \tag{11.12b}$$

and hence

$$\psi_1 + \psi_2 = \psi_3, \tag{11.13a}$$

$$\frac{1}{v_A}(\psi_1 - \psi_2) = \frac{1}{v_B}\psi_3. \tag{11.13b}$$

Elimination of ψ_2 or ψ_3 to give the reflected and transmitted amplitudes in accordance with equations (11.9) is now straightforward.

Although for a single interface it is unnecessary to limit our treatment in this way to single-frequency components, we shall see in Section 11.4 that such an analysis is helpful when we consider the interference between multiple reflections, as occurs in high-quality optical filters.

11.2.3 Conservation of energy at an interface

There is an instructive alternative to our examination of wave propagation at an interface. We again consider a travelling wave $\psi_1(u_1)$ that is incident on the interface, giving reflected and transmitted components $\psi_2(u_2)$ and $\psi_3(u_3)$, where u_1, u_2 and u_3 are as above.

Conservation of energy requires that the incident power should equal the sum of the reflected and transmitted powers, where we have seen in Section 4.2.1 that the power P transmitted by the wave motion of a string of mass per unit length ρ under a tension W and with a propagation speed v_p may be written as

$$P = W v_p \left(\frac{\partial \psi}{\partial x}\right)^2 = \rho v_p \left(\frac{\partial \psi}{\partial t}\right)^2. \tag{11.14}$$

Hence, for our frayed string under tension W, at the interface ($x = 0$),

$$W v_A \left(\frac{\partial \psi_1}{\partial x} \right)^2 = W v_A \left(\frac{\partial \psi_2}{\partial x} \right)^2 + W v_B \left(\frac{\partial \psi_3}{\partial x} \right)^2 . \tag{11.15}$$

The tension W is the same for all three components, and may be eliminated, simplifying equation (11.15) to

$$v_A \left[\left(\frac{\partial \psi_1}{\partial x} \right)^2 - \left(\frac{\partial \psi_2}{\partial x} \right)^2 \right] = v_B \left(\frac{\partial \psi_3}{\partial x} \right)^2 , \tag{11.16}$$

whence

$$v_A \left(\frac{\partial \psi_1}{\partial x} - \frac{\partial \psi_2}{\partial x} \right) \left(\frac{\partial \psi_1}{\partial x} + \frac{\partial \psi_2}{\partial x} \right) = v_B \left(\frac{\partial \psi_3}{\partial x} \right)^2 . \tag{11.17}$$

Once again, application of the chain rule to our travelling wave $\psi(x, t) = \psi(u_1)$, where $u_1 \equiv t - x/v_A$, gives

$$\frac{\partial \psi_1}{\partial x} = \frac{1}{v_A} \frac{d\psi}{du_1} \tag{11.18}$$

and

$$\frac{\partial \psi_1}{\partial t} = \frac{d\psi}{du_1}, \tag{11.19}$$

so that, on combining equations (11.18) and (11.19), we have

$$v_A \frac{\partial \psi_1}{\partial x} = \frac{\partial \psi_1}{\partial t} . \tag{11.20}$$

Similarly, taking into account the change in sign for ψ_2, we may write

$$v_A \frac{\partial \psi_2}{\partial x} = -\frac{\partial \psi_2}{\partial t}, \tag{11.21a}$$

$$v_B \frac{\partial \psi_3}{\partial x} = \frac{\partial \psi_3}{\partial t}, \tag{11.21b}$$

so that equation (11.17) becomes

$$\left(\frac{\partial \psi_1}{\partial x} + \frac{\partial \psi_2}{\partial x} \right) \left(\frac{\partial \psi_1}{\partial t} + \frac{\partial \psi_2}{\partial t} \right) = \frac{\partial \psi_3}{\partial x} \frac{\partial \psi_3}{\partial t} . \tag{11.22}$$

Because the string is continuous, the displacement at the interface must be the same on both sides, which gives the first of our continuity conditions,

Continuity condition 1: continuity of displacement

$$\psi_1 + \psi_2 = \psi_3 . \tag{11.23}$$

Thus, on differentiating with respect to time, $\partial \psi_1 / \partial t + \partial \psi_2 / \partial t = \partial \psi_3 / \partial t$, and equation (11.22) simplifies to the second continuity condition,

Continuity condition 2: continuity of spatial derivative of displacement

$$\frac{\partial \psi_1}{\partial x} + \frac{\partial \psi_2}{\partial x} = \frac{\partial \psi_3}{\partial x} . \tag{11.24}$$

At the interface, then, the displacement ψ and its temporal and spatial derivatives are all continuous. Since the simplest interface is that between two identical regions, these continuity conditions apply generally, even when there is no apparent change.

Although we have in this section considered the specific example of transverse waves on a taut string, we shall find that equivalent continuity conditions apply to the propagation of any wave disturbance.

11.3 General continuity conditions and characteristic impedance

Although the tension W of the string of Section 11.2 must be the same both sides of the interface, the equivalent property in other examples of wave motion is not always so constrained. Re-labelling the waves as *incident*, *reflected* and *transmitted*, we therefore re-derive equation (11.24) to allow for such cases:

$$W_i \frac{\partial \psi_i}{\partial x} + W_r \frac{\partial \psi_r}{\partial x} = W_t \frac{\partial \psi_t}{\partial x}. \tag{11.25}$$

Making the substitution of equations (11.20) and (11.21), and noting that $v_r = -v_i$, our continuity conditions may now be written as

$$\psi_i + \psi_r = \psi_t, \tag{11.26a}$$

$$\frac{W_i}{v_i} \frac{\partial \psi_i}{\partial t} + \frac{W_r}{v_r} \frac{\partial \psi_r}{\partial t} = \frac{W_t}{v_t} \frac{\partial \psi_t}{\partial t}. \tag{11.26b}$$

Pain [66] pp. 122–123 and 163–165

French [29] pp. 259–264

Main [60] pp. 148–156

We have used in Section 11.2.3 the trick of differentiating the first continuity equation to give the same variables as in the second. This can be used again here and, in the subsequent algebra, we shall deal with the quantities W_i/v_i etc. so often that we shall be tempted to define their moduli explicitly by the **characteristic impedance** of the medium,

$$Z \equiv \left| \frac{W}{v} \right|. \tag{11.27}$$

Equations (11.26a) and (11.26b) hence assume a general form

$$\psi_i + \psi_r = \psi_t, \tag{11.28a}$$

General continuity conditions

$$Z_1 \frac{\partial \psi_i}{\partial t} + Z_r \frac{\partial \psi_r}{\partial t} = Z_t \frac{\partial \psi_t}{\partial t}, \tag{11.28b}$$

and the amplitude reflection and transmission coefficients become

$$r = \frac{Z_1 - Z_2}{Z_1 + Z_2}, \tag{11.29a}$$

$$t = \frac{2Z_1}{Z_1 + Z_2}. \tag{11.29b}$$

Equivalent impedances can be derived for all other wave motions, so equations (11.29a) and (11.29b) are quite general. For light, the characteristic impedance $Z = \sqrt{\mu_0 \mu_r/(\varepsilon_0 \varepsilon_r)}$, given in vacuum by $Z_0 = 376.7\,\Omega$, is inversely proportional to the refractive index $\eta = \sqrt{\varepsilon_r}$; the formulae defining the corresponding reflection and transmission coefficients are known as the **Fresnel equations**.

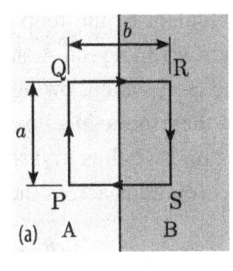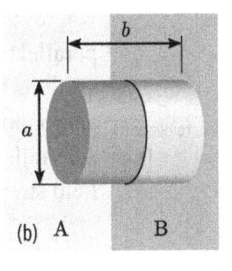

Fig. 11.3 Geometries for determining boundary conditions of components (a) parallel and (b) perpendicular to the interface between two media A and B with relative permittivities ε_A and ε_B and permeabilities μ_A and μ_B.

Although continuity conditions must in each case be derived from the physical mechanisms governing the wave propagation, they always yield expressions of the form of equations (11.5) and (11.6), suggesting a deeper meaning. When we develop our wave mechanics into *quantum mechanics*, we shall see that the continuity conditions express the **conservation of energy and momentum**. These are indeed concepts that pervade all physical wave motions, including sound and waves of electron concentration and spin, each of which can be quantized into particles such as *phonons*, *plasmons*, *polarons* and *polaritons*. *Photons* are the quanta of electromagnetic waves with, as we shall see in Section 17.2, energies and momenta given by multiplying their frequencies and wavevectors, respectively, by Planck's constant.

11.3.1 Continuity of electromagnetic waves

A recurring example of the use and importance of continuity conditions is in determining the effect of optical materials upon electromagnetic waves. **Maxwell's equations** in their integral forms allow their ready derivation by considering components of the electric and magnetic fields parallel ($E_\parallel$ etc.) and perpendicular ($E_\perp$ etc.) to the interface between the two media such as air and glass.

We first consider the *parallel* component of the electric field strength $E_\parallel$, for which we apply **Faraday's law** to a rectangular loop enclosing an element of the interface as shown in Fig. 11.3(a), where it relates the integral of the electric field E around the edge of the loop to an integral of the temporal derivative of the magnetic flux density B over the area that it encloses. We shall assume that the interface is essentially flat, and thus has translational symmetry over the scale of our analysis, and indeed we shall later ensure this by taking the limit as lengths a and b tend to zero. Breaking the loop integral into four parts, we may write Faraday's law for this case as

$$\int_P^Q E_{\parallel,A}\,dx + \int_R^S E_{\parallel,B}\,dx = \int_{PQRS} -\frac{\partial \mathbf{B}}{\partial t}\cdot d\mathbf{S}, \qquad (11.30)$$

where $\mathrm{d}\mathbf{S}$ is an element of the loop area, $E_{\parallel,\mathrm{A}}$ indicates the component of E parallel to the interface in region A, and we have omitted the integrals along the sections Q–R and S–P, which, owing to the translational symmetry, are equal and opposite and therefore cancel out. Taking the limit $a \to 0$, the right-hand side falls to zero, and we thus find that the parallel component of the electric field strength is continuous across the boundary,

Equivalently, the parallel components of the electric displacement field **D** are related by

$$\frac{D_{\parallel,\mathrm{A}}}{\varepsilon_\mathrm{A}} = \frac{D_{\parallel,\mathrm{B}}}{\varepsilon_\mathrm{B}},$$

where ε_A and ε_B are the relative permittivities in the two regions.

$$E_{\parallel,\mathrm{A}} = E_{\parallel,\mathrm{B}}. \tag{11.31}$$

Applying **Ampère's law** using an identical analysis, we find that the parallel component of the magnetic field strength is similarly continuous:

$$\int_\mathrm{P}^\mathrm{Q} H_{\parallel,\mathrm{A}}\,\mathrm{d}x + \int_\mathrm{R}^\mathrm{S} H_{\parallel,\mathrm{B}}\,\mathrm{d}x = \int_\mathrm{PQRS} \mathbf{J} + \frac{\partial\mathbf{D}}{\partial t} \cdot \mathrm{d}\mathbf{S}, \tag{11.32}$$

The parallel components of the magnetic flux density **B** are hence related by

$$\frac{B_{\parallel,\mathrm{A}}}{\mu_\mathrm{A}} = \frac{B_{\parallel,\mathrm{B}}}{\mu_\mathrm{B}},$$

where μ_A and μ_B are the relative permeabilities in the two regions.

where **J** is the current density (usually zero) parallel to the interface, and hence

$$H_{\parallel,\mathrm{A}} = H_{\parallel,\mathrm{B}}. \tag{11.33}$$

To determine the conditions relating field components *perpendicular* to the interface, we follow a similar analysis using the remaining pair of Maxwell's equations, this time referring to a closed surface containing a volume of integration, shown in Fig. 11.3(b), and once again taking the limit as its dimensions fall to zero. In the absence of surface charge, we obtain

Equivalently

$$\varepsilon_\mathrm{A} E_{\perp,\mathrm{A}} = \varepsilon_\mathrm{B} E_{\perp,\mathrm{B}},$$

$$\mu_\mathrm{A} H_{\perp,\mathrm{A}} = \mu_\mathrm{B} H_{\perp,\mathrm{B}}.$$

$$D_{\perp,\mathrm{A}} = D_{\perp,\mathrm{B}}, \tag{11.34a}$$

$$B_{\perp,\mathrm{A}} = B_{\perp,\mathrm{B}}. \tag{11.34b}$$

We may now consider a specific example of our analysis at the beginning of this section. If the incident, reflected and transmitted electric fields are E_i, E_r and E_t, equation (11.31) first gives

$$E_\mathrm{i} + E_\mathrm{r} = E_\mathrm{t}. \tag{11.35}$$

In terms of the characteristic impedance $Z = \sqrt{\mu/\varepsilon}$, which for optical media is simply inversely proportional to the refractive index, the corresponding magnetic fields H_i etc. may be written in terms of the electric field,

$$H = \pm ZE, \tag{11.36}$$

where the sign depends upon the direction of wave propagation, and hence equation (11.33) gives

$$Z_\mathrm{A}(E_\mathrm{i} - E_\mathrm{r}) = Z_\mathrm{B} E_\mathrm{t}. \tag{11.37}$$

On combining equations (11.35) and (11.37), we therefore reproduce equation (11.29a) for the case of electromagnetic waves.

We have in this section restricted ourselves to the one-dimensional case of plane waves that are normally incident upon a plane interface between two media. For the more general case in which the wave is incident at some angle ϑ

we must also consider the continuity conditions for the perpendicular components of the electric and magnetic fields (equations (11.34a) and (11.34b)). The working is rather more lengthy, and will not be covered here, but takes exactly the same path as we have followed above.

11.4 Reflection and transmission by multiple interfaces

Hecht [40] pp. 426–428
Lipson *et al.* [56] Section 5.4
Pedrotti *et al.* [68] pp. 391–404

We have seen that knowledge of the continuity conditions allows us to determine how much of an incident wave is reflected upon meeting the interface between two media. We shall now extend our analysis of a single interface to the arbitrary arrangements of such boundaries found in a number of commonly used optical technologies. We begin with an alternative to the analysis of thin-film interference given in Section 5.3.

11.4.1 Thin-film interference and blooming

Photographic lenses, when viewed in daylight, often have a blueish tint. This is the result of a thin coating of a transparent material (typically magnesium fluoride) that reduces reflections of light passing through the multiple elements of the lens and leaves only a weak residual reflectivity towards the edges of the visible spectrum. The coating of lenses in this fashion is known as **blooming**.

find travelling solutions to wave equations in each uniform, isotropic region

↓

add counter-propagating waves to form general solution in each region

↓

apply continuity conditions to match solutions at interfaces

Figure 11.4 shows schematically a typical arrangement. As in Fig. 11.2, light is incident from the left, and passes from air A, through the coating B of thickness d, into the glass or silica C from which the lens is made. The incident wave, as in Section 11.2, is taken to be $\psi_1(t - x/v_A)$, and, as before, we write the waves that are reflected and transmitted at the first interface as $\psi_2(t + x/v_A)$, and $\psi_3(t - x/v_B)$. This time, however, the transmitted beam subsequently strikes the second interface, and may itself be either reflected or transmitted; we therefore add further waves $\psi_4(t + x/v_B)$ and $\psi_5(t - x/v_C)$, corresponding to these possibilities. Because we must satisfy continuity conditions at two different positions, we may not immediately write sinusoidal solutions simply by matching the temporal dependences at the origin; we therefore leave the wave components in the form of arbitrary travelling waves.

The total wave amplitudes in the three sections are therefore

$$\psi_A(x, t) = \psi_1(t - x/v_A) + \psi_2(t + x/v_A), \qquad (11.38a)$$

$$\psi_B(x, t) = \psi_3(t - x/v_B) + \psi_4(t + x/v_B), \qquad (11.38b)$$

$$\psi_C(x, t) = \psi_5(t - x/v_C). \qquad (11.38c)$$

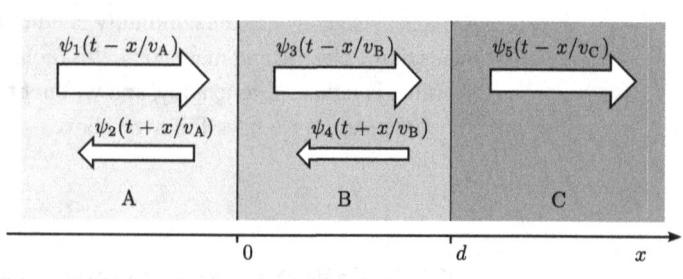

Fig. 11.4 Transmission and reflection of waves at a multiple interface.

We now apply continuity conditions (11.28a) and (11.28b) at the two boundaries, giving

$$\psi_1(t - 0) + \psi_2(t + 0) = \psi_3(t - 0) + \psi_4(t + 0), \tag{11.39a}$$

$$Z_A[\psi_1(t - 0) - \psi_2(t + 0)] = Z_B[\psi_3(t - 0) - \psi_4(t + 0)], \tag{11.39b}$$

$$\psi_3(t - d/v_B) + \psi_4(t + d/v_B) = \psi_5(t - d/v_C), \tag{11.39c}$$

$$Z_B[\psi_3(t - d/v_B) - \psi_4(t + d/v_B)] = Z_C\,\psi_5(t - d/v_C). \tag{11.39d}$$

Solution of these four equations is rather less messy if the fields are all complex exponentials $\psi_n(u) \equiv a_n \exp(i\omega u)$, so that equations (11.39) become

$$a_1 + a_2 = a_3 + a_4, \tag{11.40a}$$

$$Z_A(a_1 - a_2) = Z_B(a_3 - a_4), \tag{11.40b}$$

$$a_3 \exp(-i\omega t/v_B) + a_4 \exp(i\omega t/v_B) = a_5 \exp(-i\omega t/v_C), \tag{11.40c}$$

$$Z_B[a_3 \exp(-i\omega t/v_B) - a_4 \exp(i\omega t/v_B)] = Z_C a_5 \exp(-i\omega t/v_C). \tag{11.40d}$$

Apart from the phase factors, the second pair of equations is solved as in Section 11.2.1, giving, as in equation (11.29),

$$\frac{a_5 \exp(-i\omega t/v_C)}{a_3 \exp(-i\omega t/v_B)} = \frac{2Z_B}{Z_B + Z_C}, \tag{11.41a}$$

$$\frac{a_4 \exp(i\omega t/v_B)}{a_3 \exp(-i\omega t/v_B)} = \frac{Z_B - Z_C}{Z_B + Z_C}. \tag{11.41b}$$

Pain [66] pp. 350–353
Hecht [40] pp. 426–430
Lipson *et al.* [56]
Section 5.4

These results may now be inserted into the first pair of equations to give, after a certain amount of grief,

$$r \equiv \frac{a_2}{a_1} = \frac{Z_B(Z_C - Z_A)\cos(k_Bd) + i(Z_B^2 - Z_CZ_A)\sin(k_Bd)}{Z_B(Z_C + Z_A)\cos(k_Bd) + i(Z_B^2 + Z_CZ_A)\sin(k_Bd)}, \tag{11.42a}$$

$$t \equiv \frac{a_5}{a_1} = \frac{2Z_BZ_C \exp(ik_Cd)}{Z_B(Z_C + Z_A)\cos(k_Bd) + i(Z_B^2 + Z_CZ_A)\sin(k_Bd)}, \tag{11.42b}$$

where $k_A \equiv \omega/v_A$, etc. Note that, while this approach may in principle be extended to an arbitrary number of interfaces, the bookkeeping can be simplified by adopting matrix methods that relate the fields at adjacent interfaces.

A crucial result is that the reflection coefficient will be zero if the blooming layer is a *quarter of a wavelength* thick ($k_Bd = \pi/2$) and is made from a

Fig. 11.5   Radar-absorbing materials, which are based upon modified *quarter-wave* structures for radar
wavelengths, helped cut the radar cross-section of the F-117A Nighthawk *'stealth fighter'* to that of a
metal tea-pot.   Ⓒ U.S. Air Force photo

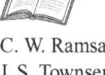

C. W. Ramsauer [73]
J. S. Townsend &
V. A. Bailey [87]

material with a refractive index (and hence *characteristic impedance*) equal
to the geometric mean of those of the adjacent lens material and air (i.e.
$Z_B^2 = Z_A Z_C$). This is the principle of optical *antireflection coatings*, and the
same principle is used to reduce reflections in other situations, such as from
the '*stealth*' aircraft of Fig. 11.5; with a quantum-mechanical wavefunction,
it is the origin of the anomalous scattering of the *Ramsauer–Townsend effect*,
whereby for certain electron energies a gas of argon or some similar element
can become relatively transparent to a beam of electrons.

The eagle-eyed will spot that, when these conditions for an antireflection
coating are met, the amplitude transmission coefficient is not equal to unity;
instead, we have

$$t \equiv \frac{a_5}{a_1} = \exp[i(k_C - k_B)d]\sqrt{\frac{Z_C}{Z_A}}, \tag{11.43}$$

where $\exp(ik_B d) = i$ and the complex factor simply accounts for the difference in phase from the case of a continuous medium A. Depending upon the
impedances Z_A and A_C, the transmitted wave amplitude a_5 can therefore exceed
the incident amplitude a_1, but only if the wave energy propagates more slowly
in C than it does in A; it is the incident and transmitted intensities, given by
equation (4.18), that will be equal.

This analysis is an alternative to the simple method we used to address thin-
film interference in Section 5.3. Although it is more complex, it does not begin
with any assumptions about the reflectivity of the interface, which we should
have to calculate as in Section 11.2. Indeed, the two approaches would then be
formally equivalent. The calculation may be extended to non-normal incidence

by allowing perpendicular field components, whose continuity conditions were given in equations (11.34).

The power of this method is that it may now be extended to arbitrary stacks of thin layers, albeit with some tidying of the algebra. Today's best mirrors, which have reflectivities differing from unity by less than a millionth part, are indeed made from multiple layers of materials that are individually highly transmitting. Metals, in contrast, typically have reflectivities no higher than 95%.

11.5 Total internal reflection

Pain [66] pp. 218–221
Hecht [40] pp. 113–127
Lipson *et al.* [56]
Section 5.5

One of the most profound consequences of the continuity conditions at the interface between two media requires us to venture beyond our largely one-dimensional treatment of wave propagation and consider what happens when light passes from one medium to another at an angle to the interface.

One method of analysis, whose details we leave as an exercise for the reader, is to repeat the treatment of Section 11.2, with incident plane wavefronts making an angle ϑ_1 to the interface in the medium of refractive index η_1 and the refracted wavefronts making an angle ϑ_2 in the medium of index η_2, where, according to Snell's law from Section 6.4,

$$\eta_1 \sin \vartheta_1 = \eta_2 \sin \vartheta_2. \tag{11.44}$$

Applying the electromagnetic boundary conditions of equations (11.31), (11.33) and (11.34), with the transverse electric and magnetic fields resolved parallel and perpendicular to the interface, then yields the **Fresnel equations**, which define the relative strengths of the reflected and transmitted fields,

$$\frac{E_r^p}{E_i^p} = -\frac{\tan(\vartheta_1 - \vartheta_2)}{\tan(\vartheta_1 + \vartheta_2)}, \tag{11.45a}$$

$$\frac{E_t^p}{E_i^p} = \frac{2 \sin \vartheta_2 \cos \vartheta_1}{\sin(\vartheta_1 + \vartheta_2) \cos(\vartheta_1 - \vartheta_2)}, \tag{11.45b}$$

$$\frac{E_r^s}{E_i^s} = -\frac{\sin(\vartheta_1 - \vartheta_2)}{\sin(\vartheta_1 + \vartheta_2)}, \tag{11.45c}$$

$$\frac{E_t^s}{E_i^s} = \frac{2 \sin \vartheta_2 \cos \vartheta_1}{\sin(\vartheta_1 + \vartheta_2)}. \tag{11.45d}$$

Fresnel equations

The sign of the p-polarization amplitude reflectivity in equation (11.45a) depends upon the convention chosen for the positive field direction of the reflected beam. Here we take all positive electric field directions to coincide at normal incidence; some authors take the incident and reflected field directions to coincide at grazing incidence.

The labels p and s refer, respectively, to the *polarization* of the electric field – which we shall meet properly in Chapter 15 – parallel and perpendicular to the **plane of incidence** that contains the ray direction and interface normal.

The behaviour of the reflected and transmitted light is now apparent. If we begin with light that is incident normally upon the interface, then we have the situation considered in Section 11.2; the transmitted light leaves undeviated but slightly attenuated, the balance being reflected back along the direction of incidence. As we tilt the incident light away from the interface normal, the

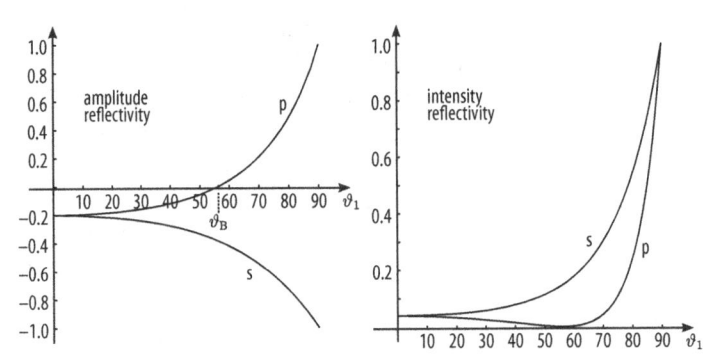

Fig. 11.6 Typical amplitude and intensity reflection coefficients at a glass–air interface for s- and p-polarizations of light as a function of the air-side angle ϑ_1. The p-polarization reflectivity is zero when light strikes the interface at Brewster's angle ϑ_B.

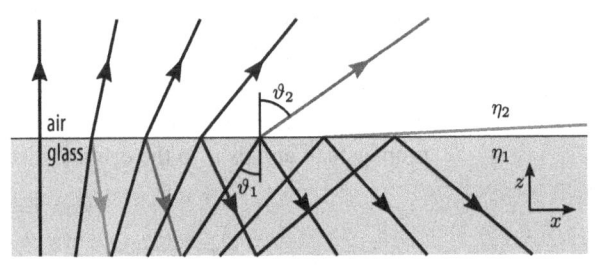

Fig. 11.7 Rays approaching a glass–air interface at different angles from within, showing internal and total internal reflection.

reflected beam leaves at an equal but opposite angle, and the transmitted beam is refracted towards or away from the normal according to equation (11.44). As shown in Fig. 11.6 for an air–glass interface with $\eta_1 < \eta_2$, the reflected fraction of light increases monotonically for the s-polarization, while for the p-polarization it first falls to a zero at **Brewster's angle** ϑ_B, which is given by

$$\vartheta_B = \tan^{-1}\left(\frac{\eta_2}{\eta_1}\right). \tag{11.46}$$

The average, experienced by unpolarized (randomly polarized) illumination, increases monotonically and, as the angle of refraction approaches 90°, the reflectivity for both polarizations approaches unity.

Equations (11.45) are valid for any two media, and hence apply equally to the situation illustrated in Fig. 11.7, when the ray path is reversed to start in the more refractive region, such as when light passes from glass or water into air. For passage into a medium of lower refractive index, a refraction angle of 90° is achieved when the incident beam is only somewhat inclined to the interface: this **critical angle** of incidence ϑ_C, from equation (11.44), will be

$$\vartheta_C = \sin^{-1}\left(\frac{\eta_2}{\eta_1}\right). \tag{11.47}$$

We may, however, incline the incident beam still further. Snell's law no longer appears to help us, for it suggests that $\sin \vartheta_2$ will be greater than unity. Experimentally, we observe that the transmitted beam, which had already vanished at the critical angle, remains absent; all the light is reflected, and we refer to **total internal reflection**. A detailed analysis of the field beyond the interface, as we shall now see, yields some important surprises.

11.5.1 The evanescent wave

evanesce/,i:vənes,,e-/v. intr. **1** fade from sight; disappear. **2** become effaced. [L *evanescere* (as E-.*vanus* empty)] **evanescent**/,i:və'nes(ə)nt,,e-/ adj. (of an impression or appearance etc.) quickly fading. [13]

If we perform the analysis outlined above, we find that it is impossible to satisfy the continuity conditions if the transmitted optical field is zero; yet no single plane-wave solution will satisfy them either. The analysis will in fact yield the true field with little difficulty, but, before discussing it in detail, we may consider how it arises by returning to our fundamental ideas of wave propagation.

Consider a diffraction grating comprising a series of narrow transmitting slits, of spacing d, in an otherwise opaque mask, illuminated at normal incidence by a plane sinusoidal wave of wavelength λ. As we have seen in Section 9.4, the waves originating in each slit add constructively to form new plane waves that propagate at angles ϑ to the original direction, where

$$d \sin \vartheta = m\lambda, \tag{11.48}$$

m being an integer. Even if the spacing d is less than the wavelength, which would require $|\sin \vartheta| > 1$ for $m \neq 0$, we shall always be left with the undeflected, *zeroth-order* beam corresponding to $m = 0$.

We may, however, eliminate even this beam by opening an additional series of slits, interleaved midway between the first, and arranging that the waves from these are delayed by a half wavelength – for example, by increasing the thickness of a glass plate in this region or, at longer wavelengths, by appropriate programming of a phased-array radar transmitter. The contributions from the two sets of slits will be equal but opposite, and will hence cancel out even the undeflected beam.

If we now zoom in on a region close to the grating, however, so that we are significantly closer to one of the slits in particular, then the wave originating from that single slit will be stronger than those from adjacent slits – and, in this regime of *Fresnel diffraction*, the relative path lengths will not show the simple dependence upon slit position to which we are accustomed in the Fraunhofer regime. The result is that, close to the grating, the waves do not cancel out completely, and some disturbance will definitely occur.

This is effectively the situation under conditions of total internal reflection. The incident wavefronts strike the interface at an angle, and hence with a phase that varies linearly along the interface, corresponding in the Huygens description to a line of secondary sources that alternate between one phase and its opposite (and also fill in all the intermediate phases as well). If there is an angle at which the waves from these secondary sources add constructively, then a transmitted beam will occur; if there is not, because the spacing between successive sources of a given phase is less than the wavelength in the second medium, then total internal reflection occurs. Just as in our imagined example with a modified diffraction grating, though, some disturbance will occur close to the interface, but its magnitude will decrease – it turns out exponentially – as we move away from the interface. This is known as the **evanescent** (or boundary) wave.

11.5.2 Motion of the evanescent wave

A little thought will show that the intersections of incident wavefronts with the interface between the two media will move along the surface with a speed $v_\parallel$ determined by the incident velocity c/η_1 and the angle of incidence ϑ_1,

$$v_\parallel = \frac{c}{\eta_1 \sin \vartheta_1}. \qquad (11.49)$$

If ϑ_1 is below the critical angle ϑ_C, then $v_\parallel$ will be *greater* than the wave speed c/η_2 in the second region (as it is also in the first), and the intersections will match the wavefronts of a wave propagating at ϑ_2 to the normal, in accordance with Snell's law (11.44) and as illustrated in Fig. 6.7. Above the critical angle, however, since $\eta_1 \sin \vartheta_1 > \eta_2$, $v_\parallel$ will be *below* the speed of free wave propagation in the second medium, and an evanescent wave results. It may be helpful to consider placing Huygens secondary sources at these intersections, and to allow them to move with them. Points of constant phase in the evanescent wave will move in step with these secondary sources but, perhaps surprisingly, there is no variation in the phase of the evanescent wave in the direction normal to the interface. The evanescent wave thus takes the form of an oscillating disturbance, decreasing rapidly in amplitude away from the surface as illustrated in Fig. 11.8, that skims across the surface more slowly than the speed of free waves. The evanescent wave does not truly propagate from point to point within itself, however, but rather represents different waves originating from the same plane-wave sources before the interface.

As we ponder the curious phenomenon of the evanescent wave, several questions eventually arise, the first being the decrease in wave magnitude with distance from the interface. For this, it is instructive to consider first the case in which the angle of the incident beam is just below critical (given by equation (11.47)), so that a refracted beam is just allowed to propagate, almost parallel to the interface, and the speed of propagation of the secondary sources just matches that of the wave in the second medium. Since the sources move at the same speed as the refracted wave, their effects all add constructively, and the broad plane wavefronts are produced as normal.

As the angle of incidence ϑ_1 is increased to and beyond the critical angle, we find that the speed of the sources slows until it can no longer match that of wavefronts in the second medium. If we are only a few wavelengths from the

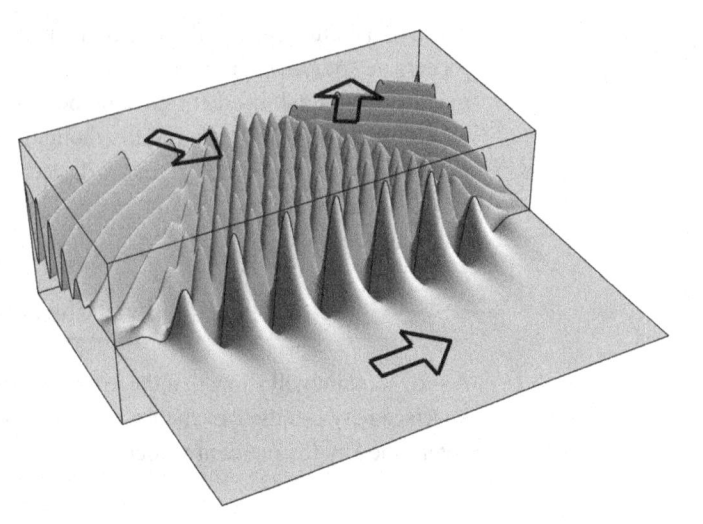

Fig. 11.8 An evanescent wave skims quickly along the surface of a refracting medium, and falls off exponentially with distance from the interface.

interface, then we see only a small number of sources and the phase mismatch at any time is small. At greater distances, however, more sources contribute a significant disturbance, and the range of phases increases, so that most of them interfere destructively. The greater the mismatch between the speed of the secondary sources and that of wave propagation in the second medium, the greater the effect of destructive interference and the shorter the range of the evanescent wave. When the angle of incidence is exactly critical, then, the range is infinite, but only a small change in angle is required in order to bring the range down to a few wavelengths.

11.5.3 Energy flow in the evanescent wave

A further question concerns the energy of the evanescent wave, for the disturbance must carry energy, yet the incident wave appears to be totally reflected. Two observations are instructive. First, the extent of the evanescent wave into the second medium, discussed above, is finite, while the plane waves are infinite in width; the fraction of the wave energy in the evanescent field is therefore infinitesimal. Secondly, the energy of the evanescent wave does not propagate away from the interface, but travels parallel to it, and is constantly exchanged with the waves within the first medium. Once the evanescent wave has been established, then, no average energy flow away from the surface occurs.

11.5.4 Characterization of the evanescent wave

We have characterized sinusoidal travelling and standing waves through their amplitude, frequency ω and wavenumber k. The evanescent wave described above results from refraction of such a travelling wave, yet we find that it does not share the usual form. For the most common example of light, a complete analysis frustratingly requires a vector treatment of waves and a little knowledge of electromagnetism; we here describe some of the important features.

Parallel to the interface, the structure of the evanescent wave is unremarkable: it follows the same travelling sinusoidal motion as the intersections of the incident wavefronts with the interface itself. The evanescent waveform ψ_e proves to be separable into functions of the coordinates x and z parallel and perpendicular to the interface (i.e. $\psi_e(x, z) = X(x)Z(z)$). As the intersections of successive wavefronts are at any time separated by a distance $\lambda_1 / \sin \vartheta_1$, where $\lambda_1 = 2\pi c/(\eta_1 \omega)$ is the wavelength in the first medium, the variation along the interface will be

$$X(x) = \cos\left(\frac{2\pi \sin \vartheta_1}{\lambda_1}x - \omega t\right)$$
$$= \cos(k_x x - \omega t), \tag{11.50}$$

where we have written an effective wavenumber in the x direction,

$$k_x \equiv \frac{\eta_1}{c} \sin \vartheta_1 \, \omega. \tag{11.51}$$

If we take a slice through our wave system that is normal to the interface, then the incident wave ψ_i will vary perpendicular to the interface according to

$$\psi_i(z) = \cos\left(\frac{2\pi \cos \vartheta_1}{\lambda_1}z - \omega t\right)$$
$$= \cos(k_z z - \omega t), \tag{11.52}$$

where

$$k_z \equiv \frac{\eta_1}{c} \omega \cos \vartheta_1. \tag{11.53}$$

The evanescent wave, in contrast, proves to vary as

$$Z(z) = \exp(-\alpha z)\cos(\omega t), \tag{11.54}$$

where it is instructive to write $\alpha \equiv -ik_z'$, with k_z' imaginary and given by

$$k_z' \equiv \frac{\omega}{c}\sqrt{\eta_2^2 - (\eta_1 \sin \vartheta_1)^2}. \tag{11.55}$$

The continuity conditions may be applied at the interface as before to determine the relative wave amplitudes. If equation (11.52) is split into complex waves $\exp[\pm i(k_z z - \omega t)]$, then these strongly resemble the evanescent form of equation (11.54), except that the imaginary argument of the exponent in the incident wave becomes real in the evanescent case. Sections of incident and evanescent waves normal to the interface are shown for various cases in Fig. 11.9. The reflected travelling wave combines with the incident field as if reflected by a mirror, except that the resulting standing wave is shifted according to the refractive indices – a phenomenon known as the *Goos–Hänchen effect* that may be used for the spectroscopic detection of surface layers.

H. F. G. Goos &
H. Hänchen [33]

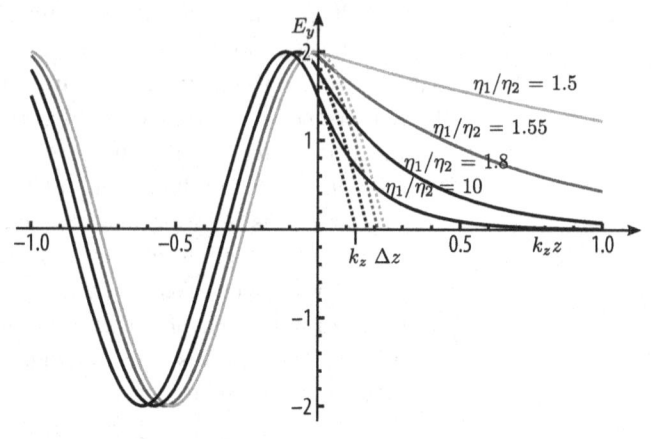

Fig. 11.9 Section through incident and evanescent wave perpendicular to the interface for $\vartheta_1 = 42°$, showing the standing sinusoidal and decaying exponential waves. The continuity conditions apply as usual. The displacement Δz of the apparent node in the extrapolated (dashed) field is the Goos–Hänchen shift.

Equation (11.55) may be re-written as

$$k_z' \equiv \sqrt{\left(\frac{\eta_2 \omega}{c}\right)^2 - k_x^2}, \tag{11.56}$$

so that

$$k_x^2 + k_z'^2 = \left(\frac{\eta_2 \omega}{c}\right)^2. \tag{11.57}$$

Since $k = \eta_2 \omega/c$ would be the wavenumber of a travelling wave of angular frequency ω in the second medium, it is tempting to regard k_x and k_z' as the components of a **wavevector**. This is indeed precisely the way that we shall analyse a range of manifestations of real and evanescent waves; it is mathematically rigorous, and an important tool in the application of wave physics to quantum mechanics. If we define ϑ_2 by Snell's law (11.44) and simply allow it to take a complex value in the evanescent case, then we may write

$$k_z' = \frac{\eta_2}{c} \omega \cos \vartheta_2, \tag{11.58}$$

which strongly resembles the form of equation (11.53).

11.6 Frustrated total internal reflection

If the change of medium, from high to low refractive index, that creates an evanescent wave is closely followed by a second change from the lower back to the higher refractive index, as illustrated in Fig. 11.10, then the evanescent wave can be converted back into an ordinary propagating wave. The continuity conditions are the same, and hence the reflection and transmission coefficients

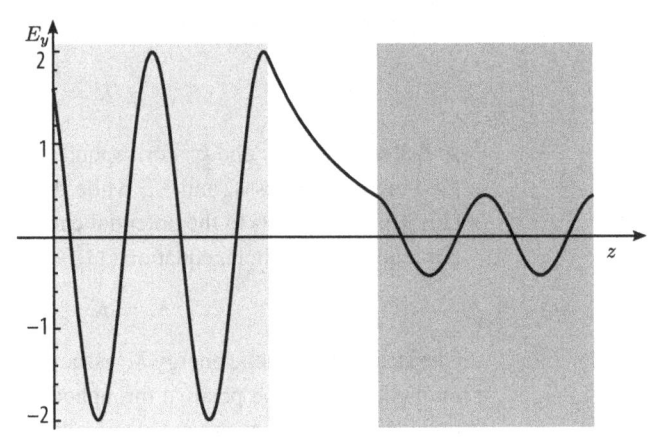

Fig. 11.10
Section through incident, evanescent and transmitted waves perpendicular to the interface, showing the travelling sinusoidal and decaying exponential waves. The continuity conditions apply as usual.

are again given by the Fresnel equations (11.45), with ϑ_1 and ϑ_2 interchanged, and we find that what remains of the evanescent field by the time it reaches the second interface is divided between a transmitted normal beam and a reflected evanescent wave that continues its exponential decay back as far as the first interface, where the wave cancels out part of the first surface reflection as required for energy conservation. An air gap between two blocks of glass, for example, will therefore act as a variable beamsplitter, with a transmission that decreases exponentially with the separation of the two pieces of glass. In most practical situations, the gap between the glass blocks must be tiny – a few wavelengths at most – if the transmission is to be appreciable.

11.6.1 Tunnelling

We have seen above that an evanescent field results when the component k_z' of the wavevector normal to the interface is imaginary. On combining equation (11.57) with the equivalent expression before the interface, we may write

$$k_z'^2 = k_z^2 - \left[\left(\frac{\eta_1 \omega}{c} \right)^2 - \left(\frac{\eta_2 \omega}{c} \right)^2 \right]. \qquad (11.59)$$

An evanescent wave may therefore be considered to occur when the normal component of the incident wavevector is insufficient to exceed the quantity $(\omega/c)\sqrt{\eta_1^2 - \eta_2^2}$. Free propagation under such circumstances is not permitted.

    The process of frustrated total internal reflection is known in quantum mechanics as **tunnelling**. The quantum wavefunction behaves exactly as the optical wavefunction sketched in Figs. 11.9 and 11.10, but the evanescent wave condition of equation (11.59) has an alternative interpretation. The kinetic energy $\mathcal{K} = mv^2/2 = p^2/(2m)$ for a particle of mass m, velocity v and normal component of the momentum $p = \hbar k_z$ may be written as

$$\mathcal{K} = \frac{\hbar^2 k_z^2}{2m}, \tag{11.60}$$

so that the terms k_z^2 and $k_z'^2$ correspond, appropriately scaled, to components of the kinetic energies $\mathcal{K}$ and $\mathcal{K}'$, while the term in square brackets in equation (11.59) corresponds to the potential energy $\mathcal{U}_{1,2}$ of the particle in each region. For a quantum particle, equation (11.59) hence becomes

$$\mathcal{K}' = \mathcal{K} - (\mathcal{U}_1 - \mathcal{U}_2). \tag{11.61}$$

Classically, the kinetic energy $\mathcal{K}'$ associated with the normal component of the momentum must be positive throughout; particles are seen to *tunnel* through regions in which they are not classically permitted.

Frustrated total internal reflection may equivalently be regarded as tunnelling of light through a region in which free propagation is not allowed. While there is no precise equivalent to the potential and kinetic energies, the quantity $-k_z^2$ mathematically plays the rôle of the kinetic energy, and $-(\eta\omega/c)^2$ plays the rôle of the potential energy.

11.7   Applications of internal reflection and evanescent fields

Evanescent fields are general wave phenomena, and can be found in a variety of situations, from the deep-water gravity waves of Section 3.3.3 to the *surface plasmons* (or *surface plasmon polaritons*) – a combination of longitudinal charge-density waves and optical evanescent waves that exist around the interface between, and penetrate with exponentially decaying strength into, a metal and a dielectric. Total internal reflection guides radio waves around the Earth's curvature, gives diamonds their sparkle, and allows retro-reflecting prisms to operate without mirror coatings; frustrated total internal reflection has been observed where ocean waves cross a seabed canyon.

11.7.1   Optical-fibre waveguides

Total internal reflection plays a crucial rôle in modern high-bandwidth communications, where it is the principle behind **optical fibres**. These thin strands of silica or other transparent refractive material guide narrow beams of light, which, even when the fibre is bent, travel at near-grazing angles to the fibre surface, and hence strike the fibre edge at an angle that comfortably exceeds the critical angle for total internal reflection. Since the beam is completely confined to the fibre, the only way in which energy can be lost from the propagating beam is by absorption, and with carefully prepared materials of great purity an optical beam can lose as little as 4% of its power per kilometre, and thereby travel tens of kilometres before it must be amplified back to its original level.

This remarkable performance is possible only if there are no losses through frustrated total internal reflection, so the fibre must be kept well away from any supporting materials or contamination. In practice, the fibre therefore forms the central *core* of a rather thicker fibre of somewhat lower refractive index, so that the evanescent field is substantially attenuated before it reaches the edge of the outer *cladding*. The presence of the cladding reduces the refractive-index difference at the edge of the fibre core, and thus increases the critical angle for total internal reflection, so the fibre cannot be bent as sharply if losses are to be avoided.

Some fibre-optic devices, including optical switches and chemical sensors, exploit frustrated total internal reflection to provide a controllable, or species-dependent, coupling of light into or out of the fibre: the cladding is reduced in thickness by chemical etching or by *drawing* to a smaller diameter. If a planar waveguide (a sheet of glass or transparent polymer) is used instead of a fibre, frustrated total internal reflection can be used to determine the position of a touching object, for fingerprint recording and optical touch-screens.

11.7.2 Photon scanning tunnelling microscopy

The sharp dependence of the transmission of frustrated total internal reflection upon the separation of the refractive surfaces provides a method of high resolution *near-field* microscopy for the study of optical materials and devices. The finely drawn tip of an optical fibre or similar structure forms one of the refractive surfaces, with the second provided by the material or structure to be studied. The tip is then mechanically scanned across the surface, and the variation in optical coupling between the tip and material is used to infer the varying distance between them and hence the profile of the studied surface.

An equivalent arrangement, measuring the current of tunnelling electrons rather than the intensity of tunnelling photons, is the basis of the *scanning tunnelling microscope*.

11.8 Evanescent-wave confusions and conundrums

Evanescent waves and frustrated total internal reflection are fertile areas for fascinating conundrums that can both lend helpful insight and cause great confusion. The interested reader is naturally encouraged to explore such intriguing curiosities, but warned to be aware that much thought may be required, and that there are some dubious explanations in circulation.

Evanescent waves are intrinsically near-field phenomena, so far-field approximations may be invalid. Pulsed evanescent waves must contain a spread of frequencies or wavelengths; and beams of finite transverse extent must correspond to a spread of wavevectors. When puzzles concern mysterious energy flow or apparently superluminal propagation velocities, it must be remembered that results for infinite, monochromatic, plane waves might not be valid.

Most theoretical analyses begin with Maxwell's equations, which are fundamentally causal and satisfy the principles of relativity and energy conservation.

Exercises

11.1 Show that the continuity conditions for the longitudinal displacement ξ of sound waves as they pass from a region 1 of density ρ_1 and elasticity E_1 into another region 2 of density ρ_2 and elasticity E_2 are

$$\xi_1(x_0, t) = \xi_2(x_0, t), \tag{11.62}$$

$$E_1 \frac{\partial \xi_1}{\partial x}(x_0, t) = E_2 \frac{\partial \xi_2}{\partial x}(x_0, t), \tag{11.63}$$

where $\xi_{1,2}(x, t)$ are the wave displacements in the two regions at position x and time t, and the boundary between the two regions occurs at $x = x_0$.

11.2 The displacements of sinusoidal sound waves in the two regions of Exercise 11.1 may be written in the form

$$\xi_{1,2}(x, t) = a_{1,2} \cos(\omega t - k_{1,2}(x - x_0)). \tag{11.64}$$

What are the possible values for the wavenumber $k_{1,2}$ in the two regions, for a given angular frequency ω?

Show that when a sound wave of amplitude a_1 passes from region 1 to region 2, it results in a reflected wave of amplitude a_r, where

$$\frac{a_r}{a_i} = \frac{Z_1 - Z_2}{Z_1 + Z_2} \tag{11.65}$$

and the acoustic impedance Z is given in terms of the density ρ and elasticity E by $Z = \sqrt{E\rho}$.

11.3 By using your results from Exercises 11.1 and 11.2, find the fraction of the wave *intensity* that is reflected when a sonar pulse encounters the boundaries between

1. freshwater from an Arctic river and the salty ocean into which it flows (both at about 0 °C), and
2. cold water from a mountain stream at 0 °C and the bottom of a warm lake at 20 °C into which it flows.

You may take the values for the density of sea water at 0 °C and 20 °C to be 1028.1 and 1024.8 kg m^{-3} respectively, and the corresponding values for freshwater to be 999.8 and 998.2 kg m^{-3}. The elasticities may be taken to be 2.32×10^9 Pa for sea water and 2.05×10^9 Pa for freshwater, and are approximately independent of temperature.

11.4 Given that the modulus of elasticity of the flesh of the Atlantic mackerel *Scomber scombrus* is about 2.58×10^9 Pa, and that the fish may be assumed homogeneous and neutrally buoyant, find the fraction of the wave intensity that is reflected when a *fish-finder* sonar pulse is normally incident upon the unsuspecting creature.

11.5 The Acme Corporation flying club has recently fitted double glazing to the windows of its airfield bar so that, after a day's aviation, members can enjoy a tipple in warmth and comfort while watching the club's more ambitious students practising their night-flying in the clear but dark winter evenings. The double glazing comprises pairs of panes of glass, each 6 mm thick (and with a refractive index of 1.52), separated by a 4-mm air gap. The panes are of reasonable optical quality, and there have been no complaints during the daytime of any noticeable distortion of the view of the airfield from the bar.

One by one, the students complete their exercises and retire to the bar to swap heroic tales of derring-do, until only one student remains airborne, practicing circuits, approaches and landings in the crisp winter night. But Pontius, the oldest pilot in the club, looking up from his double whisky, is amazed to see not one but two sets of navigation lights on the final approach to the airfield.

Is Pontius seeing double after a malt whisky too many, or can he really see a second aircraft in the black night sky? What features of the double glazing might give rise to such a phenomenon? How much more distant will the second aircraft appear to be?

12   Boundary conditions

12.1   The imposition of external constraints

Pain [66] pp. 124–126
French [29] pp. 161–181
Main [60] pp. 181–183

We have established in Chapter 11 that the motion of an unobstructed wave is generally constrained by two conditions, which, for our guitar-string example, are the continuity of the wave displacement ψ and of its first derivative $\mathrm{d}\psi/\mathrm{d}x$. For a string to follow a wave motion in which ψ were discontinuous would involve the finite extension of the arbitrarily small element containing the discontinuity, and hence require that that element be stretched beyond its linear range, causing it to snap. Similarly, were the wave motion to include a region of discontinuity in $\mathrm{d}\psi/\mathrm{d}x$, the corresponding arbitrarily small element would experience a finite force, and thus arbitrarily large acceleration. We have seen that there are equivalent continuity conditions for other wave motions; in each case, the constraints have similarly practical origins.

At boundaries, such as where the guitar string touches a fret, we have two options. If the fret has a degree of compliance, then we may regard it as a separate wave-sustaining system with different characteristics, and apply the continuity conditions as in Chapter 11 to determine how the string's motion is transmitted, for example, to the body of the guitar and hence to the air and ultimately our eardrum. If we were interested only in the motion of the string, however, then it might suffice to assume a completely rigid fret, which could in this approach be represented by a string section with infinite mass per unit length.

The alternative approach in such circumstances is simply to characterize such a boundary directly by the effect that it has upon the string motion – in this case, forcing the string displacement to be zero at that point. We might also wish to consider, however, the point at which a violin string is driven by the bow, the surface of a loudspeaker where the air meets the moving diaphragm, or the region in which the reed of a woodwind instrument vibrates the column of air. In such cases, the constraints may depend upon time, and may control some combination of the displacement and its gradient. In the general case, then, we shall have to handle the **boundary conditions** with the same formality as that which we afforded the continuity conditions in the previous chapter.

find travelling solutions
to wave equation in
unrestricted region

↓

add counter-propagating
waves to form general
solution in each region

↓

apply boundary
conditions
at limits of region to
give particular solutions

12.2 The guitar and other stringed musical instruments

In Section 4.4.2, we saw that constraining the motion of a string at its ends limits the frequencies of sinusoidal wave motion to particular values for which an integral number of half wavelengths will fit between the two ends. We found these motions by seeking specific *standing-wave* solutions to the wave equation for the guitar string and noting that particular examples satisfied the boundary conditions, namely that there be no motion at the bridge and at the fret.

The treatment presented in Chapter 11 offers an alternative approach to the effect of the boundary conditions imposed by the guitar bridge and fret at $x = 0$ and $x = l$, for we may take a general solution to the wave equation

$$\psi(x, t) = \psi_+(u_+) + \psi_-(u_-), \tag{12.1}$$

where $u_+ \equiv t - x/v_\mathrm{p}$ and $u_- \equiv t + x/v_\mathrm{p}$, and impose the boundary conditions mathematically. Our first condition is that the displacement of the guitar string at the bridge and fret must equal the displacement of the bridge or fret, which in this case we take always to be zero,

$$\psi(0, t) = \psi(l, t) = 0. \tag{12.2}$$

The second continuity condition for $\mathrm{d}\psi/\mathrm{d}x$ is lifted (more generally, modified) because the rigid fret can withstand finite point forces. On substituting equation (12.1) into equation (12.2), we find that the wave motion must therefore satisfy

$$\psi_+(t) + \psi_-(t) = 0, \tag{12.3a}$$
$$\psi_+(t - l/v_\mathrm{p}) + \psi_-(t + l/v_\mathrm{p}) = 0, \tag{12.3b}$$

so that

$$\psi_-(t) = -\psi_+(t), \tag{12.4a}$$
$$\psi_+(t - l/v_\mathrm{p}) = \psi_+(t + l/v_\mathrm{p}) \tag{12.4b}$$

and the total wave displacement is simply

$$\psi(x, t) = \psi_1(t - x/v_\mathrm{p}) - \psi_1(t + x/v_\mathrm{p}). \tag{12.5}$$

On writing $t' = t - l/v$, equation (12.4b) becomes

$$\psi_+(t') = \psi_+(t' + 2l/v_\mathrm{p}) \tag{12.6}$$

and hence requires that the function $\psi_+(t)$ be periodic, repeating every $\Delta t = 2l/v_\mathrm{p}$ – the time taken for any point of a travelling wave to make a round-trip of the string, be reflected at each end, and return to its starting point. Any periodic function satisfying this condition may, we shall see in Chapter 14, be formed from combinations of the sinusoidal standing-wave modes that we obtained in Section 4.4.

12.2.1 Musical harmony

Even amateur musicians have discovered, either by accident or by design, that an instrument configured for one note can also support other, higher notes corresponding to its overtones. A violin string touched at its mid-point will resonate at twice the fundamental frequency – an octave higher; an overblown clarinet will squeak at three times the frequency intended. Components with these frequencies do indeed accompany the fundamental when the note is played normally, as may be demonstrated by striking a piano key while another an octave higher is held depressed: the higher note is resonantly excited by the harmonic in the note played, and may be heard to persist beyond it.

A stringed instrument, as we have seen above and in Section 4.4, can support any integer multiple of the fundamental frequency. Figure 12.1 shows as an example the first eight harmonics when the fundamental (lowest) has a frequency of 220 Hz (known to musicians as A_3 or a), plotted vertically on a logarithmic scale known as a **musical stave**. Most of the harmonics lie on or close to the notes of the major chord (the third and sixth lie 0.02 semitones above, on an equally tempered scale; the fifth about 0.14 of a semitone below), and the eighth harmonic also falls on the scale, but the seventh lies about a third of a semitone below $G_{\natural 6}$ ($g''_\natural$) and is therefore the first to sound significantly discordant. It is the presence of such, albeit weak, components that in part determines the shrill sound of a strongly played instrument.

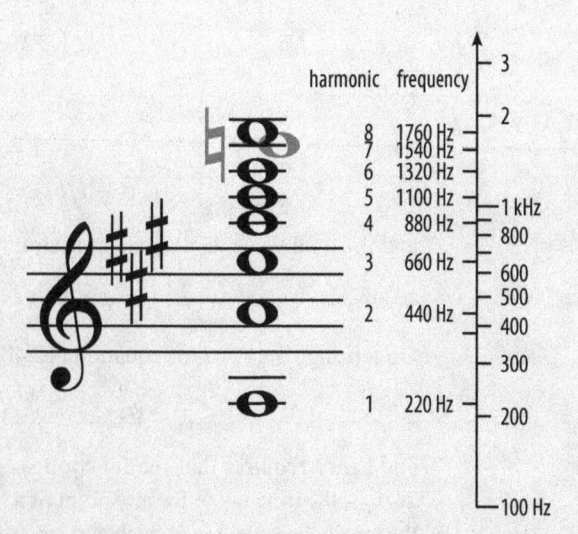

Fig. 12.1 The first eight harmonics from the notes of the A major chord, represented on the musical stave (a logarithmic plot of frequency). The seventh, shown in grey, is the first to sound significantly discordant.

Musical instruments often prove rather more complicated than presented here. The basic resonances are affected by non-ideal characteristics in the mechanism, and their relative amplitudes depend upon the means of excitation and their coupling to resonances, such as the violin soundbox, elsewhere in the instrument. Our perception of different notes and harmonics is then due to the complex physical and psychological behaviour of our hearing. Many musicians, physicists, psychologists and anatomists have therefore for years dedicated themselves to the study of musical instruments, scales and harmonies. The starting point is nonetheless an analysis such as that outlined above.

A. H. Benade [4–6]
H. Helmholtz [42]
J. H. Jeans [50]
A. Wood [93]
T. D. Rossing [75]
N. H. Fletcher & S.
Thwaites [25]
R. C. Chanaud [12]

12.3 Organ pipes and wind instruments

Organ pipes and wind instruments – both brass and woodwind – are based around long, narrow tubes or *pipes* that enclose a column of air, in which longitudinal sound waves propagate as described in Chapter 10. They vary, however, in whether the ends of the pipe are open or closed ('stopped'), and this has a profound effect upon the musical qualities of the instrument.

12.3.1 Boundary conditions in pipes

If the end of the pipe is closed – either by its construction, as with an organ pipe, or by the musician's lips over the mouthpiece of a brass instrument – then the boundary condition is that the displacement ξ of the sound wave at the closure ($x = 0$, say) must be zero,

$$\xi(0, t) = 0, \tag{12.7}$$

for otherwise the motion of the air would penetrate the barrier or create a temporary vacuum. There must therefore be a node in the standing wave, which, following the same reasoning as in Section 12.2, will therefore be of the form

$$\xi(x, t) = \xi_+(u_+) + \xi_-(u_-) = \xi_+(u_+) - \xi_+(u_-), \tag{12.8}$$

where $u_\pm \equiv t \mp x/v_{\mathrm{p}}$. The wave pressure (the difference from the static pressure P_0), or 'tension' per unit cross-sectional area, is given as in Section 10.2.1 by

$$\delta P(x, t) = P_0 \gamma \frac{\partial \xi}{\partial x} \tag{12.9a}$$

$$= P_0 \gamma \left(\frac{\mathrm{d}\xi_+}{\mathrm{d}u_+} \frac{\partial u_+}{\partial x} - \frac{\mathrm{d}\xi_+}{\mathrm{d}u_-} \frac{\partial u_-}{\partial x} \right)$$

$$= -\frac{P_0 \gamma}{v_{\mathrm{p}}} \left(\frac{\mathrm{d}\xi_+}{\mathrm{d}u_+} + \frac{\mathrm{d}\xi_+}{\mathrm{d}u_-} \right). \tag{12.9b}$$

It follows that, at $x = 0$, where $u_+ = u_- = t$ and the forward and backward propagating components are exactly in step,

$$\delta P(0, t) = -2\frac{P_0\gamma}{v_p}\frac{\partial\xi_+}{\partial x}(0, t),\tag{12.10}$$

and the wave pressure hence has its maximum variation. We may see this explicitly by differentiating equation (12.9b),

$$\frac{\partial\delta P}{\partial x}(x, t) = -\frac{P_0\gamma}{v_p}\left(\frac{\partial^2\xi_1}{\partial u_+^2}\frac{\partial u_+}{\partial x} + \frac{\partial^2\xi_1}{\partial u_-^2}\frac{\partial u_-}{\partial x}\right)$$

$$= \frac{P_0\gamma}{v_p^2}\left(\frac{\partial^2\xi_1}{\partial u_+^2} - \frac{\partial^2\xi_1}{\partial u_-^2}\right),\tag{12.11}$$

so that, at $x = 0$, $\partial\delta P/\partial x = 0$; a *node* in the wave displacement therefore corresponds to an *antinode* in the pressure variation.

The open ends of clarinets, oboes, trumpets, saxophones and flutes are characterized by a quite different boundary condition from that of equation (12.7) for a closed end. There is no constraint, indeed, upon the wave displacement, which simply causes a corresponding displacement of the surrounding air – this is the mechanism by which the musical sound is radiated. Provided that the tube is sufficiently narrow, however, the pressure will always be equal to that of the unconstrained atmosphere around it – or, at least, the pressure variation will diminish rapidly when the distance from the open end exceeds the tube diameter, which will usually be a short distance in comparison with the oscillation wavelength. The boundary condition is therefore, at $x = l$,

$$\delta P(l, t) = 0,\tag{12.12}$$

and hence, using equation (12.9a),

$$\frac{\partial\xi}{\partial x}(l, t) = 0.\tag{12.13}$$

In such cases, we therefore have a *node* in the pressure variation and an *antinode* in the wave displacement.

12.3.2 Open pipes

If the pipe were closed at both ends, we should require corresponding nodes in the wave displacement exactly as for the transverse motion of the string in Section 12.2. Generally, however, little sound can be emitted by such an arrangement, and practical instruments have at least one open end. Conversely, then, for *open pipes* that are open at both ends, the boundary conditions require nodes in the wave-pressure variation at both ends and therefore antinodes in the wave displacement. As with the guitar string, the result in both cases is that an integer number of half wavelengths must fit within the instrument, and the standing-wave modes may be any harmonic of the fundamental frequency.

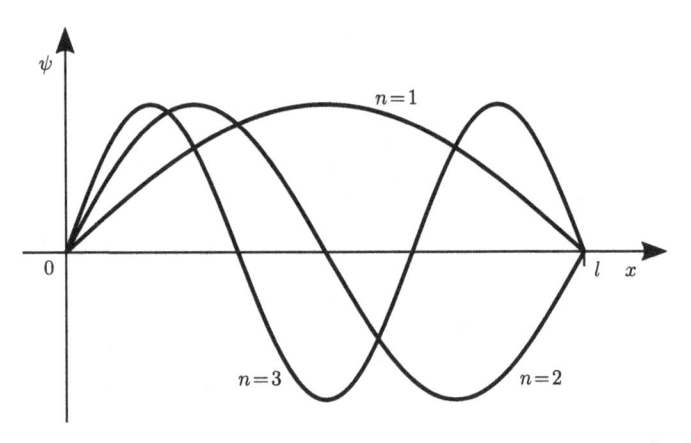

Fig. 12.2 The first three sinusoidal modes of an open pipe, with ψ representing the wave pressure δP; for a pipe that is closed at both ends, ψ would represent the displacement ξ.

The first three harmonics in these cases are illustrated in Fig. 12.2, where the wavefunction corresponds to the displacement ξ for a pipe that is closed at both ends, and the pressure δP for one whose ends are open. The gentle tone of the flute, which is open at both ends, results from the dominance in its sound of low harmonics, which, as discussed in Section 12.2.1, fall within a major chord.

12.3.3 Stopped pipes

For *stopped pipes*, which are open at one end and closed at the other, we must simultaneously satisfy equations (12.7) and (12.13), and hence set equation (12.9b) equal to zero for $u_{\pm} = t \mp l/v_{\mathrm{p}}$. This is possible only if

$$\xi_{+}(t') = -\xi_{+}(t' + 2l/v_{\mathrm{p}}), \tag{12.14}$$

where, as in Section 12.2, we have substituted $t' = t - l/v_{\mathrm{p}}$. This anti-periodic property is demonstrated by sinusoidal waves that have an odd number of quarter wavelengths between the two ends, as shown in Fig. 12.3. In contrast with an open pipe, which supports all overtones of the fundamental, the stopped pipe shows only the odd-numbered harmonics. Lacking the even harmonics, stopped pipes such as the recorder or trumpet tend to produce a harsher sound than that of stringed instruments such as the guitar or harp.

To play a piece of music, we change the effective length of the instrument by uncovering holes in the wall of the pipe. In most cases, this raises the pitch by reducing the effective length l of the instrument. If, however, we place the hole at a point that would have been an antinode in the wave displacement – and hence a node in the wave-pressure variation – then no effect will be incurred. By opening a hole near the closed, mouthpiece end of a stopped pipe, we may affect the fundamental notes more than the harmonic whose antinode roughly coincides with it. The result is that the note is determined, as usual, by the

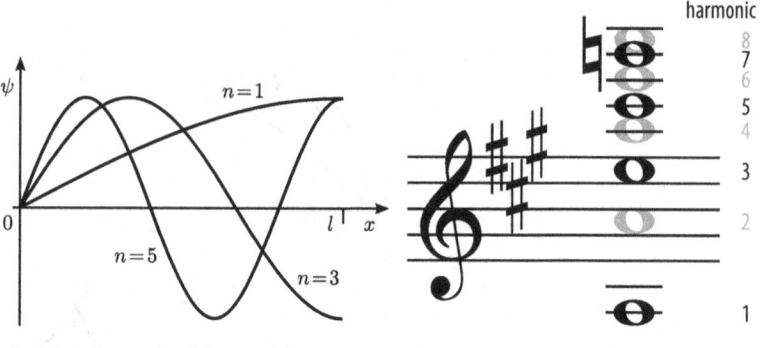

Fig. 12.3 The first three harmonics of the stopped pipe correspond to the odd harmonics of the corresponding string. The even harmonics (shown in grey) are absent.

effective length of the pipe overall, but that the predominant sound is from the first available harmonic, an octave or more higher.

12.3.4 Stopped conical pipes

An interesting and important effect occurs if, instead of having a pipe of uniform cross-section, the diameter increases from one end to the other to form a cone. In such cases, we find that the wave is distorted towards the narrow end of the cone, with an increasing intensity as the cross-section narrows, and an increase in the distance between nodes.

We may understand this roughly by considering first the travelling-wave component that moves towards the apex of the cone, which in free space would have a simple sinusoidal form. Conservation of energy requires the amplitude of the sine wave to increase according to the reciprocal of the distance r from the apex; the reflected wave shows the same dependence, but simply moves in the opposite direction. The result is part of a standing *spherical wave* – which we shall consider in Section 15.6.1 – of the form

$$\delta P(r, t) = \delta P_0 \frac{\sin(kr + \varphi)}{r}, \tag{12.15}$$

where φ depends upon the relative phase of the counter-propagating components and gives a finite amplitude at $r = 0$ only if $\varphi = 0$. The pressure of the standing wave hence follows a **sinc**-function dependence of the form shown in Fig. 9.2, in which the maximum at $kr = 0$ occurs at a *minimum* of the sinusoidal numerator. A stopped conical pipe, which requires a pressure node at its output and an antinode at its apex, hence shows a full set of harmonics of the fundamental frequency. Conically shaped instruments such as the oboe, clarinet and saxophone hence produce a rather sweeter, fuller sound than that of the tubular trumpet and trombone.

12.4 Boundary conditions in other systems

Just as a mechanical constraint upon the displacement or pressure at a boundary results in the reflection of waves from a guitar fret or the end of an organ pipe, so equivalent boundary conditions in other systems have similarly profound consequences. We consider a few examples below.

12.4.1 Electromagnetic boundaries

The ability of a good conductor to eliminate electric fields within it results in the high reflectivity of a metal surface for electromagnetic waves. When an electric field $\mathbf{E}(x, t)$ is applied to a metal of conductivity σ, a current of density $\mathbf{J}$ will flow, where Ohm's law in three dimensions gives

$$\mathbf{J} = \sigma \mathbf{E}. \tag{12.16}$$

The redistribution of charge, described by the current flow, is such as to create an additional electric field component that is equal but opposite to the applied field $\mathbf{E}$. The result is that, provided the conductivity σ is sufficiently high, the net electric field at the metal surface is zero at all times. If the surface lies at $x = 0$, then the boundary condition

$$\mathbf{E}(0, t) = 0 \tag{12.17}$$

is the direct equivalent of equations (12.7) and (12.12); formal solution follows the same procedure, and has the same qualitative result.

It may be noted that the currents that flow to accommodate the constantly changing charge distribution mean that the magnetic field is not eliminated, and indeed forms an antinode at the mirror surface where the electric field has a node. For normal incidence, a standing wave results, with the electric and magnetic field components a quarter of a wavelength out of step.

12.4.2 Boundaries for water gravity waves

Main [60] Chapter 13

The ripples across the surface of a swimming pool can propagate only if there is some horizontal flow of water to accommodate the variation in height of the water surface as the wave passes. For no water to flow into or out of the impermeable sides of the pool, the horizontal velocity v_x has to be zero at all times, and equation (3.34) shows that the water surface at the edge must at all times be level – that is, for a surface height $h(x, t)$ meeting an edge at $x = 0$,

$$\frac{\partial h}{\partial x} = 0. \tag{12.18}$$

Capillary waves experience quite different conditions that depend upon the properties of the adjoining surface, and vary from requiring the level surface of equation (12.18) to defining a fixed point of contact as at a guitar fret.

12.4.3 Thermal boundaries

The organ-pipe boundaries of zero displacement and constant pressure correspond in thermal systems to boundaries of zero heat flow and constant temperature – that is, to the presence of either a good insulator or a large thermal mass. Although, as we saw in Section 10.3, the thermal-wave equation is of a different form from that for sound waves, and consequently has different solutions, the influence of isothermal and insulating boundaries follows exactly the procedures outlined above. Indeed, when we considered longitudinal heat flow in Section 10.3, our assumption that the bar was well lagged at its sides corresponds, as in Section 12.4.2, to the level temperature profiles in the transverse direction that we assumed implicitly in our one-dimensional analysis.

12.4.4 Deep-water ocean waves

We stated in Chapter 3 that the amplitude of the motion of deep-water ocean waves decreases with the depth below the water surface, in a fashion that strongly resembles the evanescent waves examined in Section 11.5. We may now derive equations (3.52) explicitly.

We have seen in Section 3.3 that sinusoidal deep-water ocean waves may be of the form

$$\xi = \xi_0(z_0)\cos(kx - \omega t), \tag{12.19a}$$

$$\zeta = \zeta_0(z_0)\sin(kx - \omega t), \tag{12.19b}$$

where ξ and ζ represent, respectively, the horizontal and vertical displacements a distance z_0 above the seabed at a horizontal position x and time t, and the depth-dependent amplitudes ξ_0 and ζ_0 must satisfy

$$\frac{d^2\xi_0}{dz_0^2} = k^2\xi_0, \tag{12.20a}$$

$$\frac{d\xi_0}{dz_0} = k\zeta_0. \tag{12.20b}$$

Equation (12.20a) may easily be solved to give solutions of the form $\xi_0 = a\exp(\alpha z_0)$, where $\alpha = \pm k$, so that for a given value of the wavenumber k the general solution will be

$$\xi_0 = a_+\exp(kz_0) + a_-\exp(-kz_0). \tag{12.21}$$

Inserting this into equation (12.20b) hence gives

$$\zeta_0 = a_+\exp(kz_0) - a_-\exp(-kz_0). \tag{12.22}$$

There must be no vertical motion at the seabed, for water cannot flow into or out of the solid from which it is formed. We therefore have the boundary condition, if the seabed lies at $z_0 = 0$,

$$\zeta_0(0) = 0, \tag{12.23}$$

from which it follows that the coefficients a_+ and a_- must be equal. The overall wave amplitudes must therefore be of the form

$$\xi_0(z_0) = 2a \cosh(kz_0), \tag{12.24a}$$

$$\zeta_0(z_0) = 2a \sinh(kz_0), \tag{12.24b}$$

where $a = a_+ = a_-$. On setting the amplitude $\zeta_0(h_0)$ equal to h_1 at the sea surface a height h_0 above the seabed, we obtain $2a \sinh(kh_0) = h_1$ and hence, as in equations (3.52),

$$\xi_0(z_0) = \frac{\cosh(kz_0)}{\sinh(kh_0)} h_1 \tag{12.25a}$$

$$\zeta_0(z_0) = \frac{\sinh(kz_0)}{\sinh(kh_0)} h_1. \tag{12.25b}$$

12.5 Driven boundaries

There is no reason why boundaries should be limited to fixed points. The reed that closes the end of a clarinet, for example, supplies the necessary wave energy by making tiny movements that are nonetheless small enough to neglect when determining the nodes and antinodes; and we could, though it would be of little utility, excite the motion of a guitar string by attaching the bridge to a transducer driven at the appropriate frequency. We deal with such circumstances simply by adapting the appropriate boundary condition so that equation (12.7) for the clarinet might become, for example,

$$\xi(0, t) = a \cos(\omega t). \tag{12.26}$$

Although the specific form that we obtain for the wave may be rather more complex, solution then follows precisely the same sequence as before: we simply apply the boundary conditions to the generic form of solution, and solve the resulting equations to find the undefined parameters.

12.6 Cyclic boundary conditions

An interesting and significant boundary condition is that imposed not by external constraints as such, but rather through the geometry of the medium itself.

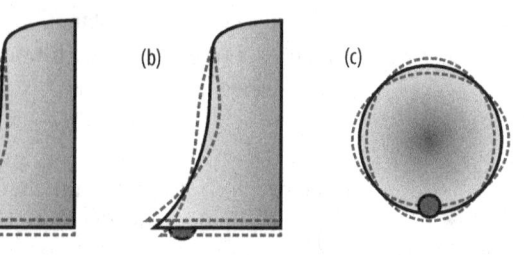

Fig. 12.4 Vibrational modes of a bell, shown from the side, (a) and (b), and from below, (c). Both the hum (a) and the fundamental (b) motions show the lowest excitation mode when viewed from below (c), although from the side the motions are of different orders, with the fundamental showing a node in its vertical structure. Higher-order motions occur from both perspectives, and give the bell its characteristically complex sound. (After Westcott [92].)

W. Westcott [92]

Such a condition occurs when a length of linear medium is bent, and the ends joined, to form a circle or other closed curve. If the circumference is l and the coordinate x, then the boundary condition is simply

$$\psi\left(t + \frac{x}{v_p}\right) = \psi\left(t + \frac{x+l}{v_p}\right) \tag{12.27}$$

and, as with the similarly confined systems described by equations (12.6) and (12.14), solutions include the sinusoidal motions whose wavelengths are integer fractions of the circumference.

Such cyclic boundary conditions are manifest in the modes of vibration of church bells, which, when viewed from below, show a standing wave around the circumference that results in a symmetrical displacement about the neutral position. (An antisymmetric motion would involve motion of the centre of mass, which is prohibited by conservation of momentum – a further constraint in this case.) The waves that propagate around a bell are *flexural*, like those of a tuning fork or ruler protruding from the edge of a desk, the restoring force being provided by the elasticity when the thickness of the material cannot be neglected.

Bells are, however, complex instruments, and each mode also has a structure with nodes in the vertical dimension. Differences between the vertical profiles of different harmonics result in significant departures in frequency from the simple geometrical relationship that would otherwise be expected, for the motions may be concentrated in different regions with different dimensions. Changing the shape of the bell allows these dimensions to be controlled, and the frequencies of the harmonics thereby adjusted to achieve the sweetest sound.

Resonant ring structures occur in many other manifestations in physics, from the electron wavefunction around an atomic nucleus to the optical field in a laser cavity. Since there is in general no other boundary condition constraining the ring motion – corresponding to the absence of any particular point on the ring – travelling-wave solutions are permitted, and the position or phase of the sinusoidal solution is unconstrained. A pulse may therefore travel repeatedly around the *waveguide*, which is precisely what happens in the case of the ring cavity of a mode-locked or Q-switched laser. The ability of the ring to support unidirectional, and not just standing, waves can be of great significance in modern laser systems.

Exercises

12.1 Establish how the pitch of an organ pipe varies with ambient pressure P_0 between different altitudes on Earth. You may take the adiabatic elasticity of air to be that of an ideal gas, given by $E = \gamma P_0$, where γ is a constant equal to the ratio of the specific heats at constant pressure and constant volume. You should consider how the density of the atmosphere varies with altitude.

12.2 How will the pitch of an organ pipe or woodwind instrument vary as the instrument warms up, if the ambient pressure remains the same?

12.3 If the bottom string of a guitar has a mass of 5.4 g m^{-1} and its length is determined by the distance of 0.648 m from the *bridge* to the *nut*, find the tension required to tune the string to the note known as E_2 (a frequency of 82.4 Hz). How far along the *neck* must the first *fret* be, to give a frequency of 87.3 Hz – i.e. a semitone higher?

12.4 The electric guitar radiates very little sound directly, but instead relies upon pick-ups (transducers) placed beneath the strings. Establish which harmonics of the open string you are likely to hear if the pick-ups are placed

(1) close to the bridge of the guitar,

(2) around the mid-point of the strings,

(3) a quarter of the way from the bridge to the nut, and

(4) 98 mm from the bridge, when the length of the open string is 648 mm.

How will the sounds reproduced by pick-ups at these positions compare? Given that the seventh harmonic causes a *dissonance*, which position will give the sweetest sound?

12.5 The classic sound used by guitarist Mark Knopfler is famously achieved by taking the *difference* between the *bridge* and *middle* pick-ups – an arrangement that originally also reduced the sensitivity of the guitar to mains-frequency magnetic fields. Assuming that the pick-ups have the same sensitivities to the displacements of the strings above them, calculate the relative sensitivities to the first eight harmonics of a guitar string (i.e. the signal produced by harmonics with the same amplitudes at their antinodes) using

(1) the bridge pick-up alone,

(2) the middle pick-up alone, and

(3) the difference between the two pick-ups.

(The *bridge, middle* and *head* pick-ups of a *Fender Stratocaster* are positioned 1.625″ (41 mm), 3.875″ (98 mm) and 6.375″ (162 mm) from the bridge; the open length of the string (the *scale length*) is 25.5″ (648 mm).)

13 Linearity and superpositions

13.1 Wave motions in linear systems

Pain [66] pp. 128–132
French [29] pp. 213–216 and 230–234
Main [60] pp. 37–72 and 213–220
Feynman [23] Vol. I, pp. 25-1–25-5

We have already on several occasions mentioned the superposition principle for linear systems. It is, indeed, reasonably intuitive that different wave motions may be combined and will propagate without interacting in the linear systems that form most of the cases we meet. The principle is nevertheless so important that we should address it directly.

By no means, though, are all physical systems linear: the breaking waves of a wind-torn ocean provide a classic counter-example, and nonlinearities prove fundamental to some of the optical phenomena used in pulsed lasers, optical modulators and frequency converters. There are, however, few general approaches to the analysis and categorization of wave motions in nonlinear systems, which therefore require careful, individual consideration and frequently demonstrate unexpected and puzzling behaviour. On the other hand, most systems of interest can be at least initially approximated to linear equivalents, to which their behaviour usually tends at small wave amplitudes. The wave phenomena that we commonly recognize and apply are therefore overwhelmingly those of linear systems, and their analysis leads naturally to the powerful principles of *normal modes* and *Fourier synthesis* that prove of very general utility.

That a system is linear does not, however, eliminate all interesting phenomena, for linear systems may still demonstrate *dispersion*, which causes wavepackets to spread as they propagate. Dispersion, at some level, proves to be intrinsic in any real material, and is a fundamental characteristic of quantum wavefunctions.

In this chapter, we therefore consider formally what is meant by linearity, how it permits superpositions and thereby the simple analysis of wavepackets. Finally, we consider how wavepackets are affected by dispersion, which leads to the introduction of yet another wave velocity: the *group velocity* – which, like evanescent waves, proves to be the source of many fascinating conundrums to do with superluminal propagation.

13.2 Linearity and the superposition principle

The wave equations that we have met so far are all *linear* in the wave displacement, ψ. This means that, if, say, the wave equation

$$\frac{\partial^2 \psi}{\partial t^2} = v_p^2 \frac{\partial^2 \psi}{\partial x^2} \qquad (13.1)$$

has two solutions given by $\psi_1(x, t)$ and $\psi_2(x, t)$, then any linear combination, or **superposition**, of the two solutions,

$$\psi(x, t) = a\psi_1(x, t) + b\psi_2(x, t), \qquad (13.2)$$

is also a solution, for all values of the constants a and b. We may see this directly, by substituting the general superposition of equation (13.2) into the wave equation:

$$a\frac{\partial^2 \psi_1}{\partial t^2} + b\frac{\partial^2 \psi_2}{\partial t^2} = v_p^2 \left(a\frac{\partial^2 \psi_1}{\partial x^2} + b\frac{\partial^2 \psi_2}{\partial x^2} \right), \qquad (13.3)$$

whence

$$a\left(\frac{\partial^2 \psi_1}{\partial t^2} - v_p^2 \frac{\partial^2 \psi_1}{\partial x^2} \right) + b\left(\frac{\partial^2 \psi_2}{\partial t^2} - v_p^2 \frac{\partial^2 \psi_2}{\partial x^2} \right) = 0, \qquad (13.4)$$

which, since ψ_1 and ψ_2 are individually solutions to the wave equation, must be true for any choice of a and b.

The principle of superposition may be applied repeatedly, so that, if the wavefunctions $\psi_n(x, t)$ are all solutions of the wave equation (13.1), then so will be

$$\psi(x, t) = \sum_n c_n \psi_n(x, t). \qquad (13.5)$$

We have used the principle of superposition to form standing waves from forward- and backward-travelling waves in Section 4.4, and to convert between sinusoidal and complex exponential oscillations in Section 5.1. In Chapter 14, we shall see that it allows the extremely powerful technique of *Fourier analysis* – which itself is just one example of the process of analysing a wavefunction into *normal modes*.

13.2.1 Nonlinear wave equations

Whether or not a wave equation is linear is defined simply by whether a general superposition of solutions is also itself a solution, as we have seen above, or whether, conversely, there are nonlinear terms in the wave equation, e.g.

$$\frac{\partial^2 \psi}{\partial t^2} = \psi \frac{\partial \psi}{\partial x}, \qquad (13.6)$$

where the term on the right will be quadrupled if ψ is doubled. A general superposition such as that of equation (13.2) will therefore *not* be a solution:

$$
\begin{aligned}
\frac{\partial^2}{\partial t^2}(a\psi_1 + b\psi_2) &= (a\psi_1 + b\psi_2)\frac{\partial}{\partial x}(a\psi_1 + b\psi_2) \\
&= a^2\psi_1\frac{\partial\psi_1}{\partial x} + b^2\psi_2\frac{\partial\psi_2}{\partial x} + ab\left(\psi_1\frac{\partial\psi_2}{\partial x} + \psi_2\frac{\partial\psi_1}{\partial x}\right) \\
&\neq a\psi_1\frac{\partial\psi_1}{\partial x} + b\psi_2\frac{\partial\psi_2}{\partial x}.
\end{aligned}
\tag{13.7}
$$

The two components of the superposition are in such cases *coupled* by the nonlinearity. We may illustrate the physical significance of this by considering a superposition of two initially separate pulses or *wavepackets* that move towards each other and overlap. Initially, then, the superposition will be a solution of the wave equation, for there is nowhere that both wavefunctions are non-zero and hence the product terms in equation (13.7) will not occur. As the wavepackets overlap, however, they interact, and will be modified as they pass through each other. In a *linear* system, in contrast, they will emerge unscathed.

Nonlinear wave phenomena can be very important, with applications and consequences for optical communications and laser light sources, the design of musical instruments, tsunami prediction and Concorde's sonic boom. In some cases, the nonlinearity may be treated as a small deviation from the linear approximation and, with care, can be straightforward to handle; in others, it may require the full nonlinear wave equation to be solved directly, with complex and sometimes chaotic solutions. Here, however, we shall limit our attention to systems that are linear, in which the principle of superposition may be used.

13.3 Wavepackets

The sinusoidal and complex exponential waves that we met in Chapters 4 and 5 are simple to handle and, as we have seen in Section 5.2, are often still valid solutions even when dispersion prevents an arbitrary pulse shape from progressing undistorted. Certainly we shall often be interested in waves of a single frequency, when we consider the note produced by a musical instrument or the propagation of a laser beam of defined wavelength: sine waves, we have already mentioned, result naturally when the source follows a free rotation or simple harmonic motion. But we are left with the question of how to handle waves of more arbitrary shapes that, in the presence of dispersion, are not travelling-wave solutions of the wave equation. Numerical integration of the wave equation will in such cases provide one approach; a more analytical method, which is the subject of Chapter 14, is to break the waveform down into sinusoidal or complex exponential components. Before embarking upon this,

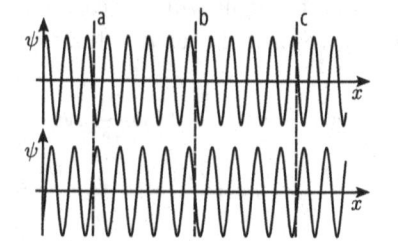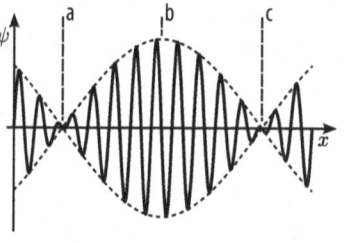

Fig. 13.1   Beating when two sinusoidal components of slightly different wavenumbers k_1 and k_2 (left) are superposed (right).

however, we first approach the problem from the opposite direction, and examine how we may create **wavepackets** of limited spatial extent by adding sine waves of a range of frequencies and wavelengths so that they add constructively only within a limited region, beyond which they tend to average to zero.

13.3.1 Beats

Although it bears little initial resemblance to a single wave pulse, the simplest prototype wavepacket is the equal superposition of just two wavelengths or frequencies, whose snapshot at a given time might be

$$\psi(x, t) = \cos(k_1 x) + \cos(k_2 x). \tag{13.8}$$

If we choose k_1 and k_2 to be nearly equal, then it is clear that the two contributions will be very similar for small values of x, but that at greater distances they will drift in and out of phase. It is helpful to rearrange equation (13.8) so that, instead of a summation, we have a product:

$$\psi(x, t)$$
$$= \cos\left[\left(\frac{k_1 + k_2}{2} + \frac{k_1 - k_2}{2}\right)x\right] + \cos\left[\left(\frac{k_1 + k_2}{2} - \frac{k_1 - k_2}{2}\right)x\right]$$
$$= \cos\left(\frac{k_1 + k_2}{2}x\right)\cos\left(\frac{k_1 - k_2}{2}x\right) - \sin\left(\frac{k_1 + k_2}{2}x\right)\sin\left(\frac{k_1 - k_2}{2}x\right)$$
$$+ \cos\left(\frac{k_1 + k_2}{2}x\right)\cos\left(\frac{k_1 - k_2}{2}x\right) + \sin\left(\frac{k_1 + k_2}{2}x\right)\sin\left(\frac{k_1 - k_2}{2}x\right)$$

so that

$$\psi(x, t) = 2\cos\left(\frac{k_1 + k_2}{2}x\right)\cos\left(\frac{k_1 - k_2}{2}x\right). \tag{13.9}$$

The superposition corresponds, therefore, to a single sinusoidal oscillation with the average wavenumber of the two components, which is multiplied by a slowly varying function that depends upon their difference, and wanes and waxes as the two components drift out of and back into phase – the phenomenon of **beating**, illustrated in Fig. 13.1.

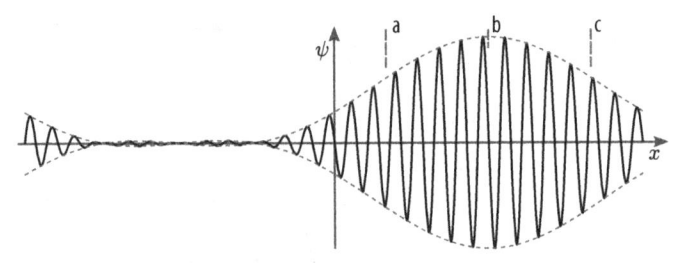

Fig. 13.2   Elimination of the first beat revival by adding two additional components of intermediate wavenumbers to the superposition of Fig. 13.1.

The section of the superposition shown in Fig. 13.1 has some of the characteristics that we might desire of a wavepacket, in that the amplitude of the wave falls from its maximum at b, where the two components are in phase, so that it has minima on each side at a and c, where the two components are exactly half a cycle out of step. Beyond a and c, however, the steadily increasing phase difference between the two components causes the wave amplitude to grow again to form a succession of maxima and minima that continues indefinitely.

These periodic revivals in the wave amplitude may be eliminated by adding further sinusoidal components with wavenumbers between k_1 and k_2 as shown in Fig. 13.2. Around b, we find that the components are all in phase: the maximum phase difference, if the wavepacket is centred on x_0, is $(k_1 - k_2)(x - x_0)$, and, provided that this remains small, there will be only a slight fall in amplitude. As we move away from b, however, this phase difference may be large enough for some components to cancel out completely the contribution of other wavenumbers. As x increases, the phases will become roughly uniformly distributed, and the cancellation will tend to become almost total. With a finite number of superposed components, the cancellation will never be complete, and there will always be some sort of revival after a great enough distance, but with sufficient components the residue after cancellation can be made to be arbitrarily small and the revival arbitrarily distant. The result is that the wave will resemble a single region of sinusoidal oscillation whose amplitude falls to zero as we move further from the origin – a true wavepacket.

The reasoning behind the spatial localization of the wavepacket in many ways resembles that behind the Huygens approach to image formation that we met in Section 7.4, whereby only at the focus did the contributions arriving via different paths arrive with the same phase and consequently add constructively. The arguments do indeed prove to be identical, and are one more manifestation of the broad equivalence of the temporal and spatial domains. Just as space and time are linked by the rules of special relativity, where they are often combined into a space-time *four-vector*, so also may be the wave frequency and wavenumbers for propagation along each of the coordinate axes.

Feynman [23] Vol. I,
pp. 34-7–34-11

13.4 Dispersion and the group velocity

The only requirement that must be met in order for the superposition principle to be valid is that which defines it: that the system be linear. There is no need for a single wave speed, so different components may travel at different speeds according to, for example, their frequency or (in three-dimensions) their direction of motion. The latter characteristic is known as *anisotropy*; the former, which was introduced in Section 5.2, is **dispersion**.

Dispersion has an important effect upon the propagation of superpositions for, when sinusoidal waves of different frequencies are superposed to form a wavepacket or other complex waveform, the individual components will travel with different speeds: their wavefronts will drift apart, and their phase relationships will therefore change as the wave progresses. A wavepacket hence tends to break up, or *disperse*, as it propagates, and indeed the motion of the wavepacket may be quite different from that of its sinusoidal components.

To illustrate the effect of dispersion, we consider the simple example of an equal superposition of two single-frequency motions, characterized by wave-numbers k_1 and k_2 that we shall assume to be nearly equal. The working is as in Section 13.3.1, so for variety we here apply it to a superposition of complex exponential waves,

$$\psi(x, t) = \psi_1(x, t) + \psi_2(x, t)$$
$$= \exp[i(k_1 x - \omega_1 t)] + \exp[i(k_2 x - \omega_2 t)]. \qquad (13.10)$$

As in Section 13.3.1, we may write the two components as

$$\psi_1(x, t)$$
$$= \exp\left[i\left(\frac{k_1 + k_2}{2}x - \frac{\omega_1 + \omega_2}{2}t\right)\right] \exp\left[i\left(\frac{k_1 - k_2}{2}x - \frac{\omega_1 - \omega_2}{2}t\right)\right],$$
$$(13.11a)$$

$$\psi_2(x, t)$$
$$= \exp\left[i\left(\frac{k_1 + k_2}{2}x - \frac{\omega_1 + \omega_2}{2}t\right)\right] \exp\left[-i\left(\frac{k_1 - k_2}{2}x - \frac{\omega_1 - \omega_2}{2}t\right)\right].$$
$$(13.11b)$$

On substituting these forms into the superposition, we have

$$\psi(x, t) = 2 \exp\left[i\left(\frac{k_1 + k_2}{2}x - \frac{\omega_1 + \omega_2}{2}t\right)\right] \cos\left(\frac{k_1 - k_2}{2}x - \frac{\omega_1 - \omega_2}{2}t\right)$$
$$= 2 \exp[i(k_0 x - \omega_0 t)]\cos(\delta k x - \delta\omega t), \qquad (13.12)$$

where $k_{1,2} = k_0 \pm \delta k$ and $\omega_{1,2} = \omega_0 \pm \delta\omega$. The superposition, illustrated in Fig. 13.3, may thus be written as the product of a rapidly oscillating complex exponential term, which travels with a phase velocity of $v_p = (\omega_1 + \omega_2)/(k_1 + $

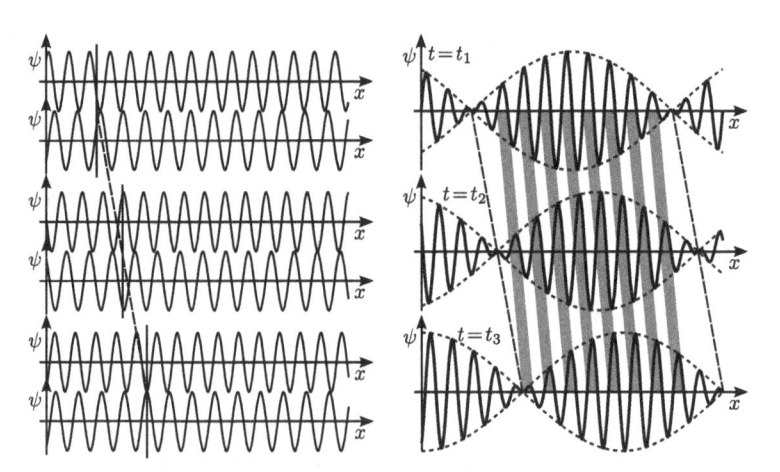

Fig. 13.3 Motion of the superposition of two sinusoidal components described by the real part of equation (13.10), shown at successive times t_1, t_2 and t_3. The group velocity is here twice the phase velocity, and hence the envelope (constant intensity) moves faster than the wavefronts (constant phase).

k_2) $\equiv \omega_0/k_0$ determined by the average frequency and wavenumber, and a slowly varying sinusoidal term that moves with the **group velocity**,

Group velocity

$$v_g = \frac{\omega_1 - \omega_2}{k_1 - k_2} = \frac{\delta\omega}{\delta k} \approx \frac{d\omega}{dk}, \tag{13.13}$$

where the final approximation is the limit as $\delta k \to 0$. The group velocity is thus the speed with which the overall wavepacket envelope progresses.

13.4.1 The dispersion relation

To analyse wave propagation in dispersive systems, we must know how the wave speed varies with the wave frequency or wavenumber. It is common to express this in the form of a **dispersion relation** between the frequency ω and the corresponding wavenumber k. For example, for a quantum particle moving freely in a region of potential energy $\mathcal{U} = 0$ and described by the time-dependent Schrödinger equation

$$\frac{\partial \psi}{\partial t} = i\frac{\hbar}{2m}\frac{\partial^2 \psi}{\partial x^2}, \tag{13.14}$$

it may be shown that

$$\omega = \frac{\hbar}{2m}k^2. \tag{13.15}$$

The *phase velocity* in this case is therefore

$$v_p \equiv \frac{\omega}{k} = \frac{\hbar}{2m}k, \tag{13.16}$$

while the *group velocity* proves to be a factor of two larger,

$$v_g \equiv \frac{d\omega}{dk} = \frac{\hbar}{m} k. \tag{13.17}$$

Since the wavefronts therefore move within the wavepacket as it progresses, solutions to dispersive wave equations such as equation (13.14) cannot be written in the simple travelling-wave form of equation (2.2).

It is useful to note the general expression

$$\frac{d}{dk} v_p \equiv \frac{d}{dk} \frac{\omega}{k} = \frac{1}{k} \frac{d\omega}{dk} - \frac{\omega}{k^2}$$
$$= \frac{1}{k}(v_g - v_p). \tag{13.18}$$

If, as in the example of a quantum particle above, the phase and group velocities are not equal, it immediately follows that the phase velocity varies with wavenumber and the system is therefore dispersive.

In most optical materials at the wavelengths of use, dv_p/dk (and hence also $d\eta/d\lambda$) is found to be negative, so that $v_p > v_g$; in such cases, the material is said to show **normal dispersion**. Near to absorption resonances within the material, however, the opposite behaviour may be found: it is apparent from equations (13.16) and (13.17) that this is also the case for the quantum particle above.

The dispersion relation must always be found either by considering the physical mechanisms within the specific system, or by experimentally measuring the dependence of the phase velocity upon wavenumber.

13.4.2 Superluminal velocities

A keystone of Einstein's *special theory of relativity* is that no physical entity can travel faster than c, the speed of light in vacuum. A wavefront – a point of constant phase – does not correspond to any physical entity, and thus the phase velocity is unconstrained. However, the group velocity describes the speed of motion of the energy of a wavepacket, or in quantum mechanics the probability density of a particle, and thus corresponds to measurable physical entities. The group velocity of any physical wave motion, therefore, should never exceed the speed of light.

There are two important caveats to this statement. The first concerns nonlinear systems, and we could as an example consider the passage of a pulse of light through a laser medium that amplifies the first part of the pulse but attenuates the second part after the laser gain has been depleted: the peak of the pulse hence advances within the wavepacket and therefore appears to move faster than the light within it. The change of waveshape here occurs through the exchange of energy with the laser medium, however, and not because energy moves forward within the pulse itself.

The second caveat is that, in our derivations above, we have assumed the wavenumbers of the sinusoidal components to be nearly equal, so with the phase and group velocities we have effectively considered just the first two terms in a Taylor-series expansion of the dispersion relation between the frequency ω and wavenumber k,

$$\omega(k) = \omega(k_0) + (k - k_0)\frac{\mathrm{d}\omega}{\mathrm{d}k} + \frac{(k - k_0)^2}{2}\frac{\mathrm{d}^2\omega}{\mathrm{d}k^2} + \cdots . \tag{13.19}$$

If the range of wavenumbers in the superposition is large, we may therefore need to revisit our analysis to take into account the higher-order terms.

Numerous theoretical analyses, and sometimes experimental observations, have been interpreted as demonstrating superluminal motion or information transfer. Most can be fairly swiftly placed into one of the two categories above.

Exercises

13.1 Show that the wave equation (in which m is a real constant)

$$\mathrm{i}m\frac{\partial\psi}{\partial t} = -\frac{\partial^2\psi}{\partial x^2} \tag{13.20}$$

is linear.

13.2 Show that equation (13.20) has complex exponential travelling-wave solutions of the form

$$\psi(x, t) = \psi_0 \exp[\mathrm{i}(kx - \omega t + \varphi)] \tag{13.21}$$

and explain the significance of the parameters k, ω and φ.

13.3 Determine the *dispersion relation* between k and ω for a system described by the wave equation (13.20).

13.4 A travelling wave has two components of the form given in equation (13.21), of equal magnitude and frequencies $\omega_0 \pm \delta\omega$, where $\delta\omega \ll \omega_0$. Show that the wave may be written in the form

$$\psi(x, t) = \psi_1 \exp[\mathrm{i}(k_0 x - \omega_0 t + \varphi_0)]\cos(k_1 x - \omega_1 t + \varphi_1) \tag{13.22}$$

and thus takes the form of a complex exponential travelling wave that is modulated by a slowly varying, real periodic function.

13.5 Explain what is meant by the *group velocity* of a wave motion, and show that, for the example of Exercises 13.1–13.4, the group velocity differs from the phase velocity by a factor of two.

14 Fourier series and transforms

14.1 Fourier synthesis and analysis

In our discussion of linearity and dispersion in Chapter 13, we have seen how simple sinusoidal components may be combined to form more complex waveforms: two frequencies produce periodic beats, and further components allow a single wavepacket to be isolated. With sufficient components spanning a great enough range of wavenumbers, the single wavepacket may also be made arbitrarily narrow. If a single narrow wavepacket can be synthesized by adding sinusoidal components then, by the principle of superposition, so can an arbitrary waveform, for we may construct the desired form from a series of consecutive pulses whose individual magnitudes correspond to the amplitude of the required waveform at that point. This, although we shall usually cast it slightly differently, is the principle of Fourier synthesis, which proves to be both an example of a more generic method of synthesis from component waveforms, and an immensely important, useful and fundamental procedure in its own right.

So much is usually written about the elegant and powerful techniques of Fourier synthesis and analysis that they acquire a reputation for complexity that is completely unmerited. Although we too shall eventually discuss some of the subtleties of Fourier techniques, we therefore begin with a straightforward statement of what they are and how they are applied.

14.1.1 The principle of Fourier synthesis

J. P. J. Fourier's principle, developed during his analysis of thermal waves, is simply that any wavefunction may be reproduced by adding together sinusoidal components with appropriate amplitudes c_n and phases φ_n – that is, we may write

$$\psi(t) \equiv \sum_n c_n \cos(\omega_n t + \varphi_n) \tag{14.1a}$$

$$\equiv \sum_n a_n \cos(\omega_n t) + b_n \sin(\omega_n t), \tag{14.1b}$$

where $a_n = c_n \cos \varphi_n$ and $b_n = -c_n \sin \varphi_n$. When the wavefunction $\psi(t)$ is represented in this way by a superposition of discrete sinusoidal components it is known as a **Fourier series**. Note that, for simplicity, we shall initially refer

James [48]
Pain [66] pp. 267–298
French [29] pp. 189–196
Main [60] pp. 192–202 and
209–210
Lipson *et al.* [56] Chapter 4
Feynman [23] Vol. I,
pp. 50-2–50-8
Boas [8] Chapter 7

J. P. J. Fourier [26]

Fourier series

to functions $\psi(t)$ of time alone; in due course, they may be trivially extended to travelling waves $\psi(u)$, where $u \equiv x - v_\mathrm{p}t$.

An alternative expression of Fourier's principle is as a **Fourier integral**,

$$\psi(t) \equiv \int_{-\infty}^{\infty} c(\omega)\cos\left(\omega t + \varphi(\omega)\right) \mathrm{d}\omega \qquad (14.2a)$$

$$\equiv \int_{-\infty}^{\infty} a(\omega)\cos(\omega t) + b(\omega)\sin(\omega t)\mathrm{d}\omega, \qquad (14.2b)$$

which means nothing more than the explicit summation of the Fourier series but often proves considerably tidier to handle. The coefficients $a(\omega)$ and $b(\omega)$, which are simply set to zero for frequencies that are not required, now represent the *amplitude per unit frequency range*, and are therefore more suitable when the superposed frequencies are so closely spaced that they form a continuum. Superpositions of individual, discrete frequency components imply an infinite amplitude per unit frequency range; while the integral representation can describe them quite satisfactorily, they require careful bookkeeping using the Dirac δ-function that we shall introduce in Section 14.4.1.

It may be shown that there is only one combination of sinusoidal components that will reproduce any given function $\psi(t)$, and the coefficients are found using

$$a(\omega) = \frac{1}{2\pi} \int_{-\infty}^{\infty} \psi(t)\cos(\omega t)\mathrm{d}t, \qquad (14.3a)$$

$$b(\omega) = \frac{1}{2\pi} \int_{-\infty}^{\infty} \psi(t)\sin(\omega t)\mathrm{d}t, \qquad (14.3b)$$

from which $c(\omega)$ and $\varphi(\omega)$ may be obtained using $c(\omega)^2 = a(\omega)^2 + b(\omega)^2$, $\tan\varphi(\omega) = -b(\omega)/a(\omega)$.

Fourier analysis is therefore a procedure for breaking an arbitrary waveform down into its single-frequency components, and **Fourier synthesis** is the reverse process of combining those components back into the original waveform. Mathematically, we shall see that they depend upon the *orthogonality* of sinusoidal waves. Physically, they correspond to the translation of a musical sound, which we hear, into the notes that we write as a chord; or of the electric field of light into the *spectrum* obtained from a spectroscope or spectrometer. Prisms, and our ears, then, perform Fourier analysis all the time.

Fourier analyses into, and syntheses from, series and integrals are together known as **Fourier transforms**, and the rest of this chapter will be concerned with their subtleties, justification, manifold applications and variations, and a particularly useful consequence known as the *convolution theorem*. The essence, however, is contained within the equations above.

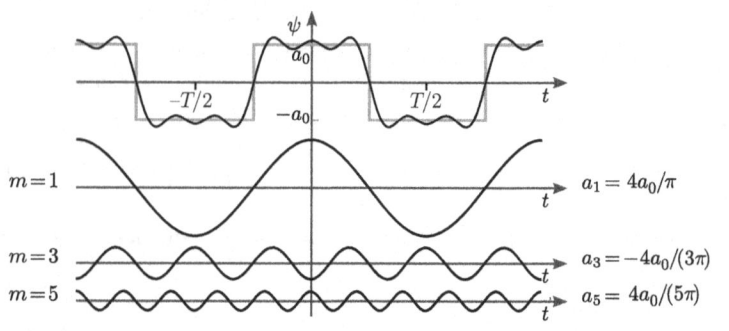

Fig. 14.1
The square wave (grey) and its approximate synthesis (black) from its first three Fourier components, shown beneath.

14.2 Fourier series and the analysis of a periodic function

The most straightforward waveform to produce electronically is the *square wave*, which is simply a periodic oscillation between two fixed displacements. This is the wave produced by the 'melody-generator' circuits in novelty greetings cards, and has a characteristically nasal buzz to its tone. We shall take it as an example with which to demonstrate the principle of Fourier analysis. Specifically, we consider the square wave illustrated in Fig. 14.1, which alternates between $\psi(t) = \pm a_0$ with a period T and is symmetrical around $t = 0$.

To simplify our analysis, we begin with some slightly subtle arguments that are based upon symmetries. The symmetry of our square wave about $t = 0$ turns out to mean that there cannot be any antisymmetric components, and the Fourier analysis will therefore find components that are cosines, rather than sines. Secondly, the periodicity of the waveform must be reflected in a periodicity of the components. Both of these conclusions may be proved explicitly by calculating the Fourier integrals of equation (14.3) and taking the symmetries into account; or they may be reached by careful non-arithmetic reasoning. The result is that we may explicitly write the square wave $\psi(t)$ as

$$\psi(t) = \sum_{n=1}^{\infty} a_n \cos\left(\frac{2\pi n}{T} t\right). \tag{14.4}$$

The expression in this way of a **periodic function** as a superposition of a series of discrete sinusoidal functions is known, rather unimaginatively, as a **Fourier series**; since its definition does not involve calculus, it is often introduced before the more general integral form for non-periodic functions.

The formula for the Fourier-series coefficients a_n may now be derived explicitly, by multiplying equation (14.4) by an arbitrary cosine function of the same periodicity and integrating with respect to time,

$$\int_{-T/2}^{T/2} \psi(t)\cos\left(\frac{2\pi m}{T}t\right) dt = \int_{-T/2}^{T/2} \sum_{n-1}^{\infty} a_n \cos\left(\frac{2\pi n}{T}t\right)\cos\left(\frac{2\pi m}{T}t\right) dt.$$
(14.5)

Note that for Fourier *series* we restrict the range of integration to a single period. This remains consistent with the principle of Fourier analysis, for the integral over all time corresponds simply to a constant multiple of the integral over the single period; since the constant multiple in this case would be infinite, however, it is convenient to avoid its introduction.

Equation (14.5) is simplified by the use of the trigonometric identity

$$2\cos\left(\frac{2\pi n}{T}t\right)\cos\left(\frac{2\pi m}{T}t\right) \equiv \cos\left(\frac{2\pi(n+m)}{T}t\right) + \cos\left(\frac{2\pi(n-m)}{T}t\right)$$
(14.6)

and its integral hence proves to be zero except when $n = \pm m$, since for all other cases the cosine terms make an integral number of complete cycles in the period of integration. When $n = m$, equation (14.5) becomes

$$\int_{-T/2}^{T/2} \psi(t)\cos\left(\frac{2\pi m}{T}t\right) dt = a_m \int_{-T/2}^{T/2} \frac{1}{2}\, dt$$
$$= a_m \frac{T}{2},$$
(14.7)

so that

$$a_m = \frac{2}{T}\int_{-T/2}^{T/2} \psi(t)\cos\left(\frac{2\pi m}{T}t\right) dt.$$
(14.8)

Had our waveform not been a symmetrical function, we should have found that the sine-wave coefficients were similarly given by

$$b_m = \frac{2}{T}\int_{-T/2}^{T/2} \psi(t)\sin\left(\frac{2\pi m}{T}t\right) dt.$$
(14.9)

The term T in the denominator in both cases ensures that the same result is obtained if the integration is extended to cover multiple periods of the waveform. This derivation, when taken to the limit $T \to \infty$ with appropriate consideration of the infinities introduced, offers an alternative route to the definitions of the integral Fourier transform given earlier in this chapter.

We may now insert our specific waveform, shown in Fig. 14.1. The integrand of equation (14.8) is symmetrical about $t = 0$, so we need only calculate one half of it; and, since the corresponding section of the waveform comprises two

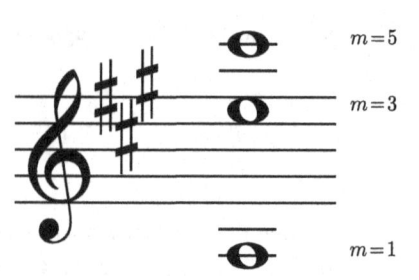

Fig. 14.2 The first three frequency components of a square or triangular wave of period $1/220$ s, shown as musical notes.

regions in which the function is constant, it makes sense to evaluate the integral separately in each of these regions:

$$
\begin{aligned}
a_m &= \frac{4}{T} \int_0^{T/2} \psi(t)\cos\left(\frac{2\pi m}{T}t\right) \, dt \\
&= \frac{4}{T} \left[\int_0^{T/4} a_0 \cos\left(\frac{2\pi m}{T}t\right) \, dt + \int_{T/4}^{T/2} (-a_0)\cos\left(\frac{2\pi m}{T}t\right) \, dt \right] \\
&= \frac{4}{T}\frac{T}{2\pi m}a_0 \left\{ \left[\sin\left(\frac{2\pi m}{T}t\right)\right]_0^{T/4} - \left[\sin\left(\frac{2\pi m}{T}t\right)\right]_{T/4}^{T/2} \right\} \\
&= \frac{4a_0}{\pi m}\sin\left(\frac{m\pi}{2}\right).
\end{aligned}
\tag{14.10}
$$

The term $\sin(m\pi/2)$ here simply cycles through the sequence $(1, 0, -1, 0, \ldots)$ as m is increased, so the first few coefficients a_m, beginning with a_1, are $4a_0/\pi$ times $(1, 0, -1/3, 0, 1/5, 0, \ldots)$. The first three components are shown schematically in Fig. 14.1.

When we calculate the Fourier transform of a waveform, we are simply asking what frequency components are present within it – which is no more than asking a musician to tell us which notes are played in a particular snatch of music, and the frequencies determined above may indeed be represented on a musical stave. If $T = 1/220$ s, so that the waveform repetition frequency is 220 Hz (corresponding to A_3, three semitones below Middle C), then the spectrum of frequencies present is as shown, using the musical stave of Chapter 12, in Fig. 14.2. While the musical notation records the component frequencies, it does not, however, indicate their relative amplitudes or phases.

14.2.1 The triangular wave

As a second example of the Fourier synthesis of a periodic function, we consider a triangular wave that rises or falls linearly between its peak values of $\pm b_0$, as shown in Fig. 14.3. For a little variety, we place the origin at a point where the wavefunction is zero: the wave is therefore an antisymmetric (odd) function of time t, and will prove to correspond to a superposition of sine, rather than cosine, functions.

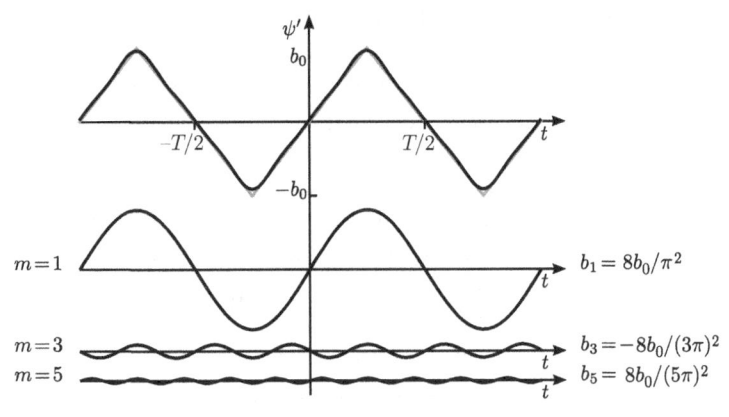

Fig. 14.3 The triangular wave (grey), just visible behind its approximate synthesis (black) from its first three Fourier components, shown beneath.

A neat mathematical trick may be used to determine the Fourier components of the triangular wave shown in Fig. 14.3 because, if its amplitude $b_0 = a_0 T/4$, the waveform is simply the integral of the square wave considered in Section 14.2,

$$\psi'(t) = \int_{-\infty}^{t} \psi(t')\mathrm{d}t', \tag{14.11}$$

where we take the constant of integration to be zero. Expressing the square wave, as earlier, as a Fourier superposition with coefficients a_m, we may write

$$
\begin{aligned}
\psi'(t) &= \int_{-\infty}^{t} \sum_{m=1}^{\infty} a_m \cos\left(\frac{2\pi m}{T}t'\right)\mathrm{d}t' \\
&= \sum_{m=1}^{\infty} a_m \int_{-\infty}^{t} \cos\left(\frac{2\pi m}{T}t'\right)\mathrm{d}t' \\
&= \sum_{m=1}^{\infty} a_m \frac{T}{2\pi m}\sin\left(\frac{2\pi m}{T}t'\right),
\end{aligned}
\tag{14.12}
$$

so that the triangle wave proves to be a Fourier superposition of sine waves

$$\psi'(t) = \sum_{m=1}^{\infty} b_m \sin\left(\frac{2\pi m}{T}t\right) \tag{14.13}$$

with coefficients given by

$$
\begin{aligned}
b_m &= \frac{T}{2\pi n}a_m \\
&= \frac{T}{2\pi m}\frac{4a_0}{\pi m}\sin\left(\frac{m\pi}{2}\right) \\
&= \frac{8b_0}{(\pi m)^2}\sin\left(\frac{m\pi}{2}\right).
\end{aligned}
\tag{14.14}
$$

As with the square wave, the term $\sin(m\pi/2)$ cycles with increasing m through the sequence $(1, 0, -1, 0, \ldots)$, so the first few coefficients b_m are $8b_0/\pi^2$ times $(1, 0, -1/9, 0, 1/25, 0, \ldots)$. Comparison with equation (14.10) shows that the same frequencies are present as in the square wave, and that both waveforms match the modes of the stopped pipe of Section 12.3.3. The higher frequencies of the triangular waveform are relatively weaker, however, falling in amplitude as $1/n^2$ rather than $1/n$. If we were to listen to such waveforms, the triangular wave would seem purer, for it is well represented by the fundamental frequency alone, and the square wave would in comparison sound harsher and more like the raucous tone of a pipe-organ.

We could have calculated the coefficients of equation (14.14) by direct integration according to equation (14.9), as in Section 14.2; the evaluation would have involved integration by parts.

14.3 Alternative forms of the Fourier transform

An important feature of the Fourier transform, and the origin of its name, is the strong similarity between the integrals of equations (14.2) and those of equations (14.3). If the function $\psi(t)$ is symmetrical about $t = 0$, for example – which, a little thought will conclude, means that it can be entirely composed of cosine, rather than sine, functions – then these equations become

$$\psi(t) = \int_{-\infty}^{\infty} a(\omega)\cos(\omega t)\,\mathrm{d}\omega, \tag{14.15a}$$

$$a(\omega) = \frac{1}{2\pi} \int_{-\infty}^{\infty} \psi(t)\cos(\omega t)\,\mathrm{d}t. \tag{14.15b}$$

The recipe for deriving the coefficients $a(\omega)$ is thus, apart from the factor of 2π, the same as that for recreating the original function. With very few caveats, then, we may consider the Fourier transform to be the transformation of the function of time $\psi(t)$ to the function of frequency $a(\omega)$ or vice versa. The symmetry can be made complete if we modify our initial definition to include a factor of $\sqrt{2\pi}$, giving

Fourier cosine transform

$$\psi(t) = \frac{1}{\sqrt{2\pi}} \int_{-\infty}^{\infty} a(\omega)\cos(\omega t)\,\mathrm{d}\omega, \tag{14.16a}$$

$$a(\omega) = \frac{1}{\sqrt{2\pi}} \int_{-\infty}^{\infty} \psi(t)\cos(\omega t)\,\mathrm{d}t. \tag{14.16b}$$

Different texts adopt different conventions for how the factor of 2π is divided between these two equations, which are sometimes explicitly labelled the **Fourier transform** (equations (14.15b) and (14.16b)) and **inverse Fourier transform** (equations (14.15a) and (14.16a)). However, the choice is arbitrary, with consequences only when we wish to calculate the absolute power of a waveform from its Fourier components, and the equations are mathematically completely equivalent mappings between what are named *conjugate variables*, from one space to another. From here on, we shall reinforce the similarity between the two by adopting the convention of equations (14.16) above.

Because the cosine is a symmetrical function, $a(\omega) \equiv a(-\omega)$, and the function $\psi(t)$ may be recreated just as well from only those components having positive frequencies, provided that the result is doubled accordingly. Beware, then, that, if you encounter other constant factors in the definitions of the Fourier transform, they are probably associated with a change in the range of integration.

14.3.1 Complex Fourier transforms

The similarity, identified above, between the Fourier transform and its inverse transform depends upon restricting ourselves to functions that are symmetrical about $t = 0$. The similarity is also apparent if the functions are completely anti-symmetrical, when the cosine terms become sines. In general, though, there is an apparent asymmetry in that we transform from one function of time, $\psi(t)$, into two functions of frequency, $a(\omega)$ and $b(\omega)$. This difficulty is resolved by combining the two real functions of frequency into one complex function,

$$g(\omega) \equiv a(\omega) - ib(\omega), \tag{14.17}$$

and, symmetrically, allowing $\psi(t)$ also to be complex. The Fourier transform and its inverse hence become

Complex Fourier transform

$$\psi(t) = \frac{1}{\sqrt{2\pi}} \int_{-\infty}^{\infty} g(\omega)\exp(i\omega t)\,d\omega, \tag{14.18a}$$

$$g(\omega) = \frac{1}{\sqrt{2\pi}} \int_{-\infty}^{\infty} \psi(t)\exp(-i\omega t)\,dt, \tag{14.18b}$$

so that the only difference between the transform and its inverse is the sign of the complex exponent. The real function ψ with which we began this chapter is simply the real part of the complex $\psi(t)$ of equations (14.18).

Instead of equation (14.17), we could have chosen $g(\omega) \equiv a(\omega) + ib(\omega)$, and the signs of the exponents in equations (14.18) would then have been interchanged. Elsewhere, both conventions will be encountered.

14.3.2 Conjugate variables

The Fourier transforms that we have presented so far allow us to transform a function of *time* into one of *frequency*, and vice versa. Time and frequency in this case form a pair of what are known as **conjugate variables**: the wave may be represented completely equivalently as a function of either, and the dimensions of the two variables are reciprocals of each other.

Fourier transforms may just as readily be used to transform a snapshot of a wave that is a function of *position* onto one of *wavenumber*, so position and wavenumber form another pair of conjugate variables for which equations (14.18) become

$$\psi(x) = \frac{1}{\sqrt{2\pi}} \int_{-\infty}^{\infty} g(k)\exp(ikx)\,\mathrm{d}k, \tag{14.19a}$$

$$g(k) = \frac{1}{\sqrt{2\pi}} \int_{-\infty}^{\infty} \psi(x)\exp(-ikx)\,\mathrm{d}x. \tag{14.19b}$$

When we later extend waves into three spatial dimensions, the three-dimensional position $\mathbf{r}$ that replaces the one-dimensional position x will be the conjugate variable of a three-dimensional *wavevector* $\mathbf{k}$. The Fourier transform definitions will, however, remain fundamentally as above: we simply pick a direction to consider and restrict ourselves at any time to that single dimension.

The implication of all of this is that whenever we can define a waveform as a function of a variable such as position or time there will exist a conjugate variable onto which the waveform is transformed, or *mapped*, by performing the Fourier transform. The Fourier transform is therefore a quite abstract, mathematical process – yet, as we shall see in due course, it corresponds to some very natural physical processes. We shall also see that conjugate variables have a particular significance in quantum mechanics, where they are linked by the *Heisenberg uncertainty principle*.

14.4   Mathematical justification of Fourier's principle

We have so far simply *assumed* the truth of Fourier's statement that any function may be formed from a unique combination of sinusoidal waves. There is, of course, a rigorous mathematical proof of this principle; but here we shall present a slightly rough but intuitively plausible justification, which is based around the δ-function introduced by Paul Dirac. In place of time and frequency, we shall consider the conjugate variables of position x and wavenumber k.

14.4.1   The Dirac δ-function

We return to the discourse of Sections 13.3 and 13.4 concerning the generation of a localized wavepacket by combining initially two, and then successively more, cosinusoidal components. Eventually, with a complete spread of wavenumbers – or *spatial frequencies* – we find that the components are all in phase only at the origin, and that away from the origin the increasingly random phases mean that as many components take a positive value as are negative, and the resultant tends to zero.

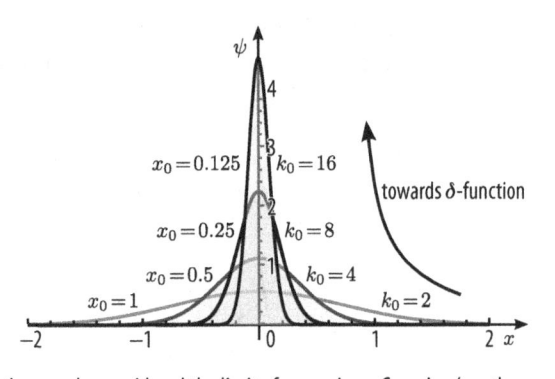

Fig. 14.4 The Dirac δ-function may be considered the limit of squeezing a Gaussian (or other monotonically peaked function) of constant area to infinitesimal width. Shown here are Gaussians $\exp(-k_0 x) \equiv \exp(-x/x_0)$, normalized with respect to unit area, with widths $x_0 = 1, 0.5$, 0.25 and 0.125 ($k_0 = 2, 4, 8$ and 16).

To treat this mathematically, we shall take components over a range of wavenumbers, but will initially concentrate them mainly within a frequency interval of k_0 around zero according to the Gaussian function

$$b(k) = \exp\left[-\left(\frac{k}{k_0}\right)^2\right]. \qquad (14.20)$$

Roughly, then, the maximum difference in wavenumber between any two components is of order k_0, and the waves will become significantly out of phase only when $k_0 x \sim 1$; the wavepacket therefore has a width of order $1/k_0$.

We may add some mathematical rigour to our argument by calculating the sum of all these components. To within an arbitrary scaling factor, which for the time being is unimportant, we have

$$\psi(x) = \int_{-\infty}^{\infty} \exp\left[-\left(\frac{k}{k_0}\right)^2\right] \cos(kx)\,\mathrm{d}k, \qquad (14.21)$$

which looks rather intractable until we express the cosine function in terms of complex exponentials,

$$\psi(x) = \frac{1}{2}\int_{-\infty}^{\infty} \exp\left[-\left(\frac{k}{k_0}\right)^2\right][\exp(\mathrm{i}kx) + \exp(-\mathrm{i}kx)]\mathrm{d}k$$

$$= \frac{1}{2}\left\{\int_{-\infty}^{\infty} \exp\left[-\left(\frac{k}{k_0}\right)^2\right]\exp(\mathrm{i}kx)\mathrm{d}k \right.$$

$$\left. + \int_{-\infty}^{\infty} \exp\left[-\left(\frac{k}{k_0}\right)^2\right]\exp(-\mathrm{i}kx)\mathrm{d}k\right\}$$

$$= \int_{-\infty}^{\infty} \exp\left[-\left(\frac{k}{k_0}\right)^2\right]\exp(-\mathrm{i}kx)\mathrm{d}k, \qquad (14.22)$$

where to achieve the final step we make the transformation $k \to -k$ in the second integral, rendering it identical to the first. We now complete the square in the exponent,

$$\psi(x) = \int_{-\infty}^{\infty} \exp\left\{-\left[\left(\frac{k}{k_0}\right)^2 + ikx + \left(\frac{ixk_0}{2}\right)^2\right]\right\} \exp\left[\left(\frac{ixk_0}{2}\right)^2\right] dk$$

$$= \int_{-\infty}^{\infty} \exp\left[-\left(\frac{k}{k_0} + \frac{ixk_0}{2}\right)^2\right] \exp\left[-\left(\frac{xk_0}{2}\right)^2\right] dk$$

$$= \exp\left[-\left(\frac{xk_0}{2}\right)^2\right] \int_{-\infty}^{\infty} \exp\left[-\left(\frac{k}{k_0} + \frac{ixk_0}{2}\right)^2\right] dk \qquad (14.23)$$

where we have taken outside the integral the term that does not depend upon k. We now write $k'/k_0 \equiv k/k_0 + ixk_0/2$, and then $z \equiv k'/k_0$, so that

$$\psi(x) = \exp\left[-\left(\frac{xk_0}{2}\right)^2\right] \int_{-\infty}^{\infty} \exp\left[-\left(\frac{k'}{k_0}\right)^2\right] dk'$$

$$= \exp\left[-\left(\frac{xk_0}{2}\right)^2\right] k_0 \int_{-\infty}^{\infty} \exp(-z^2) dz$$

$$= k_0\sqrt{\pi} \exp\left[-\left(\frac{xk_0}{2}\right)^2\right], \qquad (14.24)$$

where the integral of $\exp(-z^2)$ is a standard result, which is independent of all parameters and equal to $\sqrt{\pi}$.

The superposition therefore produces a smoothly varying **Gaussian wavepacket**, the same shape as the spectrum $b(k)$ of equation (14.20) but with a width, which we may define by the distance $x_0 \equiv 2/k_0$ at which the amplitude is a factor of e below its peak, that falls in inverse proportion to that of the spectrum, k_0. The height of the wavepacket increases with k_0, while the wavepacket narrows so as to maintain a constant area beneath it as shown in Fig. 14.4. This may be shown explicitly using the same standard integral as above,

$$\int_{-\infty}^{\infty} \psi(x) dx = k_0\sqrt{\pi} \int_{-\infty}^{\infty} \exp\left[-\left(\frac{xk_0}{2}\right)^2\right] dx$$

$$= 2\pi. \qquad (14.25)$$

These mathematical manipulations allow us to determine what happens when we add an infinite range of components of equal amplitude, for this corresponds simply to letting k_0 tend to infinity: the superposition becomes infinitely narrow and infinitely high, but at all times encloses the same area as shown by equation

(14.25). This, with appropriate scaling, is the Dirac δ-function, which may hence be written as

$$\delta(x) \equiv \lim_{k_0 \to \infty} \frac{1}{2\pi} \int_{-\infty}^{\infty} \exp\left[-\left(\frac{k}{k_0}\right)^2\right] \cos(kx) \mathrm{d}k = \frac{1}{2\pi} \int_{-\infty}^{\infty} \cos(kx) \mathrm{d}k.$$

(14.26)

The δ-function is therefore an infinitely narrow, infinitely high function with a finite area, usually taken to be unity.

An important feature of the Dirac δ-function is that it can be centred on any value of x by aligning the cosine functions around that point. Mathematically, we construct the function

$$\delta(x - x_0) \equiv \frac{1}{2\pi} \int_{-\infty}^{\infty} \cos[k(x - x_0)] \mathrm{d}k.$$

(14.27)

As we may add to equation (14.27) any odd (antisymmetric) function, whose integral is identically zero, we may also write the Dirac δ-function in terms of complex exponential components,

$$\delta(x - x_0) \equiv \frac{1}{2\pi} \int_{-\infty}^{\infty} [\cos[k(x - x_0)] + \mathrm{i}\sin[k(x - x_0)]] \mathrm{d}k$$

$$= \frac{1}{2\pi} \int_{-\infty}^{\infty} \exp[\mathrm{i}k(x - x_0)] \mathrm{d}k.$$

(14.28)

We shall find the Dirac δ-function useful both because it is the Fourier transform of a flat spectrum, as seen above, and because it allows the sampling of a single point within a continuous wavefunction.

14.4.2 The Fourier principle

Given that we may superpose sinusoidal waves to produce a δ-function centred on any value of x, it follows that any function $\psi(x)$ may be similarly produced by first writing the function as a series of δ-functions:

$$\psi(x) \equiv \int_{-\infty}^{\infty} \psi(x_0)\delta(x - x_0) \mathrm{d}x_0.$$

(14.29)

We may now proceed to obtain our expressions for the Fourier transform and its inverse. Substituting for $\delta(x - x_0)$ from equation (14.27) and expanding $\cos[k(x - x_0)]$ to $\cos(kx)\cos(kx_0) + \sin(kx)\sin(kx_0)$ gives

$$\psi(x) \equiv \frac{1}{2\pi} \int_{-\infty}^{\infty} \int_{-\infty}^{\infty} \psi(x_0)\cos[k(x - x_0)] \mathrm{d}x_0 \, \mathrm{d}k$$

$$= \int_{-\infty}^{\infty} \cos(kx) \frac{1}{2\pi} \int_{-\infty}^{\infty} \psi(x_0)\cos(kx_0) \mathrm{d}x_0 \, \mathrm{d}k$$

$$+ \int_{-\infty}^{\infty} \sin(kx) \frac{1}{2\pi} \int_{-\infty}^{\infty} \psi(x_0)\sin(kx_0) \mathrm{d}x_0 \, \mathrm{d}k,$$

(14.30)

as given by our original expressions in equations (14.2) and (14.3) for the Fourier transform and inverse transform.

The only difficulty in establishing the formal mathematical basis for Fourier analysis is in dealing with the zeros and infinities that result when k_0 is taken to infinity. This is a little messy, and troubles those of a mathematical bent. For practical physicists, however, the distinction between the extremely large and the infinite is essentially qualitative; the sketches presented here contain all the essential truths of the procedure, and the mathematics merely adds rigorous formality to the principles already outlined.

14.5 The spectrum

The square wave and triangular wave of Section 14.2 contain the same frequency components, and could therefore both be represented by the musical chord of Fig. 14.2, but this gives no indication of the relative magnitudes or phases of the components. The magnitudes may be represented pictorially in the form of a **spectrum**, which is simply a graph of the magnitudes of the components as a function of their frequencies. The *spectra* of the square and triangular waves are shown in Fig. 14.5.

The coefficients a_m and b_m that we have calculated in the preceding sections are the *amplitudes* of the sinusoidal components and, for examples such as the square wave and triangular wave that have a definite symmetry, we find that the waveform may be defined by either sine waves or cosine waves alone. It is therefore straightforward to plot the **amplitude spectra**, as in Figs. 14.5(a) and (c), where the amplitudes may be positive or negative according to their phases. More generally, however, any phase value could occur.

Often, we avoid the complications of phase by presenting the *modulus* of the amplitude for each frequency component, defined in equation (14.1) as c_n. Alternatively, we may plot the *power* or *intensity*, which, as we saw in Section 4.2, will be proportional at any frequency to the square of the modulus of the amplitude. If a spectrum is obtained experimentally by recording the currents from an array of photodiodes in the focal plane of a spectrometer, then the signal that we record is usually proportional to the intercepted power or intensity. The **power spectra** for the square and triangular waves are shown as $|a(\omega)|^2$ and $|b(\omega)|^2$ in Figs. 14.5(b) and (d). It is common in some disciplines to plot the powers (or intensities), and sometimes also the frequencies, on logarithmic axes.

We may also plot separately the **phase spectra**, which show φ_m against ω_m (or, for continuous spectra, simply $\varphi(\omega)$). Since phase is always a relative measurement with no absolute definition, it is important that, for the phase to make any sense, the basis for the representation must always be defined. When

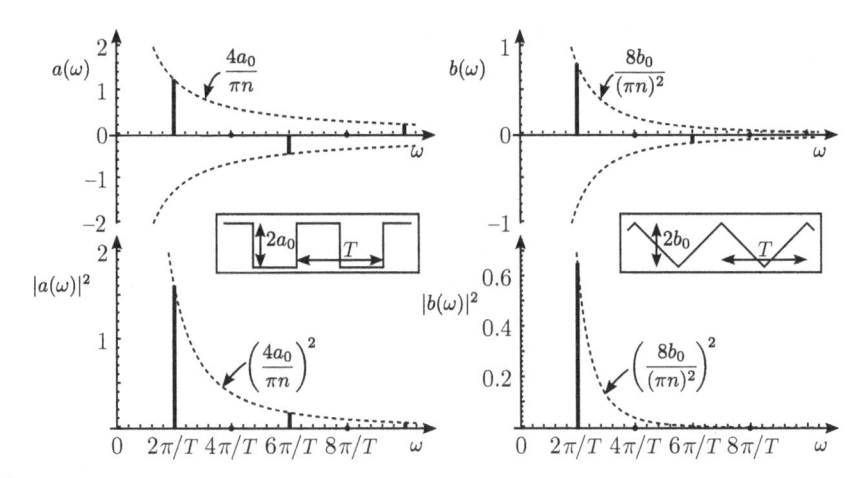

Fig. 14.5 Amplitude, (a) and (c), and power, (b) and (d), spectra of square, (a) and (b), and triangular, (c) and (d), waves. The frequencies of the components are odd multiples of $2\pi/T$; the amplitudes are proportional to $\pm 1/\omega$ and $\pm 1/\omega^2$ (shown dashed).

presenting amplitude spectra, we must also make clear whether we are using equations (14.15) or equations (14.16) as our definitions.

14.6 Orthogonality, power calculations and spectral intensities

A crucial property of sinusoidal waves, as we mentioned in Section 4.4, is their **orthogonality**. Formally, this means that the product of two sinusoidal waves averages to zero unless they share the same frequency. A practical consequence of this slightly abstruse definition involves the interpretation of power (or intensity) spectra.

The energy, and hence power and intensity, of a wave motion depend non-linearly upon the wave displacement. For transverse waves on a string, for example, we have seen in Section 4.2 that the instantaneous wave power is given by

$$\mathcal{P}(x,t) = W\upsilon_{\mathrm{p}}\left(\frac{\partial \psi(x,t)}{\partial x}\right)^2, \tag{14.31}$$

so that, for a sinusoidal wave $\psi(x,t) = a\cos(kx - \omega t)$, the power will be

$$\begin{aligned}
\mathcal{P}(x,t) &= W\upsilon_{\mathrm{p}}\left[ak\sin(kx - \omega t)\right]^2 \\
&= M\upsilon_{\mathrm{p}}\omega^2 a^2 \sin^2(kx - \omega t) \\
&= \frac{M\upsilon_{\mathrm{p}}\omega^2}{2}a^2\{1 - \cos[2(kx - \omega t)]\}.
\end{aligned} \tag{14.32}$$

The average power $\langle \mathcal{P}(x) \rangle$, which we may find by explicit integration or simply by dropping oscillatory terms, will therefore be

$$\langle \mathcal{P}(x) \rangle = \frac{M v_{\mathrm{p}} \omega^2}{2} a^2. \tag{14.33}$$

Generally, the average power or intensity of a wave motion is proportional to the product of a function of frequency and the square of the wave amplitude.

If the wave motion is represented by a Fourier superposition, such as that given at a particular point by equation (14.4), then, upon inserting the spatial dependence of a travelling wave, we may write the overall wave as

$$\psi(x, t) = \sum_{n=1}^{\infty} a_n \cos(k_n x - \omega_n t), \tag{14.34}$$

where $\omega_n \equiv 2\pi n / T$ and $k_n \equiv \omega_n / v_{\mathrm{p}}$. The power is found for convenience at $x = 0$ by applying equation (14.31) once more to obtain, instead of equation (14.33),

$$\mathcal{P}(0, t) = M v_{\mathrm{p}} \left[\sum_{n=1}^{\infty} \omega_n a_n \sin(\omega_n t) \right]^2. \tag{14.35}$$

In principle, the expansion of this expression involves a series of cross-products

$$\mathcal{P}(0, t) = M v_{\mathrm{p}} \sum_{m=1}^{\infty} \sum_{n=1}^{\infty} \omega_m \omega_n a_m a_n \, \sin(\omega_m t) \sin(\omega_n t)$$

$$= \frac{M v_{\mathrm{p}}}{2} \sum_{m=1}^{\infty} \sum_{n=1}^{\infty} \omega_m \omega_n a_m a_n \left[\cos\left(\frac{\omega_m + \omega_n}{2} t \right) + \cos\left(\frac{\omega_m - \omega_n}{2} t \right) \right]. \tag{14.36}$$

The average power is again found by dropping the oscillatory terms, leaving

$$\langle \mathcal{P}(0) \rangle = \sum_{n=1}^{\infty} \frac{M v_{\mathrm{p}} \omega_n^2}{2} a_n^2. \tag{14.37}$$

The equivalent expression for electromagnetic waves will be, from equation (4.18),

$$\langle \mathcal{I}(0) \rangle = \sum_{n=1}^{\infty} \sqrt{\frac{\varepsilon_0 \varepsilon_{\mathrm{r}}}{\mu_0 \mu_{\mathrm{r}}}} E_n^2, \tag{14.38}$$

where E_n is the amplitude of the electric field for the nth frequency component. As in Section 4.4, we therefore find that the average power of the waveform is the same as the sum of the averages for the frequency components considered individually. Given a power or intensity spectrum, then, we may calculate the total average power or intensity simply by adding the individual contributions. This result may be regarded as a generalization of that for the periodic waveforms of Section 4.4, obtained by extrapolating such results to infinite periods of repetition and integration.

In the case of a continuous spectrum, this summation involves integrating the spectrum to find the area enclosed. The quantity integrated in such a continuous spectrum must be the *power* or *intensity per unit frequency interval*, which is commonly known as the *spectral power density* or **spectral intensity**.

Note that the power (or intensity) spectrum does not allow us to determine the instantaneous power (or intensity) at any particular time, because we have discarded information about the phases with which the various frequency components add together. The amplitude spectrum (including the phase) contains all the information required to reconstruct the original waveform, and would allow us to determine the instantaneous power if required, whereas the power spectrum provides information only about average values. Nonetheless, it is in many cases the averages that are important.

14.7 Fourier analysis of dispersive propagation

An important application of Fourier analysis lies in its ability to break a complex waveform into components whose behaviour may be simpler to determine. The change in shape of a waveform as it propagates through a dispersive or dissipative medium, for example, or of an electrical signal through a network of resistors, capacitors and inductors, may in principle be determined by numerically integrating the wave equation in each case, but it is often more straightforward to break the waveform down into sinusoidal components whose propagation may be determined analytically.

To illustrate this process, we return to the thermal waves of Section 10.3 and consider their propagation along the bar shown schematically in Fig. 10.2. Here, we imagine that at the position O, which we take to be $x = 0$, a powerful heating and cooling system switches the temperature periodically between two values Θ_1 and Θ_2 so that the temperature $\Theta(0, t)$ is a square wave, with amplitude $\Theta_0 = (\Theta_2 - \Theta_1)/2$ and period T, of the form analysed in Section 14.2. We may therefore write the temperature at O as in equation (14.4) as

$$\Theta(0, t) = \sum_{n=1}^{\infty} a_n \cos(\omega_n t), \tag{14.39}$$

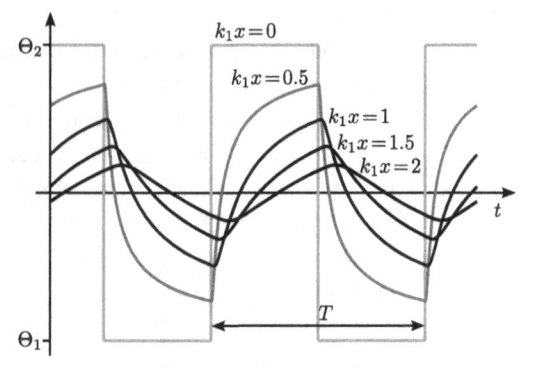

Temperature variations computed at various distances along a bar show the damped, dispersive propagation of waves that are initially square.

where $\omega_n \equiv 2\pi n/T$ and the coefficients a_n are given, after equation (14.10), as

transform waveform into single-frequency components

$$a_n = \frac{4\Theta_0}{\pi n} \sin\left(\frac{n\pi}{2}\right). \tag{14.40}$$

Each of the components of equation (14.39) will propagate as a wave of the form of equation (10.31) and, if we restrict ourselves to the region $x > 0$ and assume that the bar is long enough for reflected waves to be neglected, we may restrict ourselves to components travelling with a positive velocity, away from the origin. Matching the general form of equation (10.31) to the specific components of equation (14.39), we may therefore write the temperature at any position x as

determine propagation of individual components

combine propagated components to synthesize propagated waveform

$$\Theta(x, t) = \sum_{n=1}^{\infty} a_n \cos[(k_n x - \omega_n t)]\exp(-k_n x), \tag{14.41}$$

where, following equation (10.30), $k_n = \sqrt{\omega_n C\rho/(2\kappa)}$. The results, computed according to equation (14.41) for several points along the bar, are shown in Fig. 14.6; in addition to the increasing propagation delay, the attenuation and distortion of the waveform become more apparent as the wave propagates along the bar, until only the lowest-frequency component remains.

14.7.1 The transfer function

To determine the propagated waveform above, we used the transformation of each sinusoidal component from the summation of equation (14.39) into the corresponding component of equation (14.41). Being defined by a common single frequency, both components are, for a given propagation distance x, sinusoidal functions of infinite duration. Their relationship may therefore be fully characterized by their relative amplitude $\alpha(x)$ and phase $\varphi(x)$,

$$\alpha_n(x) = \exp(-k_n x), \tag{14.42a}$$

$$\varphi_n(x) = k_n x. \tag{14.42b}$$

Any sinusoidal component present at $x = 0$ will hence appear after a distance x to have been retarded in phase by $\varphi(x)$ and multiplied by a factor $\alpha(x)$.

If, instead of the real sinusoidal components adopted so far, we choose to work with complex exponential components of the form $\exp[i(k_n x - \omega_n t)]$, then the relative amplitude and phase shift may be combined into a single quantity known as the **transfer function**,

$$H_n(x) \equiv \alpha(x) \exp[i\varphi(x)]. \tag{14.43}$$

For the example above, this will be given by

$$H_n(x) = \exp[-(1 - i)k_n x], \tag{14.44}$$

which may be more generally written in terms of the frequency ω as

$$H(\omega) = \exp\left[-i(1 + i)\sqrt{\frac{\omega C \rho}{2\kappa}}x\right], \tag{14.45}$$

consistently with equation (10.28). The retarded wave is hence found at any position by multiplying each frequency component by the corresponding transfer function and recombining the components according to the Fourier principle.

14.8 The convolution of waveforms

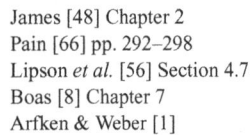
James [48] Chapter 2
Pain [66] pp. 292–298
Lipson *et al.* [56] Section 4.7
Boas [8] Chapter 7
Arfken & Weber [1] pp. 863–865

The utility of the Fourier transform lies partly in the validity of sinusoidal solutions in the presence of dispersion and partly in the ability of many instruments, including spectroscopes, resonators and even our ears, to record spectra directly. These two characteristics alone would give us ample occasions on which to require transformations between time and frequency domains, or correspondingly between physical space and the wavenumber or momentum space such as that known by solid-state physicists as the *reciprocal lattice*.

We shall shortly see that Fourier transforms are also closely linked to the phenomena of Fraunhofer and Bragg diffraction, rendering them valuable techniques in the experimental determination of molecular and crystal structures. More broadly, however, Fourier transformations are useful simply as mathematical manoeuvres, for certain operations on a mathematical function prove to be vastly simpler when the function is expressed in terms of the conjugate variable. The underlying principle in such cases is known as the *convolution theorem*, which, although an elegant example of pure calculus in its definition,

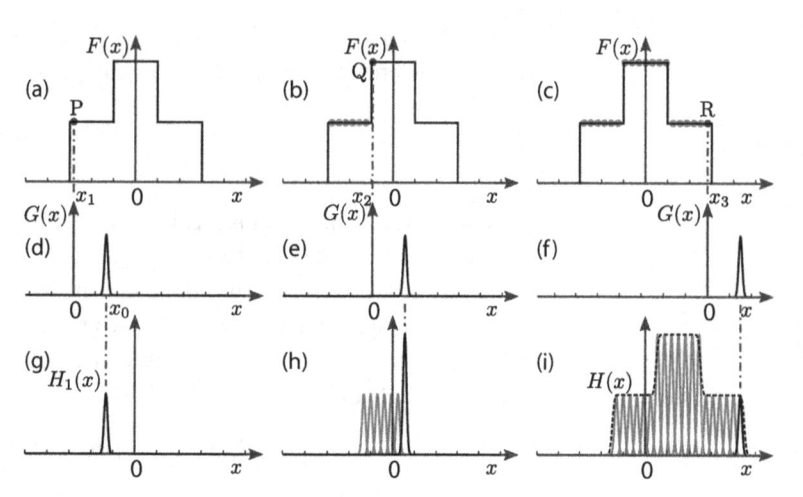

Fig. 14.7 The process of convolution of an arbitrary function $F(x)$ with a narrow Gaussian function $G(x)$ centred on x_0 to give the dashed function $H(x)$.

often allows complex mathematical operations to be performed as an essentially pictorial procedure.

14.8.1 Convolution

Before addressing the convolution theorem, we must first introduce the procedure of **convolution** itself. This is a method of combining two functions or waveforms so as to produce a third, and it is most readily explained diagrammatically. We begin by taking an arbitrary function $F(x)$, such as the stepped waveform shown in Fig. 14.7(a), and we shall consider its convolution with the narrow, peaked function $G(x)$ centred around $x = x_0$, as shown in Fig. 14.7(d).

We construct the convolution of these two functions by considering successive points on the function $F(x)$, starting at point P. We translate our function $G(x)$ until its origin lines up with the point P at $x = x_1$, as shown in Fig. 14.7(d), and then multiply $G(x)$ by the value of $F(x_1)$ at point P. The result forms the first part of our convolution, which is shown in Fig. 14.7(g).

We then move along the function $F(x)$, each time positioning the origin of $G(x)$ to coincide with the point being considered, multiply $G(x)$ by the value of $F(x)$ at that point, and add the result to what will become our convolution. Figure 14.7(h) shows some of the contributions found by working along the function $F(x)$ from P to Q; the contribution corresponding to the point Q is twice as large as that from the preceding points because $F(x_2)$ is itself twice as large as $F(x_1)$ and intervening points. Working in this fashion, we shall eventually continue along $F(x)$ until the point R, at which the convolution is complete.

The result $H(x)$ is indicated by the dotted outline in Fig. 14.7(i), and shows two significant features. First, the convoluted function roughly resembles the broader of the two initial functions, $F(x)$, but shows the rounded edges of $G(x)$; indeed, convolution may often be used as a blurring operation. Secondly, the convolution has displaced the function $F(x)$ by a distance x_0 because of the offset of the function $G(x)$ from the origin. A little thought will reveal that the same result is obtained if we exchange the functions $F(x)$ and $G(x)$. If we represent the process of convolution by the symbol $*$, then

$$H(x) = F(x) * G(x) = G(x) * F(x). \tag{14.46}$$

Mathematically, we may write the first contribution shown in Fig. 14.7(g) as

$$H_1(x_1 + x) = F(x_1)G(x), \tag{14.47}$$

where x_1 is the coordinate of the point P. On making the transformation $x \rightarrow x - x_1$, we may equivalently write

$$H_1(x) = F(x_1)G(x - x_1). \tag{14.48}$$

Adding the contributions for all values of x_1, we hence obtain

$$H(x) = \int_{-\infty}^{\infty} F(x_1)G(x - x_1)\mathrm{d}x_1. \tag{14.49}$$

The commutative nature of convolution may be demonstrated explicitly by writing $x_2 \equiv x - x_1$, giving

$$H(x) = \int_{-\infty}^{\infty} G(x_2) F(x - x_2)\mathrm{d}x_2. \tag{14.50}$$

Some examples of convolution are illustrated in Fig. 14.8. Examples (a)–(c) illustrate further the process of convolution, while example (d) shows again the blurring effect of convolution with a single, smooth peak, and example (e) shows how convolution with a closely spaced pair of sharp functions of opposite signs can perform the derivative of the function with which it is convoluted.

The convolution of two functions $F(x)$ and $G(x)$ is very similar to the process of their *cross-correlation*, which is given by

$$C(x) = \int_{-\infty}^{\infty} F(x_2)G(x_2 - x)\mathrm{d}x_2, \tag{14.51}$$

which differs from equation (14.49) only in the sign of the argument of the function G. The mathematical properties of the two processes are indeed closely linked; their practical applications, however, are usually quite different.

14.8.2 The convolution theorem

The convolution theorem provides an immensely powerful tool for the analysis of complex functions that can be assembled from more basic components: it is that the Fourier transform of the *convolution* of two functions is equal to

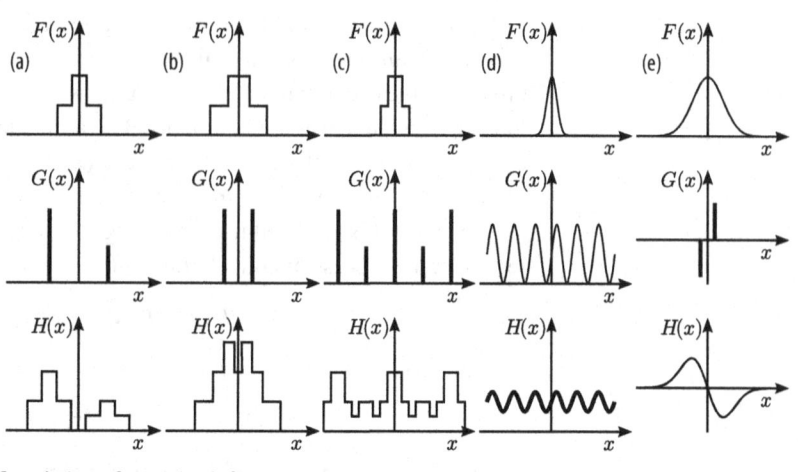

Fig. 14.8 Convolutions of some simple functions: in each case, the bottom figure $H(x)$ is the result of convoluting the two functions $F(x)$ and $G(x)$ above.

the *product* of their individual Fourier transforms. If, therefore, we write the Fourier transform of a function $F(x)$ as FT $\{F(x)\}$, then

Convolution theorem

$$FT\{F(x) * G(x)\} \equiv FT\{F(x)\} \times FT\{G(x)\}. \qquad (14.52)$$

By virtue of the symmetry of the Fourier-transform operation, it follows that the Fourier transform of the product of two functions is equal to the convolution of their individual transforms,

Convolution theorem,
alternative form

$$FT\{F(x) \times G(x)\} \equiv FT\{F(x)\} * FT\{G(x)\} . \qquad (14.53)$$

We shall see the utility of the convolution theorem in the next section, by analysing the diffraction patterns of complex apertures.

14.8.3 Proof of the convolution theorem

To derive the convolution theorem, we begin by evaluating the convolution of the function $F(x)$ in equation (14.49) with a sinusoidal or complex exponential function such as $G(x) \equiv \exp(ikx)$,

$$
\begin{aligned}
H(x) \equiv F(x) * \exp(ikx) &= \int_{-\infty}^{\infty} F(x_1)\exp[ik(x - x_1)]\mathrm{d}x_1 \\
&= \left(\int_{-\infty}^{\infty} F(x_1)\exp(-ikx_1)\mathrm{d}x_1 \right) \exp(ikx) \\
&= \sqrt{2\pi}\, f(k)\exp(ikx), \qquad (14.54)
\end{aligned}
$$

where $f(k)$ is the Fourier transform of the function $F(x)$ defined in analogy to equation (14.19b) by

$$f(k) = \frac{1}{\sqrt{2\pi}} \int_{-\infty}^{\infty} F(x)\exp(-ikx)\mathrm{d}x. \tag{14.55}$$

We repeat this for a range of k, and add the results with a k-dependent weighting factor $a(k)$:

$$\int_{-\infty}^{\infty} a(k)F(x) * \exp(ikx)\mathrm{d}k = \sqrt{2\pi} \int_{-\infty}^{\infty} a(k)f(k)\exp(ikx)\mathrm{d}k, \tag{14.56}$$

which may easily be re-written in the form

$$F(x) * \int_{-\infty}^{\infty} a(k)\exp(ikx)\mathrm{d}k = \sqrt{2\pi} \int_{-\infty}^{\infty} [a(k) \times f(k)]\exp(ikx)\mathrm{d}k. \tag{14.57}$$

Hence, writing $A(x)$ as the inverse transform of $a(k)$ from equation (14.19a),

$$A(x) \equiv \frac{1}{\sqrt{2\pi}} \int_{-\infty}^{\infty} a(k)\exp(ikx)\mathrm{d}k, \tag{14.58}$$

we see that

$$F(x) * A(x) = \mathrm{FT}^{-1}\{a(k) \times f(k)\}, \tag{14.59}$$

where FT^{-1} denotes the inverse transform of equation (14.19a). Performing the Fourier transform of both sides of this equation hence gives

$$\mathrm{FT}\{F(x) * A(x)\} = a(k) \times f(k)$$
$$= \mathrm{FT}\{F(x)\} \times \mathrm{FT}\{A(x)\}, \tag{14.60}$$

which is in the form of equation (14.52).

Similar manipulations, beginning with functions of k rather than x and omitting the final transformation, can be used to arrive at equation (14.53); equivalent expressions for sine and cosine transforms follow in the same fashion.

14.9 Fourier analysis of Fraunhofer diffraction

We have already mentioned that in many ways, when we identify a particular audible tone or distinguish the colour of an object, we are performing a Fourier transform of the acoustic or electromagnetic wave that we hear or see. There are many further examples of information-processing apparatus and algorithms by means of which we seek to identify and measure periodic components in a stream of data. There are also, however, particular wave-propagation phenomena in which the Fourier transform occurs automatically.

In Section 9.4, we showed that the amplitude of an optical wave diffracted to emerge at an angle ϑ by a mask with a transmission $t(y)$ is given by

$$a(\vartheta) = \int_{-\infty}^{\infty} t(y)\exp(-iky\sin\vartheta)\mathrm{d}y, \tag{14.61}$$

which, if we write $k' \equiv -k\sin\vartheta$, gives

$$a(k') = \int_{-\infty}^{\infty} t(y)\exp(-ik'y)\mathrm{d}y. \tag{14.62}$$

Written as a function of k', the Fraunhofer diffraction pattern is therefore the Fourier transform of the mask transmission pattern.

The variable k' is the change in path length per unit displacement y within the diffracting mask, and, for small angles ϑ, is proportional to the diffraction angle. It is also the wavenumber of sinusoidal components in the mask transmission function. If this is given, for example, by $\sin(k'y)$, then the (complex) spatial frequency components will be at $\pm k'$, and the diffraction pattern will show two diffraction orders at $\sin\vartheta = \pm k'/k$, where k is defined by the optical wavelength.

14.9.1 Diffraction by complex apertures

The correspondence between the Fraunhofer diffraction pattern of an object and the Fourier transform of the object's transmission function is the basis of many techniques of *optical information processing*, including *edge-enhancement* and *phase-contrast microscopy*. It also means that the convolution theorem may be applied to the link between an object and its diffraction pattern. If a complicated transmission function can be broken up into products and convolutions of simpler functions, then the convolution theorem provides a quick way of determining the resulting diffraction pattern.

As an example, we return to the diffraction pattern of a grating comprising a finite number of slits of non-negligible width, which we derived in Section 9.4. Such a transmission function, shown as $t(y)$ in Fig. 14.9, may be written as the convolution of a single slit $t_1(y)$ with an infinite array of δ-functions, $t_2(y)$, the result being multiplied by an overall aperture function $t_3(y)$,

$$t(y) = [t_1(y) * t_2(y)] \times t_3(y). \tag{14.63}$$

By applying the convolution theorem, we deduce that the Fraunhofer diffraction pattern must be given by

$$a(k') = [a_1(k') \times a_2(k')] * a_3(k'), \tag{14.64}$$

where $a_{1,2,3}(k')$ are the diffraction patterns of the three basic elements. This is indeed the form derived in Section 9.4, and shown in Fig. 14.9: the narrow single-slit diffraction pattern of the overall aperture appears at a regular series of points (corresponding to the diffraction pattern of an infinite comb), weighted according to the broad diffraction pattern of a single slit.

decompose object into products and convolutions of simpler elements

determine diffraction patterns of individual simpler elements

use convolution theorem to combine simple patterns into full diffraction pattern

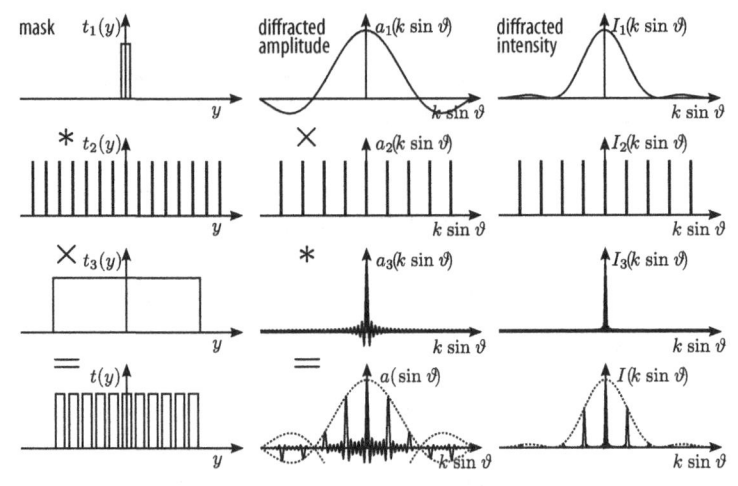

Fig. 14.9 The transmission function for a diffraction grating (left) and resulting amplitude (centre) and intensity (right) diffraction patterns may be broken down into products and convolutions of simple functions.

An alternative interpretation of Fraunhofer diffraction is that, on passing through or being reflected by the grating, an initially plane wave (with a single spatial frequency component with $k' = 0$) is converted into a wavepacket comprising a range of spatial frequencies, each of which, like the wavevector components of the evanescent waves of Section 11.5, may be regarded as the projection of a vector defining a direction of propagation. To an observer moving with the diffracted wave as it travels away from the grating, the transverse waveform spreads and disperses until, in the far field, the original wavepacket appears point-like and the diffraction pattern indicates the strengths of the various transverse wavenumber components. Since the wavenumber depends upon the illumination wavelength, Fraunhofer diffraction by a grating disperses light into its constituent colours – the principle of the grating spectrometer.

14.10 Fourier-transform spectroscopy

A further method of finding the spectrum of a light source, and another application of the Fourier transform, is the Michelson interferometer of Section 8.4, for we have seen that the transmitted intensity varies periodically with the mirror displacement according to the wavelength of illumination. On writing $\delta x \equiv 2(x_A - x_B)$ for the path difference, introducing an explicit spectral dependence to the initial intensity by writing $I_0(k)$, and dropping the constant factor $4RT$, which is unity for an ideal instrument, we may write equation (8.11) as

$$I(\delta x, k) = I_0(k)\cos^2\left(\frac{k\,\delta x}{2}\right) = \frac{I_0(k)}{2}[1 + \cos(k\,\delta x)]. \qquad (14.65)$$

The total signal is found by summing the contributions for all frequencies,

$$I(\delta x) = \int_0^\infty \frac{I_0(k)}{2}[1 + \cos(k\,\delta x)]dk$$

$$= I_1 + \frac{1}{2}\int_0^\infty I_0(k)\cos(k\,\delta x)dk, \tag{14.66}$$

where I_1 is the average transmission of the instrument. Comparing equation (14.66) with the Fourier cosine transform of equation (14.3a), and taking into account the range of integration, we may write the transmitted signal as

$$I(\delta x) = I_1 + \frac{\pi}{2}\,\mathrm{FT}\{I_0(k)\}. \tag{14.67}$$

If the average intensity is subtracted, the transmitted signal is therefore proportional to the Fourier transform of the illumination spectrum $I_0(k)$. This analysis is formalized by noting that $I(\delta x) - I_1$ is the *autocorrelation* (the cross-correlation, introduced in equation (14.51), of a function with itself) of the electric fields emerging via the two arms,

$$C(\delta x) = I(\delta x) - I_1 = \int_{-\infty}^\infty E(t)E(t + \delta x/c)dt, \tag{14.68}$$

A. Y. Khintchine [52]
Lipson *et al.* [56] Sections 11.4
and 11.5

and applying the *Wiener–Khintchine theorem*, which states that the intensity spectrum is the Fourier transform of the autocorrelation function.

If the transmitted signal $I(\delta x)$ is recorded as the path length δx is mechanically scanned, the spectrum of the incident light may therefore be found by computationally taking the inverse Fourier transform of the recorded signal. This is the principle of *Fourier-transform spectroscopy*, which is commonly used at infrared wavelengths to determine complex molecular spectra both in the laboratory and for astronomical and atmospheric studies. The limit, in any real instrument, to the range over which the path length δx may be scanned is equivalent to multiplying the recorded signal by an aperture function, so, according to the convolution theorem, the computed spectrum will appear to have been convoluted with the Fourier transform of the aperture function, blurring it slightly and hence limiting the resolution of the instrument.

Exercises

14.1   An amplitude-modulated radio signal may be written as

$$f(t) = a_0\cos(\omega_0 t)[1 + \alpha\cos(\omega_1 t)]$$

$$= a_0\cos(\omega_0 t) + \frac{\alpha}{2}a_0[\cos(\omega_0 + \omega_1)t + \cos(\omega_0 - \omega_1)t],$$

where a_0, α, ω_0 and ω_1 are constant parameters and $\omega_1 \ll \omega_0$, $\alpha \ll 1$. Show explicitly that the frequency spectrum of $f(t)$ may be written for positive frequencies ω as

$$F(\omega) = a_0 \sqrt{\frac{\pi}{2}} \left(\delta(\omega_0 - \omega) + \frac{\alpha}{2} \delta((\omega_0 + \omega_1) - \omega) \right.$$

$$\left. + \frac{\alpha}{2} \delta((\omega_0 - \omega_1) - \omega) \right). \qquad (14.69)$$

14.2 Sketch the power (or intensity) spectrum given by equation (14.69) if $\omega_0/(2\pi) = 198$ kHz, corresponding to the long-wave transmissions of BBC Radio 4, and $\alpha = 0.5$.

14.3 An experiment to investigate human hearing involves playing short bursts of duration T of a single tone with frequency $f_0 = \omega/(2\pi) \gg 1/T$,

$$a(t) = \begin{cases} \cos(\omega_0 t) & (-T/2 \le t \le T/2), \\ 0 & (t < -T/2, \ t > T/2). \end{cases} \qquad (14.70)$$

Calculate the Fourier transform of a single burst, sketch the amplitude and intensity spectra corresponding to your result, and derive the *full width at half maximum* (FWHM) of the intensity spectrum of each burst, given that $(\sin x/x)^2 = 1/2$ when $x = 1.392$.

14.4 Experimentally, it is found that, around 440 Hz, the smallest difference in tone that the ear can distinguish is about 1 Hz if the duration of the burst is 0.5 s. With reference to your answer to Exercise 14.3, comment on this result.

14.5 The diffraction pattern of an infinite regular array of narrow slits, separated by a distance d, is given by

$$a(\vartheta) = a_0 \sum_{n=0}^{\infty} \delta \left(\sin \vartheta - n \frac{\lambda}{d} \right), \qquad (14.71)$$

where ϑ is the angle through which the incident beam is diffracted and $\delta(x)$ is the Dirac δ-function. By first writing the grating transmission function as a combination of products and convolutions of functions, determine the diffraction pattern of a grating of width b, composed of narrow slits of width c spaced by a distance d, and sketch the pattern.

14.6 Use the result of your solution to Exercise 14.5 to estimate the *resolution* of a spectrograph for use at $\lambda = 589$ nm when the grating parameters are $b = 25$ mm, $d = 1/(450 \ \text{mm}^{-1})$ and $c = 0.5 \ \mu\text{m}$.

14.7 Show, by completing the square in the exponent as in the derivation of equation (14.23), that the Fourier transform of a Gaussian wavepacket $a(t)$ of width τ and centre (angular) frequency ω_0,

$$a(t) = a_0 \exp(-i\omega_0 t) \exp\left[-\left(\frac{t}{\tau} \right)^2 \right], \qquad (14.72)$$

is a Gaussian of width $2/\tau$, centred on ω_0, given by

$$a(\omega) = a_0 \tau \sqrt{\pi} \exp\left[-\left(\frac{\omega - \omega_0}{2/\tau}\right)^2\right]. \qquad (14.73)$$

14.8 Show that the convolution of any two normalized sinc functions $(\sin x / x)$ gives as its result a sinc function that is identical to the broader of the two initial functions.

Waves in three dimensions

15.1 Waves in multiple dimensions

Pain [66] Chapter 9
French [29] pp. 244–246

Our treatment of wave propagation so far has mostly been restricted to geometries that are essentially one-dimensional: absolutely, as with transverse waves on a string; approximately, as in the longitudinal waves of a narrow column of air; or implicitly, as for the propagation of periodic plane waves in three-dimensional space. We have, it is true, shown the spreading of waves (through diffraction, or simply because they emanate from a point), which implies the presence of and dependence upon additional spatial dimensions; and in our examination of ocean waves, the Huygens method of wavefront construction and its application to diffraction, we have more explicitly allowed at least a second dimension to be significant. But we have not, until now, addressed this directly.

Crucially, the principle of wave propagation in multiple dimensions is identical to that which we have described for one dimension. A disturbance propagates from one point to its neighbours, subject to a lag or retardation due to the finite speed of response, and we must take into account all of the different routes by which a given disturbance can reach the point of interest. Essentially all that remains, then, is to extend our notation from one dimension into two or three, and to consider some common examples that we shall encounter in our real, three-dimensional world; we shall also discuss the consequences of some simple geometrical features in the multipole radiation of point-like sources and the polarization of transverse waves.

15.2 Wave equations in two and three dimensions

In Section 2.2, we derived the wave equation (2.18) for transverse waves $\psi(x, t)$ on a guitar string of mass M per unit length subject to a tension W,

$$M\frac{\partial^2 \psi}{\partial t^2} = W\frac{\partial^2 \psi}{\partial x^2} \,. \tag{15.1}$$

The left-hand side of this equation describes the acceleration of an infinitesimal element of the string; the right-hand side corresponds to the elemental force (per unit length) that causes that acceleration. If the string were straight in the region of the element considered, the tension at the two ends would cancel out completely, so for there to be a force on the element the string must be curved. The second spatial derivative $\partial^2 \psi / \partial x^2$ is simply a measure of the string curvature.

Whereas the wave amplitude $\psi(x, t)$ of equation (15.1) depends at any time upon a single spatial coordinate – the position along the string, x – waves in two- or three-dimensional space will generally depend upon two or all three of the spatial coordinates x, y and z. A net force on an elemental volume will result from curvature in any of the three dimensions, and the individual contributions to the force simply add. The wave equation describing the transverse motion $\psi(x, y, t)$ of a membrane such as a drum skin, for example, proves to be

$$M \frac{\partial^2 \psi}{\partial t^2} = W \left(\frac{\partial^2 \psi}{\partial x^2} + \frac{\partial^2 \psi}{\partial y^2} \right), \tag{15.2}$$

where M is the mass per unit area and W is the tension per unit width – both being assumed here to be isotropic. Similarly, the wave equation for thermal waves in three dimensions turns out to be

$$\frac{\partial \Theta}{\partial t} = \frac{\kappa}{C} \left(\frac{\partial^2 \Theta}{\partial x^2} + \frac{\partial^2 \Theta}{\partial y^2} + \frac{\partial^2 \Theta}{\partial z^2} \right). \tag{15.3}$$

The expression describing the curvature in each case turns out not to depend at all upon the orientation of our coordinate axes, which is entirely arbitrary, and the wave equations are more compactly written using the notation of vector calculus. In three dimensions, then, the archetypal wave equation will be

In Cartesian coordinates, the *Laplacian* is (see the Appendix)

$$\nabla^2 \equiv \frac{\partial^2}{\partial x^2} + \frac{\partial^2}{\partial y^2} + \frac{\partial^2}{\partial z^2}.$$

$$\frac{\partial^2 \psi}{\partial t^2} = v_p^2 \, \nabla^2 \psi, \tag{15.4}$$

where the wave phase velocity v_p depends as usual upon a combination of material properties.

15.2.1 Separation of variables

When considering wave propagation in more than one dimension, it is sometimes simplest and most helpful to separate the variables as in Section 4.4.1 by looking for waves that are products of pure functions of the wave coordinates. We may attempt to solve equation (15.2), for example, with waves of the form

$$\psi(x, y, t) = X(x) Y(y) T(t), \tag{15.5}$$

which, upon insertion into equation (15.2), gives the equations

$$\frac{\partial^2 T}{\partial t^2} = \frac{W(a+b)}{M} T, \tag{15.6a}$$

$$\frac{\partial^2 X}{\partial x^2} = aX, \tag{15.6b}$$

$$\frac{\partial^2 Y}{\partial y^2} = bY, \tag{15.6c}$$

where a and b are the constants of separation. These equations, as in one dimension, have both sinusoidal and complex exponential solutions, corresponding to the vibrational standing-wave modes of the membrane and to travelling waves. The method of the separation of variables therefore works just as well in systems of more than one spatial dimension.

15.3 Plane waves and the wavevector

In Cartesian coordinates,

$$\mathbf{r} \equiv x\hat{\mathbf{i}} + y\hat{\mathbf{j}} + z\hat{\mathbf{k}}$$

and

$$\mathbf{k} \equiv k_x\hat{\mathbf{i}} + k_y\hat{\mathbf{j}} + k_z\hat{\mathbf{k}},$$

where $\hat{\mathbf{i}}, \hat{\mathbf{j}}$ and $\hat{\mathbf{k}}$ are unit vectors along the x, y and z axes.

Most of the one-dimensional waves that we have encountered already have in truth been in three-dimensional systems, such as the deep ocean or a coaxial cable, and their one-dimensional nature has resulted from our consideration of plane waves that have translational symmetry perpendicular to the propagation direction. Plane waves are indeed solutions to three-dimensional wave equations such as equation (15.4), and may generally be written as

$$\psi(\mathbf{r}, t) = A \cos(\omega t - \mathbf{k} \cdot \mathbf{r} + \varphi), \tag{15.7}$$

where $\mathbf{k}$ is the three-dimensional wavevector, of magnitude $k = 2\pi/\lambda$, which points in the direction of propagation, normal to the wavefront. If the x axis is aligned parallel to k, for example, then equation (15.7) reduces to the familiar one-dimensional form

$$\psi(x, t) = A \cos(\omega t - k_x x + \varphi), \tag{15.8}$$

where k_x is the component of the wavevector parallel to the x axis; in this example, of course, $k_y = k_z = 0$.

It is instructive to split equation (15.7) into its complex components,

$$\psi(\mathbf{r}, t) = \frac{A}{2}\{\exp[\mathrm{i}(\omega t - \mathbf{k} \cdot \mathbf{r} + \varphi)] + \exp[-\mathrm{i}(\omega t - \mathbf{k} \cdot \mathbf{r} + \varphi)]\}, \tag{15.9}$$

for these may themselves be separated into pure functions of the wave coordinates, as in Section 15.2.1:

$$\frac{A}{2}\exp[\mathrm{i}(\omega t - \mathbf{k} \cdot \mathbf{r} + \varphi)] \equiv \frac{A\exp(\mathrm{i}\varphi)}{2} \exp(\mathrm{i}\omega t)\exp(-\mathrm{i}k_x x)\exp(-\mathrm{i}k_y y)$$
$$\times \exp(-\mathrm{i}k_z z). \tag{15.10}$$

The individual complex exponential terms correspond to the separated functions $T(t)$, $X(x)$, $Y(y)$ and $Z(z)$ of equation (15.5), where the final term is naturally absent if we have just two spatial dimensions. Inserting these functions into equations (15.6) shows that the constants are related by

$$\omega^2 = \frac{W}{M}\left(k_x^2 + k_y^2\right) = \frac{W}{M}k^2 \tag{15.11}$$

in exactly the way that we have introduced the wavevector above. Plane waves in three-dimensional space form a complete set of orthogonal functions from which any wave motion may be composed; they are also approximations to the radiation from distant, and hence point-like, sources.

15.3.1 The three-dimensional electromagnetic wave equation

The one-dimensional analysis of electromagnetic wave propagation presented in Section 3.2 is already valid for the propagation of plane waves in three-dimensional space, and can readily be extended to the more general case. We begin with Maxwell's equations in their three-dimensional differential forms,

Fleisch [24]

$$\nabla \cdot \mathbf{D} = \rho, \tag{15.12a}$$

$$\nabla \cdot \mathbf{B} = 0, \tag{15.12b}$$

$$\nabla \times \mathbf{E} = -\frac{\partial \mathbf{B}}{\partial t}, \tag{15.12c}$$

$$\nabla \times \mathbf{H} = \mathbf{J} + \frac{\partial \mathbf{D}}{\partial t}, \tag{15.12d}$$

where $\mathbf{E}$, $\mathbf{B}$, $\mathbf{D}$ and $\mathbf{H}$ are, respectively, the electric field strength, magnetic flux density, electric displacement and magnetic field strength, $\mathbf{J}$ is the current density, ρ is the charge density and the operator ∇ is given by

$$\nabla \equiv \hat{\mathbf{i}}\frac{\partial}{\partial x} + \hat{\mathbf{j}}\frac{\partial}{\partial y} + \hat{\mathbf{k}}\frac{\partial}{\partial z}. \tag{15.13}$$

In a linear, isotropic medium, $\mathbf{D}$ and $\mathbf{E}$ are related by the dielectric constant or relative permittivity of the medium $\varepsilon = \varepsilon_0 \varepsilon_r$, and $\mathbf{B}$ and $\mathbf{H}$ by its magnetic permeability $\mu = \mu_0 \mu_r$, through

$$\mathbf{D} = \varepsilon \mathbf{E}, \tag{15.14a}$$

$$\mathbf{B} = \mu \mathbf{H}, \tag{15.14b}$$

and hence, on differentiating equation (15.12c) with respect to position and equation (15.12d) with respect to time, we may write

$$\nabla \times \nabla \times \mathbf{E} = -\nabla \times \frac{\partial \mathbf{B}}{\partial t}$$
$$= -\frac{\partial}{\partial t}(\nabla \times (\mu \mathbf{H})) = -\mu \left(\frac{\partial \mathbf{J}}{\partial t} + \frac{\partial^2 (\varepsilon \mathbf{E})}{\partial t^2} \right). \tag{15.15}$$

We may now use the vector identity

$$\nabla \times \nabla \times \mathbf{X} \equiv \nabla(\nabla \cdot \mathbf{X}) - \nabla^2 \mathbf{X} \tag{15.16}$$

to yield

$$\nabla(\nabla \cdot \mathbf{E}) - \nabla^2 \mathbf{E} = -\mu \left(\frac{\partial \mathbf{J}}{\partial t} + \frac{\partial^2 (\varepsilon \mathbf{E})}{\partial t^2} \right). \tag{15.17}$$

If there is no free charge, so that from equation (15.12a) $\nabla \cdot \mathbf{E} = 0$, and the current density $\mathbf{J}$ is everywhere constant or zero, we therefore obtain

Equation for electromagnetic waves in vacuum

Validity: no free charges

$$\nabla^2 \mathbf{E} = \mu\varepsilon \frac{\partial^2 \mathbf{E}}{\partial t^2}, \tag{15.18}$$

which is the three-dimensional equivalent of equation (3.27).

Note that, because of the similarity between equations (15.12a) and (15.12b), and between equations (15.12c) and (15.12d), we could equally have differentiated equation (15.12c) with respect to time and (15.12d) with respect to position to yield a wave equation for the magnetic flux density $\mathbf{B}$. Indeed, for harmonic plane waves with angular frequency ω, the two variables $\mathbf{E}$ and $\mathbf{B}$ are simply related, as may be found by substituting $\mathbf{E} = \mathbf{E}_0 \exp[i(\mathbf{k} \cdot \mathbf{r} - \omega t)]$ and so on into equations (15.12c) and (15.12d), giving

$$\mathbf{B} = \frac{1}{\omega}\mathbf{k} \times \mathbf{E}, \tag{15.19a}$$

$$\mathbf{D} = \frac{1}{\omega}\mathbf{k} \times \mathbf{H}. \tag{15.19b}$$

It follows that both $\mathbf{B}$ and $\mathbf{D}$ must be perpendicular to the wavevector $\mathbf{k}$ and that, in a linear isotropic medium, the same is also true of $\mathbf{E}$ and $\mathbf{H}$ – confirming the transverse nature of electromagnetic plane waves.

The energy density for electromagnetic waves in linear, non-conducting media continues to be given, as in Section 4.2, by $\epsilon = \varepsilon_0 \varepsilon_r E^2$, and the scalar intensity may be combined with a vector in the direction of energy flow to give the **Poynting vector**,

J. H. Poynting [71]

$$\mathbf{S} = \mathbf{E} \times \mathbf{H}. \tag{15.20}$$

In uniform, isotropic media, $\mathbf{S}$ and $\mathbf{k}$ will always be parallel.

15.4 Fourier transforms in two and three dimensions

The extension of Fourier transformation to higher dimensions proves to be very straightforward: the one-dimensional basis set of functions with wavenumber k simply becomes the set of two- or three-dimensional plane waves with wavevectors $\mathbf{k}$. Since functions with different $\mathbf{k}$ are orthogonal, the component of each is found as in one dimension, by multiplying the wavefunction by the complex conjugate of the basis function such as $\exp(-i\mathbf{k} \cdot \mathbf{r})$, and integrating over all space. Equations (14.19) hence become

$$\psi(\mathbf{r}) = \frac{1}{\sqrt{2\pi}} \int_{-\infty}^{\infty} g(\mathbf{k})\exp(i\mathbf{k} \cdot \mathbf{r})d\mathbf{k}, \tag{15.21a}$$

$$g(\mathbf{k}) = \frac{1}{\sqrt{2\pi}} \int_{-\infty}^{\infty} \psi(\mathbf{r})\exp(-i\mathbf{k} \cdot \mathbf{r})d\mathbf{r}. \tag{15.21b}$$

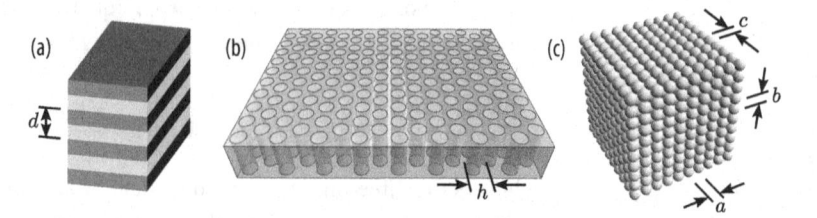

Fig. 15.1 Diffractive structures in three dimensions, shown schematically. (a) Bragg stack, (b) planar photonic crystal and (c) volume photonic crystal or opal.

The graphical representation of the three-dimensional Fourier transform can be tricky, since it in principle occupies four or five dimensions, but the transform itself is straightforward.

15.5 Diffraction in three dimensions

Just as wavefunctions can be extended into three dimensions, so can the objects that affect them. An instructive example is the diffraction grating, whose cross-section of Fig. 9.7 could also be stretched or extruded at right angles to the plane of its rulings, so that the absorption or refractive index of the rulings continues parallel to the axis of incidence to form an alternating stack of absorptive or refractive media as shown in Fig. 15.1(a); the absorption here should be kept small enough for light still to penetrate throughout the grating.

By applying either the Huygens construction or an equivalent summation of wave amplitudes, we find that, while a beam of light continues to be diffracted by the structure of periodicity d, there is an additional condition that the diffracted order exists only if it is also a mirror-like reflection from the extruded surfaces; even with a range of illumination directions, the diffracted orders continue to form individual narrow rays. The resulting pattern, which may still be obtained by detailed calculation, has an astonishingly simple property: if the incident and diffracted beams have wavevectors $\mathbf{k}_i$ and $\mathbf{k}_d$, then

$$\mathbf{k}_d = \mathbf{k}_i \pm \mathbf{k}_g, \qquad (15.22)$$

M. von Laue [88]
W. L. Bragg [10]

where $\mathbf{k}_g$ may be regarded as a wavevector of the extruded grating: a vector of length $2\pi/d$ that is normal to the extruded surfaces. This process is known as *Bragg scattering*, and equation (15.22) is the *Laue condition* defining it.

15.5.1 Crystal and lattice diffraction

If the translational symmetries of the extruded grating of Fig. 15.1(a) are broken, we may produce truly three-dimensional diffractive structures. Figure 15.1(b) shows a planar *photonic crystal*, which may be formed as an array of holes in

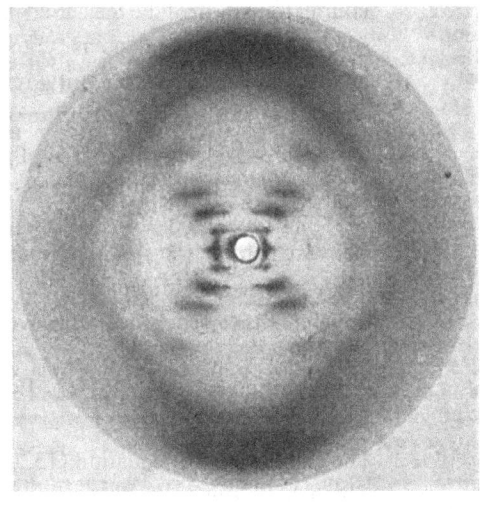

Fig. 15.2   X-ray diffraction pattern of Na^+ DNA fibre, showing features characteristic of helical structures.
Reprinted by permission from Macmillan Publishers Ltd:
R. E. Franklin & R. G. Gosling, *Nature* **171**, 740–741, copyright (1953) [28]

a refractive plate; grating wavevectors $\mathbf{k}_g$ representing the periodicities of the patterned structure lie in the plane shown, normal to the extrusion direction. For the volume photonic crystal of Fig. 15.1(c), such as a natural *opal* consisting of regularly packed silica particles, the grating wavevectors themselves map out a three-dimensional array of points known as the *reciprocal lattice*.

For a general three-dimensional structure whose optical characteristics are described by a complex position-dependent density $\psi(\mathbf{r})$, the grating wavevectors are those with non-zero amplitudes $g(\mathbf{k})$, defined by equation (15.21b); each corresponds to a component, resembling a sinusoidal or complex exponential version of the square wave of Fig. 15.1(a), with a single periodicity, and hence a single diffraction direction given by equation (15.22).

The diffraction pattern for a complex three-dimensional object is therefore found by breaking the structure down into individual periodic components with grating wavevectors $\mathbf{k}_g$, each of which then produces corresponding features in the diffraction pattern defined by equation (15.22) above. Just as with a one-dimensional grating, the diffraction pattern proves to correspond to the (now three-dimensional) Fourier transform of the diffracting object, given by equation (15.21b). With fabricated structures, the diffraction patterns may be observed with visible wavelengths, while the analysis of X-ray- and electron-diffraction patterns, such as the historic image for DNA shown in Fig. 15.2, has become the principal method of determining molecular and crystal structures.

If both the intensity and phase of the diffraction pattern were recorded, then the diffracting structure could be determined by simply performing an inverse

Fourier transform of the measured pattern. Unfortunately, most techniques measure only the diffracted intensity, and the interpretation of the diffraction pattern involves a degree of ambiguity, rendering the analysis of diffraction patterns something of an art. It was to resolve this ambiguity in electron microscopy that Dennis Gabor originally proposed what was to become the technique of *optical holography*. Although electron holography is now used as an analytical technique, notably for the study of magnetic field structures to which the electrons are particularly sensitive, sources of coherent electrons remain difficult to produce. The invention of the laser, however, has allowed holography at optical wavelengths to become a well-developed technique.

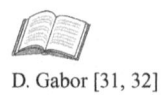

D. Gabor [31, 32]

15.5.2 Pseudo-momentum

The simple form of equation (15.22) hints at a profound underlying principle, which turns out to be that of *wave–particle duality*. The *de Broglie* waves of quantum particles are found, through experimental phenomena such as the *Compton effect*, to have a wavevector $\mathbf{k}$ related to their momentum $\mathbf{p}$ via Planck's constant $h \equiv 2\pi\hbar$ through

$$\mathbf{p} = \hbar\mathbf{k}. \tag{15.23}$$

Equation (15.22) may hence be written, on multiplying through by $\hbar$, as

$$\mathbf{p}_d = \mathbf{p}_i \pm \mathbf{p}_g, \tag{15.24}$$

where $\mathbf{p}_i$ and $\mathbf{p}_d$ are the momenta of the incident and diffracted photons (or, in the case of electron diffraction, electrons). The vector quantity $\mathbf{p}_g = \hbar\mathbf{k}_g$ is interpreted as the **pseudo-momentum** of the diffracting object and corresponds to the wavevector of a Fourier component in the device or crystal structure. The various angles through which light or electrons are diffracted are therefore those allowed if the combined momentum of the photon or electron and diffracting object is conserved – assuming, it turns out, that the incident and diffracted particles are of the same energy. Diffraction, then, corresponds to the elastic scattering of particles from a periodic structure whose pseudo-momenta are constrained to discrete values.

15.5.3 Phonons and acousto-optic modulation

While Bragg diffraction conserves momentum and energy in an elastic process, there are related phenomena in which the scattering process is inelastic, and energy is transferred to or from the scattering object. One of the most beautiful demonstrations of such processes is the *acousto-optic modulator*, shown schematically in Fig. 15.3(a). The device comprises a transparent crystal, to which is attached an acoustic transducer. Sound waves – typically in the frequency range from 100 kHz to 10 GHz – travel along the crystal, across the direction of a collimated laser beam, and the compressions and rarefactions

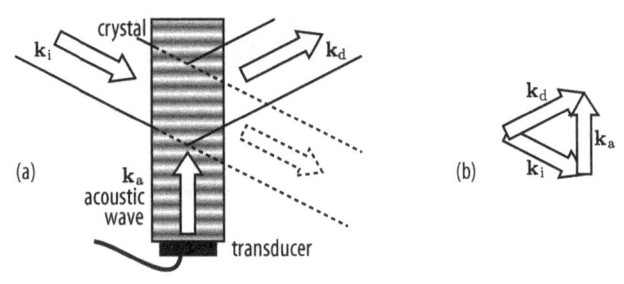

Fig. 15.3 (a) Acousto-optic modulation. (b) Triangle of wavevectors for Bragg elastic scattering of photons by the absorption or stimulated emission of phonons.

of the crystal cause a periodic modulation of the refractive index. This acts like a diffraction grating, and deflects part of the laser beam through an angle that depends upon the acoustic frequency. By varying the amplitude and frequency of the acoustic signal applied to the transducer, the angle and intensity of the diffracted beam may be controlled. The crystal is usually, however, thick enough for each acoustic wavefront to act as a mirror, and strong diffraction occurs only when the incident and deflected beams make equal angles to the acoustic wavevector, as for *Bragg diffraction*. The range of angular deviation for practical devices is determined both by the thickness of the crystal and by the curvature of the acoustic wavefronts within the crystal.

The process of reflection from a moving mirror introduces a *Doppler shift*, which we examine further in Chapter 18, into the diffracted beam, just as a moving car induces an incriminating shift of frequency in the radar beam of a traffic speed camera. This lifts the frequency degeneracy between the choices of sign in equation (15.24) and introduces a directionality to the scattering process. Acousto-optic modulators may thus be used as beam deflectors, modulators and optical frequency shifters.

The Bragg diffraction angle and the Doppler frequency shift may be calculated by straightforward and rather lengthy integration of the electromagnetic field emerging after passing through the moving acoustic wave, and such analyses will be found in the usual texts, but the results again prove to have a particularly simple form. If the incident, diffracted and acoustic frequencies and wavevectors within the crystal are ω_i, ω_d, ω_a and $\mathbf{k}_i$, $\mathbf{k}_d$, $\mathbf{k}_a$, then

$$\omega_i \pm \omega_a = \omega_d, \tag{15.25a}$$

$$\mathbf{k}_i \pm \mathbf{k}_a = \mathbf{k}_d. \tag{15.25b}$$

The momentum-conservation condition of equation (15.25b) is represented graphically in Fig. 15.1(b). The deflection process thus corresponds to the absorption or emission of an acoustic **phonon** (vibrational quantum) of frequency ω_a and wavevector $\mathbf{k}_a$, where energy and momentum are conserved. Equations

(15.25), together with the relations $\omega_d = (c/\eta)|\mathbf{k}_d|$ and $\omega_a = v_a|\mathbf{k}_a|$ that characterize the optical and acoustic propagation within the crystal, where v_a is the acoustic velocity, suffice to determine the process completely.

15.6 Wave radiation in three dimensions

Although the decomposition of wave motions into plane-wave components is perhaps the most powerful technique for their analysis, there are occasions – when we examine the wave motion close to a source, or over dimensions comparable to the distance from it – when it may be more straightforward to consider alternative solutions to the wave equation that correspond more directly to the disturbance observed. Generally, we arrive at these alternative families of solutions by changing from the Cartesian coordinate system to a different geometry that reflects the particular symmetries of the situation. For point sources that show spherical or at least cylindrical symmetry, then, we may choose to adopt a polar system of coordinates. Since the Huygens description of wave propagation is cast in terms of the radiation from point sources, this is a particularly useful and widely applicable case.

15.6.1 Spherical waves

Taking for our example the wave equation (15.4), we may write the Laplacian ∇^2 (see the Appendix) in terms of the spherical polar coordinates r, ϑ and φ,

$$\nabla^2 \equiv \frac{1}{r^2}\frac{\partial}{\partial r}r^2\frac{\partial}{\partial r} + \frac{1}{r^2\sin\vartheta}\frac{\partial}{\partial\vartheta}\sin\vartheta\frac{\partial}{\partial\vartheta} + \frac{1}{r^2\sin^2\vartheta}\frac{\partial^2}{\partial\varphi^2}, \qquad (15.26)$$

where ϑ is the angle measured from the z axis and φ is the azimuthal angle measured about it. The wave equation can then be solved formally by the technique of the separation of variables, giving solutions that multiply a function of radius by **scalar spherical harmonics** that depend upon the angular coordinates φ and ϑ alone. The scalar spherical harmonics Y_l^m, where l and m are integer parameters, have the form

$$Y_l^m(\vartheta, \varphi) \equiv \exp(im\varphi)P_l^m(\cos\vartheta), \qquad (15.27)$$

where the functions $P_l^m(u)$ are *associated Legendre polynomials* and, like the spherical harmonics to which they contribute, form a complete orthogonal set just as sinusoidal functions do. The simplest of the scalar spherical harmonics, $Y_0^0(\vartheta, \varphi) = 1$, is independent of ϑ and φ, and this proves appropriate for the radiation of isotropic, longitudinal waves (*monopole radiation*) from a point source.

If the wave displacement at a distance $r_0 \ll \lambda \equiv 2\pi v_p/\omega$ from the source varies as $\mu(\omega t)$, then after travelling outwards to a radius r we would expect the retarded wave to take the form of a radially travelling wave $\psi(u)$, where $u \equiv \omega t - kr$ and k is as usual the wavenumber. If the wave energy is conserved, however, the intensity must obey an inverse-square law, and the amplitude will hence fall with the reciprocal of the radius. From this reasoning, we might propose that the wave displacement be of the form

$$\psi(r, t) = \frac{\mu(\omega t - kr)}{r/r_0}. \tag{15.28}$$

Our exercise is hence to demonstrate that such functions are solutions to the three-dimensional wave equation (15.4).

To simplify later working, we first write

$$\begin{aligned}
\frac{\partial \mu}{\partial t} &= \omega \frac{d\mu}{du}, \\
\frac{\partial^2 \mu}{\partial t^2} &= \omega^2 \frac{d^2\mu}{du^2}, \\
\frac{\partial \mu}{\partial r} &= -k \frac{d\mu}{du}, \\
\frac{\partial^2 \mu}{\partial r^2} &= k^2 \frac{d^2\mu}{du^2}.
\end{aligned} \tag{15.29}$$

Substitution of the trial form of equation (15.28) into the wave equation, with the Laplacian as given above, now yields

$$\begin{aligned}
\frac{\omega^2}{r/r_0} \frac{d^2\psi}{du^2} &= v_p^2 \frac{1}{r^2} \frac{\partial}{\partial r} r^2 \frac{\partial}{\partial r} \frac{\mu}{r/r_0} \\
&= v_p^2 r_0 \frac{1}{r^2} \frac{\partial}{\partial r} r^2 \left(-\frac{\mu}{r^2} - \frac{k}{r} \frac{d\mu}{du} \right) \\
&= v_p^2 r_0 \left(\frac{k\mu}{r^2} - \frac{k}{r^2} + \frac{k^2}{r} \frac{d^2\mu}{du^2} \right) \\
&= \frac{v_p^2 k^2}{r/r_0} \frac{d^2\mu}{du^2}. \tag{15.30}
\end{aligned}$$

The trial form of equation (15.28) is therefore a solution to the wave equation for all times and positions, provided that $v_p = \pm\omega/k$.

15.6.2 Multipole radiation

The spherical wave of equation (15.28) is indeed typical of sound waves emanating from explosive sources and, modified by an exponential decay term, could describe thermal waves in a nuclear reactor. It cannot, however, describe transverse waves that originate from the transverse motion of a source, for no transverse force will be experienced along the extended axis of the motion, and the radiated intensity in that direction is therefore zero. This proves to be a general property of transverse waves, summarized by what is sometimes known as the *hairy-ball theorem*, which is that the hair on, for example, a tennis ball

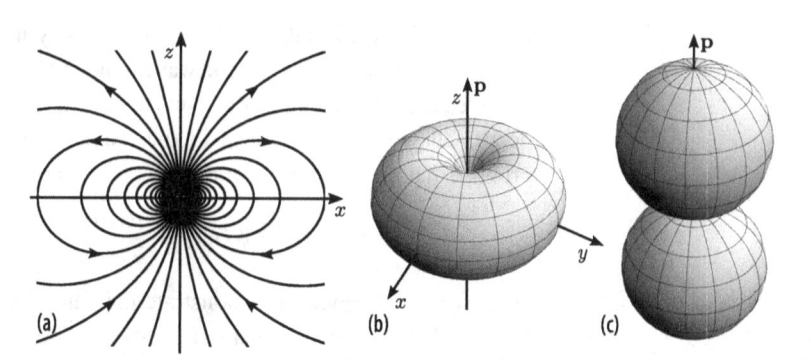

Fig. 15.4   The field of a static dipole: (a) field lines; (b) and (c) polar plots of field amplitudes for transverse and longitudinal fields, respectively.

H. Poincaré [70]

cannot be combed smooth over the entire surface but must leave at least two tufts towards and away from which the hair locally points. More prosaically, the *Poincaré–Hopf theorem* states that it is impossible to arrange a tangential vector field over the surface of a sphere without including at least two singularities at which the field is zero.

Radiated transverse waves correspond to higher orders of spherical harmonics $Y_l^m(\vartheta, \varphi)$, which successively form the terms of a *multipole* expansion. The first of these, known as the *dipole* field, corresponds for a scalar wave disturbance to the scalar spherical harmonics $Y_1^0(\vartheta, \varphi) \equiv \cos\vartheta$ and $Y_1^{\pm1}(\vartheta, \varphi) \equiv \sin\vartheta\,\exp(\pm i\varphi)$, and is the principal component radiated by an oscillating *dipole*, such as the electric potential of a pair of equal but opposite charges or the oscillating current element that they create. Since mass is always positive, the lowest-order gravitational waves correspond to *quadrupole* terms, described by the spherical harmonics Y_2^m. Most real sources produce a superposition of various multimode fields, although often one term firmly dominates the others.

There are various ways of writing the vector spherical harmonics. In spherical polar coordinates, longitudinal components will be given directly by the scalar harmonics $Y_l^m(\vartheta, \varphi)\hat{\mathbf{r}}$, while the orthogonal transverse components may be written as $r\,\nabla Y_l^m(\vartheta, \varphi)$ and $\mathbf{r} \times \nabla Y_l^m(\vartheta, \varphi)$.

Vector fields may be expressed as superpositions of *vector spherical harmonics*, whose components may be written in terms of the scalar spherical harmonics. Figure 15.4(a) shows the electric field around a static pair of equal but opposite charges that are equally spaced above and below the origin. If the charges have values $\pm q$ and their vector separation (measured from the negative to the positive charge) is $\mathbf{d}$ then, for distances $r \gg |\mathbf{d}|$, the field corresponds purely to the dipole terms and depends only upon the *electric dipole moment* $\mathbf{p} \equiv q\mathbf{d}$. It is common to consider a dipole to be the abstraction formed by reducing the charge separation $\mathbf{d}$ to zero while adjusting q so as to maintain a constant dipole moment $\mathbf{p}$. It is apparent that the field along the dipole (z) axis is parallel to it and therefore has no transverse component, whereas in the x–y plane it is only transverse. We shall address the extension from the static electric dipole field to radiated waves in Chapter 19.

Arfken & Weber [1]
Section 12.11
Jackson [47] Section 16.2

Dipole sources are not limited to electromagnetic waves: a loudspeaker, for example, is a dipole source of sound waves, compressing the air to one side as it allows it to expand on the other. In each case, the dipole radiation varies as $\sin \vartheta$ for transverse waves (Fig. 15.4(b)) and as $\cos \vartheta$ for longitudinal waves (Fig. 15.4(c)). Dipole sources of longitudinal waves, therefore, produce a pair of beams that emerge in opposite directions along the dipole axis, whereas transverse waves spread out sideways as a ring.

The reversibility of wave mathematics means that many systems used as wave sources can also be used as detectors, and the preceding analysis can therefore be applied to determine the detection sensitivity and its angular dependence for radio antennæ, microphones and so on. An interesting example is the heart-shaped *cardioid* response that is common for directional microphones, which are designed so that the acoustic paths to opposite sides of the (dipole) microphone element produce a net response that is equivalent to a superposition of monopole (spherical wave) and dipole patterns, which add for sound from in front of the microphone and cancel out for sound from behind.

15.7 Polarization

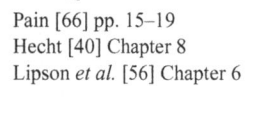

Pain [66] pp. 15–19
Hecht [40] Chapter 8
Lipson *et al.* [56] Chapter 6

The term *polarization* regrettably has two distinct meanings in electromagnetism. Here it refers to the electric field orientation in transverse electromagnetic waves; it may also describe the charge separation in the electric dipole from which they radiate or by which they are detected.

In contrast to the spherical sound waves that we considered in Section 15.6, transverse waves in three dimensions have an additional degree of freedom, for their motion may be associated with displacements anywhere within the plane perpendicular to the direction of propagation. If we label this plane with the x and y axes, then the displacement may have any combination of x and y components: the guitar string, in principle at least, may similarly oscillate parallel to the face of the instrument, or perpendicular to it. The orientation of the displacement is known as the wave **polarization**, and the wave properties may depend upon it. The most common manifestations and applications of wave polarization are with electromagnetic waves, and the difference in behaviour between opposite polarizations is crucial to the operation of many devices, from sunglasses to electronic displays. There are many fascinating manifestations in nature, including the remarkable ability of the octopus to change the polarizing properties of its body at will. For the rest of this section we shall therefore consider the polarization of a beam of light.

15.7.1 The polarization of a monochromatic field

We have already encountered some polarization-dependent phenomena in the reflectivity of an angled interface in Section 11.5, described by the Fresnel equations and illustrated in Fig. 15.5(a). The electric field vector of the reflected ray may lie in any direction within the plane normal to the wavevector **k**, and it is convenient to characterize the vector by its components parallel and

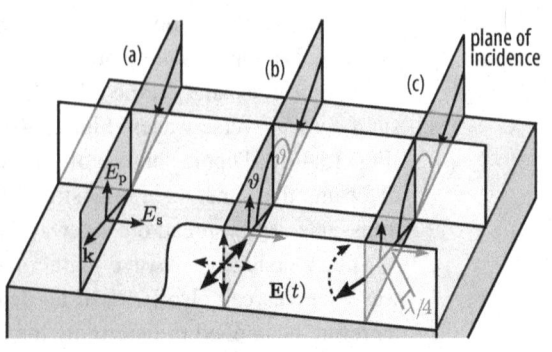

Fig. 15.5   Polarization components illustrated in the reflections of differently polarized rays from an
air–dielectric interface.

perpendicular to the *plane of incidence*, which contains both the ray and the
normal to the reflecting surface. By convention, we label these the p and s
components, from the German terms *parallel* and *senkrecht*. If the propagating
wave is monochromatic, the p and s components of its electric field will maintain
a fixed relationship in amplitude and phase, and the wave is said to be *polarized*.

The optical electric field exerts a force upon the charges in any material, and
their resulting motion is responsible for the refractive and absorptive properties
of the medium. If the material is *anisotropic*, and the response of the charges
within it depends upon the direction of the applied field, then the p and s polarization components may experience different refractive indices or absorption
coefficients, allowing an incident beam to be split along two paths (*birefringence*) or filtered by attenuating one of the two components, or the components
may be shifted in phase. This is the basis for a range of optical components,
such as polarizing filters, waveplates, optical diodes and polarization-dependent
beamsplitters, some of which may be electrically or even optically controlled.
The polarization state of the light is modified by passage through such optical
elements, which may be combined to form a variety of technological devices,
as well as by angled reflections and passage through naturally polarizing, birefringent or optically active materials.

To design and analyse such optical instruments, it is helpful to be able to
characterize the polarization state at each stage of the device. In the following
sections, we consider two commonly used approaches that are convenient for
the cases of monochromatic and broadband fields.

15.7.2 Jones vectors and Jones calculus

The polarization state of the light at any point may be characterized by the
relative amplitudes of any pair of orthogonal components; for example, we
may write the transverse electric field in terms of its components E_s and E_p in

the s and p directions as

$$E_s(z, t) = e_s E_0 \exp[i(kz - \omega t)], \tag{15.31a}$$

$$E_p(z, t) = e_p E_0 \exp[i(kz - \omega t)], \tag{15.31b}$$

where z is measured along the ray direction and the parameters e_s and e_p are in general complex and normalized to satisfy

$$|e_s|^2 + |e_p|^2 = 1. \tag{15.32}$$

The relative amplitude of the two components will then be given by the modulus of their ratio, $|e_p/e_s|$, and their relative phase by its argument, $\tan \varphi = \text{Arg}(e_p/e_s)$. While the usual parameters E_0, ω and k define the overall magnitude, frequency and wavenumber, the important polarization information is contained within the complex vector

$$\mathbf{e}(\omega) \equiv \begin{pmatrix} e_s(\omega) \\ e_p(\omega) \end{pmatrix}, \tag{15.33}$$

R. C. Jones [51]

which is known as the **Jones vector**. The vector field $\mathbf{E}(z, t)$ is then given by

$$\mathbf{E}(\omega, z, t) = \mathbf{e}(\omega) E_0 \exp[i(kz - \omega t)], \tag{15.34}$$

where we have introduced an explicit frequency dependence for the polarization properties. The real part of the electric field may be found by adding the complex electric field vector to its complex conjugate so that, if $e_s(\omega)$ is real,

$$\mathbf{E}(\omega, z, t) + \mathbf{E}^*(\omega, z, t) \equiv \begin{pmatrix} E_s(\omega, z, t) + E_s^*(\omega, z, t) \\ E_p(\omega, z, t) + E_p^*(\omega, z, t) \end{pmatrix}$$

$$= 2|E_0(\omega)| \begin{pmatrix} |e_s(\omega)| \cos(kz - \omega t) \\ |e_p(\omega)| \cos(kz - \omega t + \varphi) \end{pmatrix}. \tag{15.35}$$

The electric field vector for any superposition of waves with the same frequency and wavevector thus traces out an ellipse around the propagation axis. In general, the ellipse may have any ellipticity and orientation.

Any linear optical element will attenuate, project or shift in phase the field components so as to transform an incident Jones vector $\mathbf{e}_1(\omega)$ into an emergent vector $\mathbf{e}_2(\omega)$ according to

$$\mathbf{e}_2(\omega) = \begin{pmatrix} M_{11} & M_{12} \\ M_{21} & M_{22} \end{pmatrix} \mathbf{e}_1(\omega)$$

$$= \mathbf{M} \mathbf{e}_1(\omega), \tag{15.36}$$

where $\mathbf{M}$, whose elements M_{ij} depend upon the specific optical components, is known as the **Jones matrix**. Subsequent optical components will map the Jones vector $\mathbf{e}_2$ onto successive vectors $\mathbf{e}_3$, $\mathbf{e}_4$ and so on, each time related to the preceding vector by a Jones matrix describing the optical element. Complex arrangements of polarizing components may therefore be described, and their effects calculated, by multiplying together the Jones matrices of the individual elements in a process known as **Jones calculus**.

15.7.3   Classifications of polarized light

While the polarization state of any monochromatic electromagnetic wave can be described by a Jones vector, two specific cases are commonly noted as significant. These are illustrated in Figs. 15.5(b) and (c), which in practice would require an appropriate source or additional polarizing components.

If the elements of the Jones vector differ only by a real factor, then the s and p components of the field are in phase, as shown in Fig. 15.5(b); the ellipse is compressed to a line, and the wave is said to be **plane polarized** or **linearly polarized**. A particular example, which occurs regardless of the incident polarization if ϑ is equal to Brewster's angle ϑ_B, is when $E_p = 0$: in the geometry shown, the light is unimaginatively said to be *horizontally polarized*.

If the elements of the Jones vector instead differ only by a factor of $\pm i$, so that the s and p components are a quarter of a wavelength out of step as shown in Fig. 15.5(c), then the ellipse becomes a circle and the field is said to be **circularly polarized**. If the electric field moves clockwise around the wavevector when viewed, as illustrated, from front (i.e. looking *towards* the source), then the beam is *right-circularly polarized*; if it moves anticlockwise, then its polarization is *left-circular*. The field vector of a right-circularly polarized beam is therefore a spiral, like the thread of a right-handed screw, which moves forwards at the phase velocity of the beam.

It is clear that circularly polarized light comprises a superposition of the s and p linear polarizations, with a phase difference of $\pm i$ according to the handedness of the circular polarization. Conversely, linearly polarized light of any orientation may be expressed as a superposition of appropriately phased circular polarizations. Just as the s and p linear polarizations form the basis set for the Jones vector, from which any polarization state may be achieved, so the right- and left-circular polarizations could in principle be used.

15.7.4   The polarization of a non-monochromatic field

The single frequency component of a monochromatic field has, by definition, a constant magnitude; the s and p components of the Jones vector do not change, and the polarization state of a monochromatic field is therefore fixed. For the field magnitudes, and thereby the polarization state, to vary, we have seen in Chapter 14 that multiple frequency components are required.

Practically occurring optical fields are never completely monochromatic, however, and all but the most precisely controlled laser beams have frequency components spanning a MHz or more. Beating between these components allows the wave magnitude, and hence the polarization state, to vary at similar frequencies. While lasers usually contain polarization-defining components, many other sources of still greater bandwidth do not, and the polarization states of lamps, sunlight and so on can change more rapidly than is easily measured.

By defining the Jones vector $\mathbf{e}(\omega)$ to be frequency-dependent, we have allowed for such changes; in principle, with a range of Jones vectors, we could describe the polarization states of all frequency components present, and hence define the complete electric field. In practice, however, the number of vectors required would be huge, and the rapidly varying polarization of even a narrow-bandwidth source is often a fundamentally random process, so the Jones vectors are impossible to predict.

We choose in such cases to represent the polarization properties by a frequency-independent set of four **Stokes parameters** S_0–S_3, which describe correlations between the Jones-vector components and themselves form a four-component **Stokes vector**, $\mathbf{S}$:

G. G. Stokes [80]

$$\mathbf{S} \equiv \begin{pmatrix} S_0 \\ S_1 \\ S_2 \\ S_3 \end{pmatrix} = \begin{pmatrix} \langle E_s^2 \rangle + \langle E_p^2 \rangle \\ \langle E_s^2 \rangle - \langle E_p^2 \rangle \\ \langle 2 E_s E_p \cos\varphi \rangle \\ \langle 2 E_s E_p \sin\varphi \rangle \end{pmatrix} = \begin{pmatrix} \langle E_s^* E_s \rangle + \langle E_p^* E_p \rangle \\ \langle E_s^* E_s \rangle - \langle E_p^* E_p \rangle \\ \langle E_s^* E_p \rangle + \langle E_p^* E_s \rangle \\ \mathrm{i}(\langle E_s^* E_p \rangle - \langle E_p^* E_s \rangle) \end{pmatrix}. \tag{15.37}$$

A convenient feature of the Stokes parameters is that they may readily be determined experimentally by measuring the total intensity, the intensity transmitted by a linear polarizer oriented horizontally and at $45°$, and the intensity transmitted by a polarizer that passes right-circularly polarized light.

As with the Jones-vector representation, the effects of optical components upon the Stokes vectors may be represented by matrices, which may simply be multiplied to determine the effect of a complete instrument. The Jones and Stokes vectors for some common polarization states are listed in Table 15.1.

It may be shown that

$$S_0^2 \geq S_1^2 + S_2^2 + S_3^2, \tag{15.38}$$

where equality occurs if the Jones vector is the same for all wavelengths and the beam is therefore completely polarized. It is sometimes useful to introduce a measure of the degree of polarization, P, given by

$$P^2 \equiv \frac{S_1^2 + S_2^2 + S_3^2}{S_0^2}. \tag{15.39}$$

If $P = 0$ then the wave is said to be **randomly polarized** or **unpolarized**: the field components are uncorrelated, so the orientation of the total instantaneous field is random. Almost all naturally occurring light sources are unpolarized, although this does not mean that the light that we see is also: scattering and reflection vary with polarization, and therefore filter unpolarized light to render it at least partly polarized. The sometimes strong polarization of the scattered sunlight of a blue sky is found to be exploited by honeybees and other insects as a means of orientation and hence navigation.

Table 15.1 Jones and Stokes vectors for various polarization states

Polarization state	Jones vector	Stokes vector
Linearly (plane) polarized		
Horizontal	$\begin{pmatrix} 1 \\ 0 \end{pmatrix}$	$\begin{pmatrix} 1 & 1 & 0 & 0 \end{pmatrix}^{\mathrm{T}}$
Vertical	$\begin{pmatrix} 0 \\ 1 \end{pmatrix}$	$\begin{pmatrix} 1 & -1 & 0 & 0 \end{pmatrix}^{\mathrm{T}}$
+45°	$\frac{1}{\sqrt{2}}\begin{pmatrix} 1 \\ 1 \end{pmatrix}$	$\begin{pmatrix} 1 & 0 & 1 & 0 \end{pmatrix}^{\mathrm{T}}$
−45°	$\frac{1}{\sqrt{2}}\begin{pmatrix} 1 \\ -1 \end{pmatrix}$	$\begin{pmatrix} 1 & 0 & -1 & 0 \end{pmatrix}^{\mathrm{T}}$
ϑ	$\begin{pmatrix} \cos\vartheta \\ \sin\vartheta \end{pmatrix}$	$\begin{pmatrix} 1 & \cos(2\vartheta) & \sin(2\vartheta) & 0 \end{pmatrix}^{\mathrm{T}}$
Circularly polarized		
Right-circular	$\frac{1}{\sqrt{2}}\begin{pmatrix} 1 \\ -i \end{pmatrix}$	$\begin{pmatrix} 1 & 0 & 0 & 1 \end{pmatrix}^{\mathrm{T}}$
Left-circular	$\frac{1}{\sqrt{2}}\begin{pmatrix} 1 \\ i \end{pmatrix}$	$\begin{pmatrix} 1 & 0 & 0 & -1 \end{pmatrix}^{\mathrm{T}}$

Operators for wave motions

16.1 The mathematical operator

The *argument* here refers to the variable upon which a function depends. The same term is also used specifically to describe the phase of a complex number. For the function $f(\varphi) \equiv \exp(i\varphi)$, the variable φ will be the argument in both senses.

We are used to the concept of a mathematical *function*: a procedure that, when applied to a given number (known as its *argument*), generates another number as its result. We routinely use the sine and cosine functions, the logarithm and exponent, factorial, powers, roots and so on; they are so fundamental to the language of mathematics that we rarely give them a second thought.

Many of the functions with which we are already familiar map a single variable, such as x, onto another single variable such as y, e.g. $y(x) = \sin x$. We are also used to functions of two variables such as the sinusoidal travelling wave $\psi(x, t) \equiv \psi_0 \cos(kx - \omega t)$, while its equivalent in three dimensions $\psi(\mathbf{r}, t) \equiv \psi_0 \cos(k_x x + k_y y + k_z z - \omega t)$ is a function of the four variables x, y, z and t – indeed, if they are not fixed, it is also a function of the parameters k_x, k_y, k_z, ω and ψ_0. Functions may thus depend upon several variables, and we could in principle devise functions of arbitrarily many variables $y(x_1, x_2, x_3, \dots)$.

However, whereas functions depend upon individual values of x, t and so on, **operators** are procedures that are applied to continuous variables such as a wavefunction ψ, and the recipe described by an operator can take into account not just the instantaneous value of its argument but also how quickly that variable is changing. An example of an operator would therefore be the spatial derivative $\partial/\partial x$, which, applied to a wavefunction ψ, would yield

$$y(\psi) \equiv \partial \psi / \partial x. \tag{16.1}$$

Evaluation hence requires knowledge not only of the wavefunction at a given point but also of its values at adjacent points. A *function* may therefore be regarded as a sub-category of operator that depends only upon the instantaneous and local value of its argument. Since wave motions fundamentally describe variations with time that result from variations with position, operators prove to be of immense use in the analysis and characterization of wave motions.

16.1.1 Operators for wave motions

An operator, which we shall identify by a 'hat' (*circumflex* or *caret*, ˆ) over the symbol representing it, may hence be regarded as a recipe or operation for determining a wave property from the wavefunction to which it refers. For example, an operator $\hat{\mathbf{O}}$ might be constructed so as to yield the parameter o associated with the wavefunction $\psi(x, t)$ via the equation

$$o = \hat{\mathbf{O}}\big(\psi(x, t)\big). \tag{16.2}$$

Matthews [62] Chapter 2
Feynman [23] Vol. III, pp. 8-5
and 20-1*ff*

Just like a function, then, the operator $\hat{\mathbf{O}}$ *operates* on the wavefunction $\psi(x, t)$ to yield the required parameter or property o – which could be the wave frequency, wavenumber, wavevector, intensity, Poynting vector and so on. Generally, although not exclusively, the parameters yielded by the application of an operator are *observable* (experimentally measurable) properties of a wave motion.

We commonly make two changes to the definition sketched out in equation (16.2). First, while the result of any linear operator will increase in proportion to the magnitude of the wavefunction, this is not in general true of the observable properties in which we shall be interested and, while we could take this into account within the recipe to which the operator corresponds, it is convenient to write the normalization explicitly and define the normalized operator by

$$o(x, t) = \frac{\hat{\mathbf{O}}\big(\psi(x, t)\big)}{\psi(x, t)}. \tag{16.3}$$

Secondly, it is common to omit the brackets around the wavefunction, and let it be implicit that an operator is to operate on the following argument rather than simply multiply it. Multiplying through by the denominator hence yields the usual form in which the operator is defined,

Operator equation

$$o(x, t)\, \psi(x, t) = \hat{\mathbf{O}}\, \psi(x, t). \tag{16.4}$$

We shall see that, although the result $o(x, t)$ of the operator $\hat{\mathbf{O}}$ is defined, and may be calculated, at any position x and time t, it does not necessarily quite correspond to the intended, experimentally measurable property; indeed, in quantum physics, it will often not be possible to measure it experimentally at a single point and instant in time. The result $o(x, t)$ nonetheless provides the basis for determining such a property in a relatively straightforward manner by calculating the average value for the wavefunction, as we shall see in the next section.

16.2 Operators for frequency and wavenumber

To demonstrate their use, we shall in this section find operators for the frequency and wavenumber of a monochromatic wavefunction given by the complex exponential $\psi(x, t) = A \exp[\mathrm{i}(kx - \omega t)]$. We begin by considering the derivatives of $\psi(x, t)$ with respect to x and t, which are, respectively,

$$\frac{\partial \psi}{\partial x} = \mathrm{i}kA \exp[\mathrm{i}(kx - \omega t)] = \mathrm{i}k\psi(x, t), \qquad (16.5a)$$

In three dimensions, the wavefunction will be

$$\frac{\partial \psi}{\partial t} = -\mathrm{i}\omega A \exp[\mathrm{i}(kx - \omega t)] = -\mathrm{i}\omega \psi(x, t). \qquad (16.5b)$$

$\psi(\mathbf{r}, t) = A \exp[\mathrm{i}(\mathbf{k} \cdot \mathbf{r} - \omega t)]$ We thus find that operators for frequency and wavenumber will be given by

and the wavenumber operator will be replaced by the wavevector operator

$$\hat{\omega} = \mathrm{i}\frac{\partial}{\partial t}, \qquad (16.6a)$$

$$\hat{\mathbf{k}} = -\mathrm{i}\,\nabla.$$

$$\hat{k} = -\mathrm{i}\frac{\partial}{\partial x}. \qquad (16.6b)$$

When applied to our monochromatic wavefunction according to equation (16.4), these operators correctly give the parameters ω and k, respectively. We shall see in the next chapter that these operators have a particular significance in the context of quantum mechanics, but they are also of considerable use in themselves.

It is important to note that the instantaneous frequency or local wavenumber obtained by applying an operator describes the wave only at a particular point or time: it is simply the ratio of the wave gradient or derivative to the wave amplitude, and therefore tells us the frequency or wavenumber of a complex exponential that would give the same ratio at that place and instant. If the wave is indeed a complex exponential $\psi(x, t) = A \exp[\mathrm{i}(kx - \omega t)]$, as above, then our recipe will give the same answer for all points and times, and we may regard the frequency or wavenumber as well defined. For an arbitrary function, however, the result will vary from place to place and from one time to another; we must then conclude that the wave cannot be described by a unique frequency or wavenumber and that it corresponds to a superposition of such functions. This should not greatly surprise us, for to measure the frequency or wavenumber (and hence wavelength) of a wave in practice we should count the number of oscillations over an extended period or distance; any value that we obtain from a single point must be limited in its significance.

We may, however, use our operators to calculate the average frequency or wavenumber for a wave motion, and (with a little care) the variance or standard deviation from that average. If the variance proves to be zero, then we may deduce that the frequency or wavenumber is the same everywhere, and is given by the value whose average we have calculated.

16.3 The expectation value: the mean value of an observable

By applying an operator to a wavefunction according to equation (16.3) or (16.4), we may obtain values for the corresponding observable parameter at different points and times. To find the average value of the parameter, we simply calculate its average over all positions, weighting the values according to the magnitude of the wavefunction. To prevent regions of negative wave displacement from subtracting from the average, we represent the magnitude of the wavefunction by the square (or, for complex waves, the modulus squared) of the wave amplitude. We shall see shortly that this reasonable but apparently arbitrary choice makes the result consistent with that defined in a more conventional manner. The mean value of the observable o, which we denote $\langle o \rangle$, will hence be

$$
\begin{aligned}
\langle o \rangle &= \frac{\int_{-\infty}^{\infty} o(x)|\psi(x)|^2 \, dx}{\int_{-\infty}^{\infty} |\psi(x)|^2 \, dx} \\
&= \frac{\int_{-\infty}^{\infty} o(x)\psi^*(x)\psi(x) \, dx}{\int_{-\infty}^{\infty} |\psi(x)|^2 \, dx}.
\end{aligned}
\tag{16.7}
$$

Since the observable o is obtained from the wavefunction through the operator $\hat{O}$ as in equation (16.4), our expression for its mean becomes

Expectation value
In quantum mechanics, the wavefunction is already normalized, and the denominator is omitted.

$$
\langle o \rangle = \frac{\int_{-\infty}^{\infty} \psi^*(x)\hat{O}\psi(x) \, dx}{\int_{-\infty}^{\infty} |\psi(x)|^2 \, dx}.
\tag{16.8}
$$

Although, in the language of wave mechanics, $\langle o \rangle$ is known as the **expectation value**, it is *precisely* the same as the mean of o, as we have shown above.

16.3.1 Formal justification of the expectation value

In deciding to weight the values of the observable o at various positions according to the square of the wave amplitude at each point, we made what seemed to be an arbitrary, if reasonable, choice, which we now examine further.

Suppose that there is a set of wave motions $\psi_o(x, t)$ for each of which the observable o is well defined – that is, independent of position and time – so that

$$
\hat{O}\psi_o(x, t) = o\psi_o(x, t).
\tag{16.9}
$$

For the specific examples of frequency and wavenumber, these will be the complex exponentials $\psi_o(x, t) = A \exp[i(kx - \omega t)]$ given earlier. The principles of superposition and Fourier synthesis allow us to write any arbitrary wavefunction in terms of such components, as

$$
\psi(x, t) = \int_{-\infty}^{\infty} a(o)\psi_o(x, t)\,do,
\tag{16.10}
$$

which, for the example of frequency, corresponds with a factor $\sqrt{2\pi}$ to the Fourier integral of equation (14.18a). Substituting this into equation (16.8) gives

$$\langle o \rangle = \frac{\int_{-\infty}^{\infty} \int_{-\infty}^{\infty} a^*(o')\psi_{o'}^*(x,t)\mathrm{d}o' \ \hat{\mathbf{O}} \int_{-\infty}^{\infty} a(o)\psi_o(x,t)\mathrm{d}o\,\mathrm{d}x}{\int_{-\infty}^{\infty} \int_{-\infty}^{\infty} a^*(o')\psi_{o'}^*(x,t)\mathrm{d}o' \int_{-\infty}^{\infty} a(o)\psi_o(x,t)\mathrm{d}o\,\mathrm{d}x}$$

$$= \frac{\int_{-\infty}^{\infty} \int_{-\infty}^{\infty} a^*(o')\psi_{o'}^*(x,t)\mathrm{d}o' \int_{-\infty}^{\infty} oa(o)\psi_o(x,t)\mathrm{d}o\,\mathrm{d}x}{\int_{-\infty}^{\infty} \int_{-\infty}^{\infty} a^*(o')\psi_{o'}^*(x,t)\mathrm{d}o' \int_{-\infty}^{\infty} a(o)\psi_o(x,t)\mathrm{d}o\,\mathrm{d}x} . \quad (16.11)$$

The integrals are, of course, just summations, and we are free to change the order in which we evaluate them:

$$\langle o \rangle = \frac{\int_{-\infty}^{\infty} a^*(o') \int_{-\infty}^{\infty} oa(o) \int_{-\infty}^{\infty} \psi_{o'}^*(x,t)\psi_o(x,t)\mathrm{d}x\,\mathrm{d}o\,\mathrm{d}o'}{\int_{-\infty}^{\infty} a^*(o') \int_{-\infty}^{\infty} a(o) \int_{-\infty}^{\infty} \psi_{o'}^*(x,t)\psi_o(x,t)\mathrm{d}x\,\mathrm{d}o\,\mathrm{d}o'} . \quad (16.12)$$

If, as is the case for our complex exponentials, the functions $\psi_o(x,t)$ are orthogonal, then the integrals over x will be identically zero unless $o' = o$, and may be replaced by the Dirac δ-function $\delta(o - o')$, giving

$$\langle o \rangle = \frac{\int_{-\infty}^{\infty} a^*(o') \int_{-\infty}^{\infty} oa(o)\delta(o - o')\mathrm{d}o\,\mathrm{d}o'}{\int_{-\infty}^{\infty} a^*(o') \int_{-\infty}^{\infty} a(o)\delta(o - o')\mathrm{d}o\,\mathrm{d}o'} . \quad (16.13)$$

The δ-function effectively reduces the double integral to a single integral, for the only combinations to contribute are those for which o' and o coincide. The expectation value may therefore be written as

$$\langle o \rangle = \frac{\int_{-\infty}^{\infty} a^*(o)oa(o)\mathrm{d}o}{\int_{-\infty}^{\infty} a^*(o)a(o)\mathrm{d}o} = \frac{\int_{-\infty}^{\infty} o\,|a(o)|^2\,\mathrm{d}o}{\int_{-\infty}^{\infty} |a(o)|^2\,\mathrm{d}o} . \quad (16.14)$$

We have seen in Section 14.6 that for electromagnetic waves, $|a(o)|^2$ is proportional to the spectral energy density; for the specific example of frequency, then, the expectation value $\langle \omega \rangle$ gives the same result as we would obtain by calculating the energy-weighted mean of the frequency spectrum that we might record from an optical spectrometer or radiofrequency spectrum analyser. As we discuss in Section 17.3.3, $|\psi^2|$ corresponds for quantum wavefunctions to a particle's probability density, so that the weighting factor in the expectation value is directly the probability of the corresponding outcome.

16.4 The uncertainty: the standard deviation of an observable

Statistical measures of the width of a distribution are commonly given in terms of the **variance**, $\mathrm{var}(o)$ or Δ_o^2, which is the mean squared deviation from the mean, and its square root, known as the **standard deviation**, $\sigma(o)$. If, as in Section 16.3, we find that the spectrum of frequencies (or values of any other observable, o) present in a wave motion contains a spread of values, then the variance will be given by

$$\Delta_o^2 \equiv \sigma(o)^2 = \frac{\int_{-\infty}^{\infty} \left(o - \langle o \rangle\right)^2 a^*(o)a(o)\mathrm{d}o}{\int_{-\infty}^{\infty} a^*(o)a(o)\mathrm{d}o}. \qquad (16.15)$$

This is therefore explicitly the expectation value of the square of the deviation, $(o - \langle o \rangle)^2$, according to equation (16.14). By expanding the square, noting that the expectation value does not itself depend upon the variable of integration o, and recalling the definition of the expectation value from equation (16.14), we may re-write equation (16.15) as

$$\begin{aligned}
\Delta_o^2 \equiv \sigma(o)^2 &= \frac{\int_{-\infty}^{\infty} \left(o^2 - 2o\langle o \rangle + \langle o \rangle^2\right) a^*(o)a(o)\mathrm{d}o\,\mathrm{d}o}{\int_{-\infty}^{\infty} a^*(o)a(o)\mathrm{d}o} \\
&= \frac{\int_{-\infty}^{\infty} o^2 a^*(o)a(o)\mathrm{d}o}{\int_{-\infty}^{\infty} a^*(o)a(o)\mathrm{d}o} - 2\langle o \rangle \frac{\int_{-\infty}^{\infty} o a^*(o)a(o)\mathrm{d}o}{\int_{-\infty}^{\infty} a^*(o)a(o)\mathrm{d}o} + \langle o \rangle^2 \\
&= \langle o^2 \rangle - 2\langle o \rangle \langle o \rangle + \langle o \rangle^2 \\
&= \langle o^2 \rangle - \langle o \rangle^2. \qquad (16.16)
\end{aligned}$$

The expectation value of o^2 is simply given by applying the operator $\hat{\mathbf{O}}$ twice:

$$\begin{aligned}
o^2(x, t)\psi_o(x, t) &= o(x, t)o(x, t)\psi_o(x, t) \\
&= o(x, t)\hat{\mathbf{O}}\psi_o(x, t) \\
&= \hat{\mathbf{O}}o(x, t)\psi_o(x, t) \\
&= \hat{\mathbf{O}}\hat{\mathbf{O}}\,\psi_o(x, t). \qquad (16.17)
\end{aligned}$$

The standard deviation $\sigma(o)$ is known in wave mechanics as the **uncertainty** and represented by the symbol Δ_o.

16.5 Operator analysis of a Gaussian wavepacket

As an illustration of the use of the frequency operator, we consider a Gaussian wavepacket, shown in Fig. 16.1(a) and represented, with the omission of its spatial dependence, by

$$\psi(t) = \psi_0 \exp(-\mathrm{i}\omega_0 t)\exp\left[-\left(\frac{t}{t_0}\right)^2\right]. \qquad (16.18)$$

We expect this, from Chapter 14, to show a Gaussian spectrum centred on ω_0 with a width inversely proportional to t_0. Application of the frequency operator

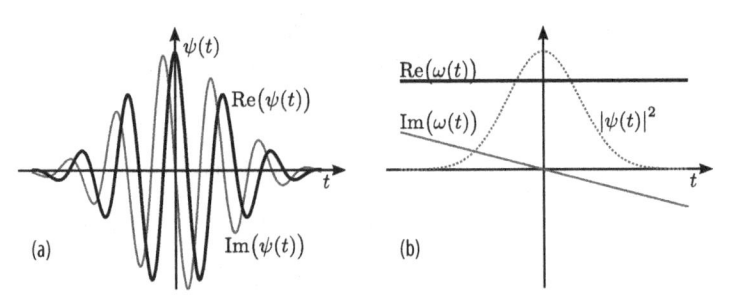

Fig. 16.1 A Gaussian wavepacket (a) and the result (b) of applying the frequency operator to obtain $\omega(t)$, shown together with the weighting function $|\psi(t)|^2$.

$\hat{\omega}$ of equation (16.6a) gives

$$
\begin{aligned}
\omega(t)\psi(t) = \hat{\omega}\psi(t) &= \psi_0 \mathrm{i} \frac{\mathrm{d}}{\mathrm{d}t} \left\{ \exp(-\mathrm{i}\omega_0 t)\exp\left[-\left(\frac{t}{t_0}\right)^2 \right] \right\} \\
&= \psi_0 \left(\omega_0 - \frac{2\mathrm{i}t}{t_0^2} \right) \exp(-\mathrm{i}\omega_0 t)\exp\left[-\left(\frac{t}{t_0}\right)^2 \right] \\
&= \left(\omega_0 - \frac{2\mathrm{i}t}{t_0^2} \right) \psi(t).
\end{aligned}
\tag{16.19}
$$

Since the weighted average of the antisymmetric t-dependent term will be zero, the expectation value $\langle\omega\rangle$ defined by equation (16.8) will be equal to ω_0. The real and imaginary parts of $\omega(t)$ are shown in Fig. 16.1(b).

Application of the frequency operator a second time gives

$$
\begin{aligned}
\omega^2(t)\psi(t) &= \hat{\omega}^2\psi(t) \\
&= \psi_0 \left(-\frac{\mathrm{d}^2}{\mathrm{d}t^2} \right) \left\{ \exp(-\mathrm{i}\omega_0 t)\exp\left[-\left(\frac{t}{t_0}\right)^2 \right] \right\} \\
&= \psi_0 \left[\omega_0^2 - \frac{4\mathrm{i}\omega_0 t}{t_0^2} - \left(\frac{2t}{t_0^2}\right)^2 + \frac{2}{t_0^2} \right] \exp(-\mathrm{i}\omega_0 t)\exp\left[-\left(\frac{t}{t_0}\right)^2 \right] \\
&= \left\{ \omega_0^2 - \frac{4\mathrm{i}\omega_0 t}{t_0^2} + \frac{2}{t_0^2} \left[1 - 2\left(\frac{t}{t_0}\right)^2 \right] \right\} \psi(t).
\end{aligned}
\tag{16.20}
$$

While we may again omit the asymmetric t-dependent term to determine the mean $\langle\omega^2\rangle$, we must explicitly evaluate the integral of the t^2 term, which proves to cancel out half of the term in $1/t_0^2$. We may then find the mean according to equation (16.8), yielding

$$
\langle\omega^2\rangle = \omega_0^2 + \frac{1}{t_0^2}.
\tag{16.21}
$$

The standard deviation or uncertainty in frequency, defined by equation (16.16) and reflecting the width of the spectrum of the wavepacket, hence proves to be

$$\Delta_\omega \equiv \sigma(\omega) = 1/t_0. \tag{16.22}$$

With a real wavefunction, such as the wavepacket

$$\psi(t) = \psi_0 \cos(\omega_0 t)\exp[-(t/t_0)^2]$$
$$\equiv \frac{\psi_0}{2}[\exp(i\omega_0 t) + \exp(-i\omega_0 t)]\exp[-(t/t_0)^2], \tag{16.23}$$

we shall always obtain the value $\omega = 0$, reflecting the equivalence to a superposition of two complex exponential waves with equal but opposite frequencies. The two components have the same value of ω^2, however, so the mean squared frequency proves to be as in equation (16.21). For the real wavepacket we therefore obtain an rms (*root mean squared*) frequency of

$$\sqrt{\langle \omega^2 \rangle} = \sqrt{\omega_0^2 + \frac{1}{t_0^2}}$$
$$\approx \omega_0 \left(1 + \frac{1}{2(\omega_0 t_0)^2}\right). \tag{16.24}$$

For wavepackets lasting several cycles, which comprise components with a narrow range of frequencies and for which $\omega_0 t_0 \gg 1$, the rms frequency will therefore tend to ω_0.

While Fig. 16.1 and equation (16.19) show that the instantaneous value of $\omega(t)$ is in general complex and should not be mistaken for the frequency itself, its statistical properties therefore prove to be identical to those which we would obtain by calculating the mean, variance and higher moments of the spectra of the wavepacket.

16.6 Complex electrical impedances

A further example of the utility of the frequency operator is in the a.c. analysis of electrical circuits. Here we find, for example, that the voltage V across an inductor L is defined in terms of the rate of change of current I flowing through it by

$$V(t) = -L\frac{dI(t)}{dt}, \tag{16.25}$$

and we might therefore expect that it would be possible to determine a circuit's performance only by resorting to differential calculus. However, the frequency operator $\hat{\omega}$ allows us to write

$$V(t) = iL\hat{\omega} I(t), \tag{16.26}$$

where, if $I(t)$ is taken to be a single complex exponential component, the operator may be replaced by the frequency ω of the component. By comparison with Ohm's law ($V = IR$) for a resistor R, it is apparent that we may consider the inductor to behave as a **complex impedance** Z, which for a resistor would be simply R, that depends upon frequency according to

$$Z_L(\omega) = i\omega L. \tag{16.27}$$

The corresponding expression for the impedance of a capacitor C is

$$Z_C(\omega) = \frac{1}{i\omega C}. \tag{16.28}$$

Networks of resistors, capacitors and inductors may therefore be analysed as if, at a given frequency, they were simple resistors with a complex impedance Z; an example is shown in Fig. 16.2. The modulus of the overall impedance, once determined, relates the magnitudes of the corresponding voltages and currents, while the argument determines their relative phase. Fourier analysis then allows the behaviour for arbitrary waveforms to be determined from that for individual frequency components as in Section 14.7.

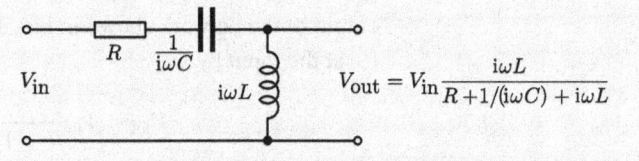

Horowitz & Hill [44] pp. 30–33

Fig. 16.2   The use of complex impedances in the analysis of an electrical circuit.

Note that, strictly, the complex impedances are operators; they act as simple multipliers only for single sinusoidal components, when ω is defined.

Exercises

16.1 With small modifications to Exercise 14.1, we may define an amplitude-modulated complex wave,

$$f(t) = a_0 \exp(i\omega_0 t) + \frac{\alpha}{2} a_0 \{\exp[i(\omega_0 + \omega_1)t] + \exp[i(\omega_0 - \omega_1)t]\}, \tag{16.29}$$

whose (amplitude) frequency spectrum will be given by

$$F(\omega) = \frac{a_0}{2}\left[\delta(\omega_0 - \omega) + \frac{\alpha}{2}\delta((\omega_0 + \omega_1) - \omega)\right.$$
$$\left. + \frac{\alpha}{2}\delta((\omega_0 - \omega_1) - \omega)\right]. \tag{16.30}$$

In contrast to Exercise 14.1, there are no corresponding terms for negative frequencies.

Given that the *power* (or intensity) spectrum is proportional to the square of the modulus of the amplitude spectrum above, find the mean frequency of the *power* spectrum, which is defined by

$$\bar{\omega} = \frac{\int_0^\infty \omega F^*(\omega)F(\omega)\,d\omega}{\int_0^\infty F^*(\omega)F(\omega)\,d\omega}. \tag{16.31}$$

Note that you should not need to evaluate the integral of the Dirac δ-function, or its square, if it occurs in both the numerator and denominator of equation (16.31).

16.2 Show that the result obtained in Exercise 16.1 may alternatively be obtained by calculating the *expectation value* of the *operator* $\hat{\omega} \equiv -i(d/dt)$,

$$\langle \hat{\omega} \rangle = \frac{\int_{-\infty}^\infty f^*(t)\hat{\omega}f(t)\,dt}{\int_{-\infty}^\infty f^*(t)f(t)\,dt}. \tag{16.32}$$

16.3 Show that, if the complex impedances Z of a resistor R, capacitor C and inductor L at angular frequency ω are given, respectively, by R, $1/(i\omega C)$ and $i\omega L$, then the amplitude V_{out} of the component of frequency ω at the output of the network shown in Fig. 16.2 will be related to the amplitude V_{in} at the input by

$$V_{out} = V_{in}\frac{i\omega L}{R + 1/(i\omega C) + i\omega L}. \tag{16.33}$$

17 Uncertainty and quantum mechanics

17.1 The bandwidth theorem

Pain [66] pp. 376–377
Main [60] pp. 206–211
Feynman [23] Vol. I,
Chapter 38; and Vol. III,
Chapter 2

We have already noted that a wave with a single sinusoidal or complex exponential component extends over all space and time, and that, if we wish to limit its extent, further sinusoidal components spanning a range of frequencies must, as in Section 13.3, be superposed. As with the Gaussian wavepacket of Sections 14.4.1 and 16.5, the tighter the wavepacket is to be confined, the greater the range of frequencies needed. This reciprocal dependence of the frequency range upon the spatial or temporal extent is known as the **bandwidth theorem**, and is a crucial concept for the analysis of wavepackets.

We have also met the bandwidth theorem in our calculation of the diffraction pattern of a single slit in Section 9.2, where we saw that a slit of width d has a sinc-function diffraction pattern with its first minimum at $\sin \vartheta = 2\pi/(kd) = \lambda/d$. Reducing the slit width d hence broadens the diffraction pattern, which, as we saw in Sections 9.7 and 14.9, directly reflects the spectrum of spatial frequencies needed to form the slit itself. If we take the width of the diffraction pattern to be the distance from the peak to the first minimum, $\Delta k' = \Delta(k \sin \vartheta) = 2\pi/d$, then we see that $d \, \Delta k' = 2\pi$. The product of the width of the aperture with the width of the spectrum is hence a constant of the order of unity.

A more precise statement requires more careful definitions of the widths of the slit and the diffraction pattern, and we conventionally adopt the *uncertainty* or *standard deviation* that we met in Section 16.4. This is the square root of the average of the square of the deviation from the mean, or the 'root mean square (rms) deviation', $\sigma(o)$, defined in equation (16.15),

$$\Delta_o^2 \equiv \sigma(o)^2 = \frac{\int_{-\infty}^{\infty} \left(o - \langle o \rangle \right)^2 a^*(o)a(o)\mathrm{d}o}{\int_{-\infty}^{\infty} a^*(o)a(o)\mathrm{d}o}, \tag{17.1}$$

where $a(o)$ is the distribution as a function of the variable o, which could be position x, time t, wavenumber k or angular frequency ω. With this definition, we may write the bandwidth theorem explicitly through relations such as

$$\Delta_x \Delta_k \geq \frac{1}{2}, \tag{17.2a}$$

$$\Delta_t \Delta_\omega \geq \frac{1}{2}. \tag{17.2b}$$

The bandwidth theorem indeed applies to any pair of *conjugate variables*, whose product is dimensionless, and reflects the Fourier-transform relationship between the conjugate representations.

It is important to note that the bandwidth theorem provides a lower limit on the product of the uncertainties; the actual product depends upon the particular shape of the distribution and its spectrum. The shape that gives the minimum value of $\frac{1}{2}$ for the product is the Gaussian, $\exp[-(x/x_0)^2]$, of Sections 14.4.1 and 16.5; its spectrum, as in Section 14.4.1, is another Gaussian, and it is the only function that is identical in shape to its Fourier transform. It is not hard to broaden the bandwidth without reducing the pulse duration, however, and for other waveforms the inequality applies.

For the square aperture considered initially, the standard deviation in wave-number $\Delta k'$ proves, rather surprisingly, always to be infinite, for the sinc function decays only as $1/k$. The standard deviation, although usually convenient, is just one of many ways of characterizing the width of a distribution.

17.1.1   The frequency resolution of human hearing

An intriguing example of the everyday significance of the bandwidth theorem emerges in our ability to distinguish the difference in pitch between two musical notes.

To characterize and gain a better understanding of human hearing, experimental psychologists measure the 'just-noticeable' difference between two single-frequency bursts of a given duration. Subjects listen to, for example, 200 ms at 249 Hz followed, after a short pause, by 200 ms at 251 Hz; if they

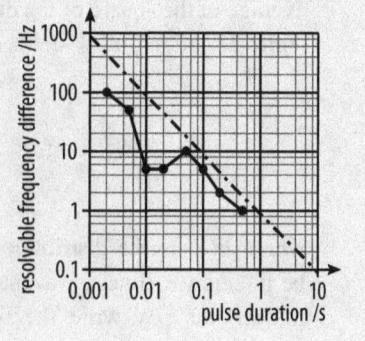

Fig. 17.1 Just-noticeable frequency difference as a function of pulse duration at 440 Hz for the 2005 Wave Physics class in Southampton. The spectral width (FWHM) is shown dotted.

can consistently identify which note is higher, then it is deduced that the 2 Hz difference is or exceeds the just-noticeable difference. Repeating the experiment with various frequency differences and pulse durations allows the resolvable frequency difference to be mapped. Typical results are shown in Fig. 17.1, together with the *full width at half maximum* (FWHM) of the pulse spectrum. It is apparent that the ear can distinguish between pulses whose spectra are displaced by only a fraction of their width.

17.2 Wave–particle duality

Astonishing and now classic experiments in the first decades of the twentieth century showed that the tiniest particles then known – electrons, protons and atoms – displayed many of the characteristics of waves, and could, for example, undergo the tunnelling and diffraction discussed in Sections 11.6.1 and 15.5.1. Conversely, as we shall see in Section 17.4.2, light could only be properly understood if considered to comprise particle-like photons. This duality of wave and particle characteristics is the fundamental origin of quantum mechanics, and it turns out that the kinetic properties of energy $\mathcal{E}$ and momentum p are related to the wave properties of angular frequency ω and wavenumber k by Planck's constant $h \equiv 2\pi\hbar$ through

In three dimensions,

$$\mathcal{E} = \hbar\omega, \tag{17.3a}$$

$$\mathbf{p} = \hbar\mathbf{k}.$$

$$p = \hbar k \tag{17.3b}$$

for any particle, body or wave motion. We have already met the three-dimensional extension of equation (17.3b) in our discussion of pseudo-momentum in Section 15.5.

If a particle is described by, or associated with, a wavefunction $\psi(x, t)$, then its kinetic properties may be determined by multiplying by $\hbar$ the expectation value of ω or k derived from the wavefunction as in Section 16.3. The corresponding value obtained by multiplying the uncertainty in ω or k by $\hbar$ gives the uncertainty in energy or momentum, and the bandwidth theorem tells us that these will be non-zero if the particle's wavefunction is localized. Interpreting the physical significance of an uncertainty in energy or momentum is one of the perpetual challenges of quantum mechanics; fundamentally, it means that, if we know something about a particle's position or time of arrival, then measurements of its energy or position will fall in a distribution. On combining equations (17.3) with the bandwidth theorem of equation (17.2) we obtain

$$\Delta_x \Delta_p \geq \frac{\hbar}{2}, \tag{17.4a}$$

$$\Delta_t \Delta_E \geq \frac{\hbar}{2}. \tag{17.4b}$$

These are the mathematical expressions of Heisenberg's **uncertainty principle**, and relate the achievable precisions of simultaneous measurements of the pairs of parameters.

17.3 The quantum-mechanical wavefunction

Given that electrons, protons and atoms show wave-like behaviour and share many of the characteristics of photons of light, then it seems reasonable – indeed, it is necessary – to associate each particle with some sort of *wave-function* ψ representing the amplitude of the particle's wave. The wavefunction proves to contain all the information that we can have about the particle.

In the following sections we shall summarize the important fundamentals of quantum mechanics, namely the quantum wavefunction and the quantum wave equation. It is best to regard each of these as a pure hypothesis; indeed, although there are strong similarities with Hamiltonian approaches to classical mechanics and electromagnetism that suggest the quantum-mechanical approach, there is – in contrast to the examples in Chapter 3 – no way in which quantum mechanics may be absolutely derived from any previous theoretical framework. So quantum mechanics is a self-contained theory, which proves to be self-consistent and which we can therefore explore to investigate the behaviour of a hypothetical system obeying its rules. Remarkably, we find that its predictions agree perfectly with the behaviour that we observe for electrons, photons and so on – and the comparison has in some cases been made to a precision of parts in 10^{15}. While we therefore conclude quantum theory to be accurate and valid, and while we can make many observations of similarity to and consistency with aspects of classical theory, we cannot derive it *a priori* from more fundamental or established rules. Classical mechanics, which we would similarly regard as an empirically justified hypothesis were it not that our everyday intuition happens to agree with it, can, however, be derived from quantum mechanics as the macroscopic limit.

17.3.1 Quantum-mechanical operators

A fundamental principle of quantum mechanics is that physically observable or measurable properties are defined entirely by the particle wavefunction, and can be determined from it by the application of the appropriate operators. We therefore begin by combining the wave–particle mappings of equations (17.3) to the wave operators of equations (16.6), and propose that the operators for the total energy $\mathcal{E}$ and momentum component p_x be given by

$$\hat{\mathcal{H}} \equiv i\hbar \frac{\partial}{\partial t}, \qquad\qquad (17.5a)$$

$$\hat{p}_x \equiv -\mathrm{i}\hbar\frac{\partial}{\partial x}, \tag{17.5b}$$

where the energy operator $\hat{\mathcal{H}}$ associated with the observable $\mathcal{E}$ is known as the **Hamiltonian**. To obtain the total energy of a particle represented by the wavefunction $\psi(x,t)$, therefore, we calculate, in accordance with equation (16.8),

$$\langle \mathcal{E}(t) \rangle = \frac{\displaystyle\int_{-\infty}^{\infty} \psi^*(x,t)\,\mathrm{i}\hbar\frac{\partial}{\partial t}\psi(x,t)\mathrm{d}x}{\displaystyle\int_{-\infty}^{\infty} \left|\psi(x,t)\right|^2 \mathrm{d}x}. \tag{17.6}$$

With the addition of the operators for position x and time t, which we take to be given simply by

$$\hat{x} \equiv x, \tag{17.7a}$$

$$\hat{t} \equiv t, \tag{17.7b}$$

we may propose further operators. For example, the angular-momentum component L_z could be

$$\begin{aligned} \hat{L}_z &\equiv \hat{x}\hat{p}_y - \hat{y}\hat{p}_x \\ &= -\mathrm{i}\left(x\frac{\partial}{\partial y} - y\frac{\partial}{\partial x}\right), \end{aligned} \tag{17.8}$$

and hence the squared total angular momentum $\hat{L}^2$ might be

$$\hat{L}^2 \equiv \hat{L}_x^2 + \hat{L}_y^2 + \hat{L}_z^2. \tag{17.9}$$

The operators of equations (17.5), (17.7), (17.8) and (17.9) prove to be some of the fundamental operators of quantum mechanics and are therefore the essential basis for modern atomic, molecular, condensed-matter and particle physics.

17.3.2 Schrödinger's wave equation

E. Schrödinger [77]

The equation governing the non-relativistic evolution of the quantum wavefunction was proposed in 1926 by the Austrian physicist Erwin Schrödinger, who considered that the operator-derived expression for the total energy should equal the sum of the potential energy $\mathcal{U}(x)$ and the kinetic energy $p^2/(2m)$, where m is the particle mass and the momentum p is again determined by applying its operator to the wavefunction ψ. Making these substitutions, he thereby arrived at the now famous equation

$$-\mathrm{i}\hbar\frac{\partial}{\partial t}\psi = \frac{\hbar^2}{2m}\frac{\partial^2}{\partial x^2}\psi + \mathcal{U}(x)\psi, \tag{17.10}$$

which, as we saw in Section 13.4.1, is a linear, dispersive wave equation, with a group velocity twice the phase velocity. In three dimensions,

$$-i\hbar\frac{\partial}{\partial t}\psi = \frac{\hbar^2}{2m}\nabla^2\psi + \mathcal{U}(\mathbf{r})\psi. \qquad (17.11)$$

It should again be stressed that, while building upon analogous approaches to classical mechanics and electromagnetism, Schrödinger's equation is still essentially a hypothesis. Nonetheless, apart from modification for relativistic situations and the inclusion of additional components of the energy, it is one of the most precisely investigated equations in physics, and 85 years of testing have found no inconsistency with experimental observation.

17.3.3   The nature of the quantum wavefunction

There are perhaps three mysteries of quantum mechanics. The first, the wave–particle duality of Section 17.2, is to some extent resolved if one accepts that the particle properties of energy and momentum are, even in classical mechanics, always expressions of wave characteristics. The second, the quantum measurement problem, we shall describe shortly. There always remains, however, the question of what is meant by the quantum wavefunction ψ.

Several characteristics of the wavefunction are clear. First, it is the function whose evolution is described by the Schrödinger equation that we can apply successfully to any physical situation. It is the function from which we can determine, using appropriate operators, any observable property of the particle, so it somehow contains all the information that is available. The *squared amplitude* of the wavefunction $|\psi|^2$ corresponds to the probability density of finding the particle at a given point or with given properties; and the *phase* of the wavefunction determines whether superpositions are added constructively or destructively. What we are unable to explain is the amplitude ψ itself.

The quantum wavefunction ψ does not correspond to a physical displacement or field strength; it is neither transverse nor longitudinal, and does not imply any 'wiggling' of the particle. Nor can ψ be fully measured. The wavefunction proves to be almost always complex, and, since physical properties are invariably real, it turns out that we may measure only a projection of the wavefunction at any time. Depending upon how it is projected by the measurement, we may determine its real part or its imaginary part, or choose between its magnitude and its phase; but we can never arrange to measure both and therefore know the wavefunction completely.

17.4 Measurement of the quantum wavefunction

The most troubling aspect of quantum mechanics is *measurement* for, while the Schrödinger equation defines the evolution of a quantum wavefunction *before* a measurement, we are utterly unable to explain in detail what happens *during* the measurement process. While we can outline clear rules that link the outcomes of a measurement to the wavefunction upon which it is performed, these rules differ from those for a classical system, and we cannot truly explain why this is so or how the outcomes occur.

We illustrate the quantum measurement process, and contrast it with its classical analogue, with a pair of examples. Our quantum system will be an electron within a *quantum well*, such as the gain region of a semiconductor laser; as a classical analogue, we choose the string of a guitar or violin. Both one-dimensional systems support freely propagating waves that are bounded at the ends by nodes; wave motions in both cases can hence be represented as superpositions of standing-wave modes like the harmonics of a musical instrument.

We have seen in Chapters 14 and 16 two methods by which we can determine the frequency of the classical motion of the string. If only a single mode is excited, then the measurement will always give the frequency of that mode; if the motion is instead a superposition, then both the mean of the Fourier transform and the expectation value of the frequency operator will give some sort of weighted average of the different mode frequencies according to their relative strengths within the superposition.

If the quantum wavefunction within the quantum well comprises a single mode and therefore well-defined frequency, then measurement of the frequency will, as with the string, yield that single frequency. The case of a superposition, however, is rather different. Quantum measurements always yield one of the discrete mode frequencies, with relative likelihoods that depend upon their amplitudes within the superposition.

We may examine our two examples in more detail. Suppose that in both cases the allowed modes may be numbered with the index i and have corresponding frequencies ω_i, so that the modes are of the form $\psi_i = \cos(k_i x)\cos(\omega_i t)$ in the classical case and $\psi_i = \cos(k_i x)\exp(\mathrm{i}\omega_i t)$ for the quantum mechanical wave. We may therefore write a general superposition Ψ in each case as

$$\Psi = \sum_i a_i \psi_i, \qquad (17.12)$$

where the coefficients a_i indicate the relative contributions of the different modes to the superposition. A classical measurement of the frequency would yield the mean value for the superposition

$$\langle \omega \rangle = \frac{\sum\limits_i |a_i|^2 \omega_i}{\sum\limits_i |a_i|^2} \,, \qquad (17.13)$$

in accordance with our expression for the expectation value of the $\hat{\omega}$ operator from equation (16.8), and therefore with the mean value of the frequency of the superposition before the measurement.

While a quantum superposition would have the same expectation value $\langle \omega \rangle$ for the frequency, quantum measurements do not simply give this value: instead, they are found to yield *any* of the values ω_i for the components of the superposition; and *only* these mode frequencies are possible. If the entire experiment is repeated, with the superposition created anew before each measurement, then the outcomes are found to occur randomly with probabilities according to the contributions of the corresponding modes to the superposition – i.e. the probability p_i of a measurement yielding the frequency ω_i is

$$p_i = \frac{|a_i|^2}{\sum\limits_i |a_i|^2}. \qquad (17.14)$$

The average result of a large number of experiments, each beginning with the same wavefunction superposition, will yield the classical result, since

$$\langle \omega \rangle = \sum_i p_i \omega_i, \qquad (17.15)$$

but, unless the initial wavefunction contains only a single frequency, the outcomes of individual experiments *cannot be predicted*.

17.4.1 Collapse of the wavefunction

A crucial difference between quantum and classical measurements is that measurement in quantum mechanics fundamentally affects the system being measured. While the result of a quantum measurement does not match the expectation value of the measured quantity *before* the measurement, it does correspond to that for the wavefunction *after* the experiment, for if the measurement yields the value ω_i the wavefunction will be found to have changed from the initial superposition into the single mode ψ_i. If the experiment is repeated without reconstructing the initial superposition, every subsequent measurement will yield the same value ω_i, even though this was a random result from the initial measurement upon the superposition.

In our classical analogue, the motion of the string is unaffected by the measurement, so all subsequent measurements will yield the same result $\langle \omega \rangle$ as was obtained from the first measurement.

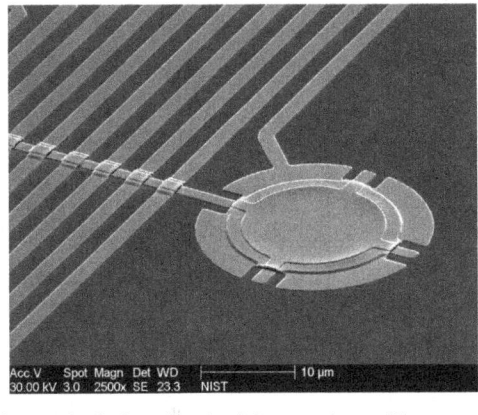

Fig. 17.2 Scanning electron micrograph of a drum-like aluminium membrane, 15 μm across by 100 nm thick, whose thermal motion can be cooled down to that of the lowest vibrational quantum state [82].

Courtesy of A. Sanders / NIST

Quantum measurement is thereby found to affect the state of the system – a phenomenon known as the **collapse of the wavefunction**. If the wavefunction before a measurement contains components corresponding to different measurement outcomes, then the measurement will randomly yield one of these outcomes, and the wavefunction thereafter will correspond to that value alone. Subsequent measurements will yield the same value with unity probability, and the wavefunction is deduced to have *collapsed* into the single mode ψ_i.

17.4.2   Quantum measurement of classical systems

If we make classical measurements with sufficient precision, we shall always find quantum behaviour to be apparent. The vibrations of our guitar or violin string, for example, could in principle be resolved into a finite number of *phonons* – vibrational quanta – and the measured energy in each oscillation frequency would be found to correspond to an integer number of phonons. In practice, we are unable to reach the precision required to observe such phenomena with musical instruments, but such experiments have been performed on microscopic mechanical oscillators, such as that shown in Fig. 17.2, fabricated using techniques developed for the processing of semiconductor devices. Such objects, which can measure as little as a few tens or hundreds of nanometres in each dimension, nonetheless contain many millions of atoms – yet clear manifestations of quantum behaviour are observed.

The earliest evidence for the quantization of classical energies was one of the seminal discoveries in the development of quantum theory: the observation that the spectral intensity of *black-body radiation*, illustrated in Fig. 17.3, does not increase indefinitely at shorter wavelengths (the *ultraviolet catastrophe*), but instead peaks at a characteristic wavelength at which the photon energy is

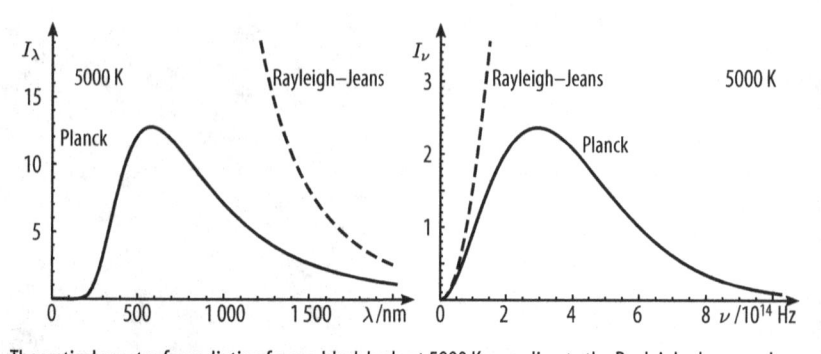

Fig. 17.3 Theoretical spectra for radiation from a black body at 5000 K according to the Rayleigh–Jeans and Planck laws (continuously variable and quantized mode energies); experimental measurements follow the Planck law. Spectral intensity (left) as a function of wavelength λ in kW m^{-2} nm^{-1} sr^{-1} and (right) as a function of frequency ν in W cm^{-2} THz^{-1} sr^{-1}.

comparable to the thermal energy $k_B \Theta$, where k_B is the Boltzmann constant and Θ the black-body temperature. The explanation considers the standing-wave modes of the electromagnetic radiation field within an imaginary box – just like the modes of our guitar string – and relies upon the assumption that the energy of each mode be quantized rather than continuously variable.

17.4.3 Uncertainty in quantum measurement

Since the wavefunction ψ_i after the measurement will in general have a different spatial distribution, energy, momentum etc. from those of the initial superposition Ψ, the measurement process must have disturbed the quantum system not only in the abstract sense, but also by redistributing the spatial probability distribution of the particle and modifying its energy and/or momentum. Given that classical measurements can in principle be instantaneous, this would imply that the system being measured would be forced to move at infinite speed and undergo infinite acceleration.

The resolution of this conundrum lies in the uncertainty principle, for in order to make a frequency measurement with resolution $\delta\omega$ the measurement must have a duration of at least $\delta t = 1/(2\,\delta\omega)$. For example, to distinguish be-

The *harmonic oscillator* describes the motion of a particle subject to a restoring force that depends linearly upon its displacement, and hence of a particle in a potential that varies quadratically with position. The ith quantum-mechanical mode of such a potential proves to have a frequency $(i + 1/2)\omega_0$, where ω_0 is the classical oscillation frequency.

tween the modes of a *harmonic oscillator*, with frequencies $\omega_i = (i+1/2)\omega_0$, requires $\delta\omega \leq \omega_0/2$, so our measurement duration must be at least $\delta t = 1/\omega_0$. The measurement must therefore last approximately as long as the classical oscillation, allowing ample time for the wavefunction to be redistributed from one part of the superposition to another.

The combination of the uncertainty principle with the collapse of the wavefunction results in the *non-commutativity* of conjugate quantum measurements and thereby of the operators that describe them. The measurement, to a given precision, of the position of a wavepacket, for example, will result in the collapse of the wavefunction to one whose position uncertainty is small, and hence whose momentum uncertainty is large; subsequent measurement of the momentum will collapse the wavefunction a second time into a state of well-defined momentum but position uncertainty, with the momentum taking any of a wide range of values. Reversing the order of the measurements will result in a well-defined position and large momentum uncertainty. This occurs only with *conjugate* pairs of properties, such as position and momentum, energy and time, or orthogonal components of angular momentum; many other pairs of properties are far better behaved. For the measurement of conjugate properties, however, the measured values and final wavefunction depend upon the order of the measurements – quite unlike in the classical case.

Exercises

17.1 The time-dependent Schrödinger equation of equation (17.10) describes the evolution of the quantum-mechanical wavefunction $\psi(x, t)$ for a particle of mass m moving in one dimension and subject to a spatially dependent potential energy $\mathcal{U}(x)$, which could for example be the electrostatic potential energy of an atomic electron in the field of the nucleus, or the gravitational potential energy that varies with height above the ground. The wavefunction ψ is complex and proves to contain all the information about the quantum particle; its (real) modulus-squared corresponds to the *probability density* of the particle – the probability per unit length of finding the particle at a given position.

Given that $\mathcal{U}(x)$ and $\psi(x, t)$ are both finite, show that, if the rate of change of ψ is to be finite, then, at a boundary between region A and region B at $x = x_0$,

$$\frac{\partial \psi_A}{\partial x}(x_0, t) = \frac{\partial \psi_B}{\partial x}(x_0, t), \tag{17.16}$$

where ψ_A and ψ_B are the wavefunctions in the adjacent regions.

17.2 By substituting wavefunctions of the form $\psi(x, t) = \psi_0 \exp[-i(\omega t - kx)]$ into the Schrödinger wave equation (17.10), show that the angular frequency ω and wavenumber k for quantum waves in a region of potential energy $\mathcal{U}$ are related by

$$k = \frac{\sqrt{2m(\hbar\omega - \mathcal{U})}}{\hbar}. \tag{17.17}$$

17.3 Using the result of Exercise 17.2, and following working similar to that in Section 11.3, show that the impedance Z for quantum particles of angular frequency ω will be proportional to the wavenumber k.

17.4 Quantum tunnelling occurs when particles enter and pass through a region in which their potential energy $\mathcal{U}$ exceeds their total energy $\hbar\omega$.

Consider a particle, with angular frequency ω in region A ($x < 0$), where the potential $\mathcal{U} = 0$, which enters a region B ($0 \le x \le d$) in which $\mathcal{U} = \mathcal{U}_0$ before emerging into a region C ($x > d$), where $\mathcal{U} = 0$ once more. By reference to equations (11.42), show that, if $2md^2(\mathcal{U}_0 - \hbar\omega)/\hbar^2 \gg 1$, the fraction of the wavefunction transmitted will vary as $\exp(-d\sqrt{2m(\mathcal{U}_0 - \hbar\omega)/\hbar^2})$, and hence that the probability of the particle tunnelling through the barrier will vary as $\exp(-2d\sqrt{2m(\mathcal{U}_0 - \hbar\omega)/\hbar^2})$. (The sine and cosine of imaginary numbers can be determined by recalling that $\cos\vartheta = [\exp(i\vartheta) + \exp(-i\vartheta)]/2$ and $\sin\vartheta = [\exp(i\vartheta) - \exp(-i\vartheta)]/(2i)$.)

18   Waves from moving sources

18.1   Waves from slowly moving sources

In our analysis so far, we have assumed all detectors, interfaces and boundaries, the centres of oscillation of sources and the medium in which waves propagate to be stationary. While such situations may indeed be considered the most common, or at least the most fundamental, they miss some important and fascinating physics, from the Doppler effect and shock waves to the wedged shape of a ship's wake and the Čerenkov radiation from quickly moving charged particles. To explore some of these phenomena we draw together Huygens' general treatment of wave propagation and some specific details of propagation mechanisms and characteristics.

18.1.1   The Doppler effect

C. A. Doppler [18]

We are all familiar with everyday manifestations of the Doppler effect, for it accounts for the change in pitch of a fire engine's siren or a sports car's engine as the vehicle passes. The underlying principle is that, if we move towards a periodic wave source, then each successive wavefront will have to travel a slightly shorter distance than its predecessor before reaching us, and its journey will therefore take a shorter time. A simple demonstration can readily be performed by climbing or descending the steps of a deserted escalator, against its motion, at exactly the right pace to remain stationary, matching the frequency at which the steps arrive; if instead we wish to move towards the 'source' of the steps, a faster pace is needed.

    Our analysis for the case of a moving source differs slightly from that for a moving observer so, although the results are identical at low speeds, we consider both. Figure 18.1 shows successive wavefronts that have been emitted by a moving source of frequency f_0, with the positions of the source at the corresponding times of emission shown in grey. The source velocity is taken to be v_1, and the speed of wave propagation to be v_0. In the period $\tau_1 = 1/f_0$ between the emission of successive wavefronts, the source will therefore move a distance $v_1\tau_1$ to the right, and each wavefront will therefore be emitted a distance λ' behind the preceding wavefront, where

$$\lambda' = (v_0 - v_1)\tau_1. \tag{18.1}$$

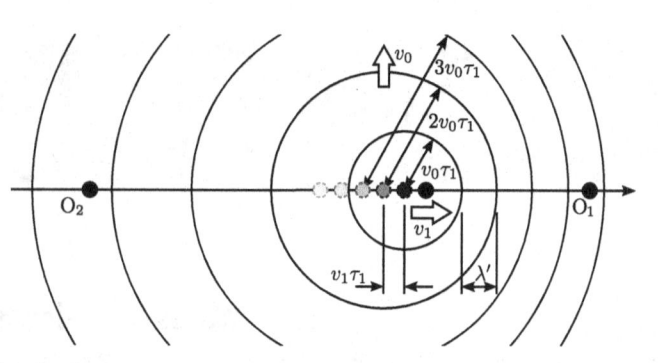

The Doppler effect for a source moving with speed v_1. In the time τ_1 between emitting successive wavefronts, the source has moved a distance $v_1\tau_1$, so that $\lambda' = (v_0 - v_1)\tau_1$.

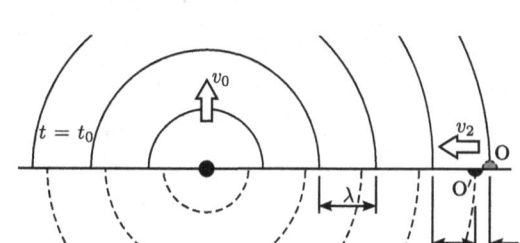

The Doppler effect recorded by an observer moving with speed v_2. In the time τ_2 between encountering successive wavefronts, the observer moves a distance $v_2\tau_2$, so that $\lambda = (v_0 + v_2)\tau_2$.

The apparent frequency of the emitted wave, according to a stationary observer O_1 to the right of the source along its line of motion, will therefore be

$$f' = \frac{c}{\lambda'} = \frac{f_0}{1 - v_1/v_0}. \tag{18.2}$$

For an approaching source, the observer at O_1 hence measures a higher frequency, whereas for a receding source, as measured at O_2, the frequency is reduced.

The situation is slightly modified if the observer O lies to the side of the track of the source, in which case the apparent frequency will be

$$f' = \frac{f_0}{1 - v_1 \cos \vartheta / v_0}, \tag{18.3}$$

where ϑ is the angle between the track of the source and its position relative to the observer. If a fire engine passes us as we stand on the pavement, $\cos \vartheta$ will vary from $+1$ to -1, and the perceived pitch will vary from high to low.

The case in which the observer moves towards a stationary source is only slightly different. The upper part of Fig. 18.2 shows a series of wavefronts at the instant when the first one reaches the observer at O, and the lower part shows the situation a little later when the observer, now at O', encounters the next wavefront. In the time τ_2, the wave will have moved a distance $v_0\tau_2$, while the observer, moving towards the source with a velocity v_2, will have moved a distance $v_2\tau_2$. The distance between wavefronts is, of course, the wavelength, $\lambda = v_0/f_1$, where f_1 is the frequency with which the waves are emitted. The apparent frequency, f'', recorded by the observer, will hence be

$$f'' = \frac{1}{\tau_2} = \frac{v_0 + v_2}{\lambda} = \left(1 + \frac{v_2}{v_0}\right)f_1. \tag{18.4}$$

Again, if the source lies at an angle ϑ to the side of the track of the observer, then the apparent frequency becomes

$$f'' = \left(1 + \frac{v_2 \cos \vartheta}{v_0}\right)f_1. \tag{18.5}$$

If waves from a stationary source are reflected by a moving target and detected by a stationary observer coincident with the source (as, for example, with active radar or sonar), then the apparent frequency may be determined by combining equations (18.3) and (18.5) with $f_1 = f'$ and $v_1 = v_2 = v$, giving

$$f'' = \frac{1 + (v/v_0)\cos \vartheta}{1 - (v/v_0)\cos \vartheta} f_0. \tag{18.6}$$

This gives a **Doppler shift** in frequency that, for $v \ll v_0$, will be approximately

$$\Delta f \equiv f'' - f_0 \approx \frac{2v \cos \vartheta}{v_0} f_0. \tag{18.7}$$

18.1.2 The relativistic Doppler effect

Although fire engines and ambulances confine themselves to velocities far below that of light, the Doppler effect is of great utility in relativistic situations, when it may account for the apparent shift of the spectral emission or absorption lines of atoms in a distant star, or in the gamma rays emitted by the sub-atomic particles within a particle accelerator. Indeed, the simple Lorentz transformation between different inertial reference frames may appear a straightforward and rigorous way in which to approach the Doppler effect in general.

If the emitted wave, in the (x, t) frame of the source, is given, for example, by

$$\psi(x, t) = A \cos(kx - \omega t) \tag{18.8}$$

then in the frame (x', t') of an observer moving with a relative velocity v_x, we simply substitute the Lorentz-transformed coordinates

$$x' = \gamma(x - v_x t), \tag{18.9a}$$

$$t' = \gamma(t - xv_x/c^2), \tag{18.9b}$$

where $\gamma \equiv 1/\sqrt{1-(v/c)^2}$, giving

$$\psi'(x',t') = A\cos\gamma\left[k(x-v_xt)-\omega(t-xv_x/c^2)\right]$$
$$= A\cos\gamma\left[(k+\omega v_x/c^2)x-(\omega+kv_x)t\right], \quad (18.10)$$

so that the apparent angular frequency and wavenumber are

$$\omega' = \gamma(\omega+kv_x), \quad (18.11\text{a})$$
$$k' = \gamma(k+\omega v_x/c^2). \quad (18.11\text{b})$$

Substituting $\omega/k = v_0$ hence gives

$$\omega' = \gamma(1+v_x/v_0)\omega, \quad (18.12)$$

and we recall that the wave speed v_0 is defined in the frame of the source. For small relative velocities of the two inertial frames ($\gamma \approx 1$), equation (18.12) correctly reduces to the Doppler shift for a moving observer, equation (18.4). The Doppler shift is therefore consistent with, and entirely contained within, the Lorentz transformations of special relativity.

18.1.3 The quantum-mechanical Doppler effect

The Doppler effect is readily apparent as a shift in the frequency of light absorbed and emitted by a moving atom. Although it is well described by the classical approach of Section 18.1.1, the Doppler shift also emerges naturally from a quantum-mechanical analysis in which energy and momentum are conserved.

The combined energy of the atom and photon after emission must equal that of the excited atom alone beforehand – and vice versa for the process of absorption. The total momentum of the atom and photon must be similarly conserved. For a photon of angular frequency ω and wavenumber k, the energy and momentum are given by equations (17.3) as $\hbar\omega$ and $\hbar k$, where $\omega/k = c$, the speed of light. The energy of an atom of mass m with momentum p is equal to the sum of its electronic energy $\mathcal{E}_{1,2}$ (depending upon whether it is in the lower state 1 or upper state 2) and its kinetic energy, $p^2/(2m)$. These energies and momenta are shown schematically in Fig. 18.3; the photon, represented by the line of slope c, must exactly bridge the two parabolas.

Let us then consider the emission of a photon of energy $\hbar\omega$ and momentum $\hbar k$, so that the momentum of the atom changes from $p_x + \hbar k/2$ to $p_x - \hbar k/2$, where $p_x \equiv mv_x$ is the mean component in the photon direction of its momenta before and after absorption and v_x is the corresponding velocity component. The change in the total energy of the atom must

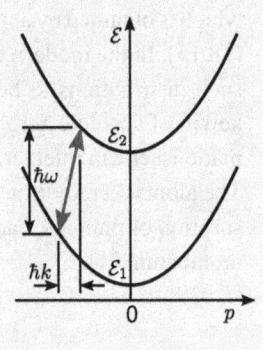

Fig. 18.3

The energy of an atom comprises both internal, electronic energy $\mathcal{E}_{1,2}$ and kinetic energy, $p^2/(2m)$. Radiative transitions must conserve the total energy and momentum of the atom–photon combination.

therefore be

$$\hbar\omega = \left[\mathcal{E}_2 + \frac{(p_x + \hbar k/2)^2}{2m}\right] - \left[\mathcal{E}_1 + \frac{(p_x - \hbar k/2)^2}{2m}\right]$$

$$= (\mathcal{E}_2 - \mathcal{E}_1) + \frac{p_x \hbar k}{m}. \tag{18.13}$$

The optical frequency hence depends upon the velocity of the atom, and is given by

$$\omega = \frac{\mathcal{E}_2 - \mathcal{E}_1}{\hbar} + \frac{p_x k}{m}$$

$$= \omega_0 + \frac{v_x}{c}\omega, \tag{18.14}$$

where $\omega_0 = (\mathcal{E}_2 - \mathcal{E}_1)/\hbar$ is the resonance frequency of a stationary atom. If, as is usually the case, $\omega_0 \approx \omega$, equation (18.14) is just an alternative expression of the Doppler shift of equation (18.4). In the quantum-mechanical description of the absorption and emission of light, then, the Doppler shift results directly from the conservation of total energy and momentum.

In an atomic sample in thermal equilibrium at temperature Θ, there will be a number of atoms $N(v_x)\delta v_x$ with velocity components between v_x and $v_x + \delta v_x$ given, according to the *Maxwell–Boltzmann distribution*, by

$$N(v_x) \propto \exp\left(-\frac{m v_x^2}{2k_B \Theta}\right), \tag{18.15}$$

where m is the atom mass. The narrow natural lineshape of a single stationary atom will therefore be *convoluted* with the temperature-dependent Gaussian

$$I(\omega) \propto \exp\left[-\frac{mc^2}{2k_B \Theta}\left(\frac{\omega}{\omega_0}\right)^2\right], \tag{18.16}$$

T. W. Hänsch &
A. L. Schawlow [38]

which is obtained by applying equation (18.14) to the distribution of equation (18.15). In the modern technique of *laser cooling*, the Doppler effect allows such distributions to be cooled from room temperatures to within a milli-kelvin of absolute zero by providing the mechanism by which a monochromatic laser can interact selectively with a narrow class of atomic velocities. The atoms thereby experience a velocity-dependent impulse through the absorption of photons, each of which, according to equation (17.3b), carries a momentum $\hbar k$.

18.2 Waves from quickly moving sources

Our derivations of the Doppler shift in Section 18.1 generally assumed the motion of the wave emitter or observer to be slow in comparison with the propagation of the wave itself. While this is commonly so, there are important cases in which the propagation speed is approached or exceeded by the speed with which the source travels through the propagation medium. Although such cases often reveal nonlinearities that must be taken into account in a full analysis of the wave-propagation mechanisms, important phenomena may nonetheless be distinguished by applying the Huygens description once more.

18.2.1 Shock waves

The momentary disturbance of a non-dispersive medium causes a single ripple to propagate, like that from a raindrop falling into a puddle, as a growing ring around the original disturbance. The ripple passes through any given point only once, and any later disturbance from the same origin will arrive at the point at a correspondingly later time. There is therefore a unique relationship between the time of the wave's arrival at a given point and the time of the initial disturbance.

French [29] pp. 274–280
Pain [66] pp. 505–513
Feynman [23] Vol. I, pp. 51-1*ff*

The same is true for the emission from a slowly moving source shown in Fig. 18.1 except that, as we have seen in Section 18.1.1, the time-frame of the source disturbance is compressed or expanded by the Doppler effect onto that observed. We may write this explicitly by considering a source whose disturbance at time t and position $\mathbf{r}_s$ is $\psi_s(t)$, from which waves propagate at a speed v_0 and are observed at a position $\mathbf{r}$. If geometrical factors such as the $1/r$ attenuation of the spherical wave of Section 15.6.1 are accounted for by a propagation factor $g(|\mathbf{r} - \mathbf{r}_s|)$, the disturbance observed will be

$$\psi(\mathbf{r}, t) = g(|\mathbf{r} - \mathbf{r}_s|)\psi_s\left(t - \frac{|\mathbf{r} - \mathbf{r}_s|}{v_0}\right). \qquad (18.17)$$

If the source moves with a constant velocity $\mathbf{v}$, we may set the source position at time t to $\mathbf{r}_s = \mathbf{r}_0 + \mathbf{v}t$, and, assuming that $|\mathbf{v}t| \ll |\mathbf{r} - \mathbf{r}_0|$, write

$$
\begin{aligned}
|\mathbf{r} - \mathbf{r}_s| = |\mathbf{r} - \mathbf{r}_0 - \mathbf{v}t| &\approx \left| \mathbf{r} - \mathbf{r}_0 - \mathbf{v}t \cdot \left(\frac{\mathbf{r} - \mathbf{r}_0}{|\mathbf{r} - \mathbf{r}_0|} \frac{\mathbf{r} - \mathbf{r}_0}{|\mathbf{r} - \mathbf{r}_0|} \right) \right| \\
&= \left| (\mathbf{r} - \mathbf{r}_0) \left(1 - \mathbf{v}t \cdot \frac{\mathbf{r} - \mathbf{r}_0}{|\mathbf{r} - \mathbf{r}_0|^2} \right) \right| \\
&= |\mathbf{r} - \mathbf{r}_0| - \mathbf{v}t \cdot \frac{\mathbf{r} - \mathbf{r}_0}{|\mathbf{r} - \mathbf{r}_0|},
\end{aligned} \tag{18.18}
$$

so that the disturbance observed may be written as

$$
\begin{aligned}
\psi(\mathbf{r}, t) &= g(|\mathbf{r} - \mathbf{r}_0 - \mathbf{v}t|) \psi_0 \left(t - \frac{|\mathbf{r} - \mathbf{r}_0 - \mathbf{v}t|}{v_0} \right) \\
&\approx g(|\mathbf{r} - \mathbf{r}_0|) \psi_0 \left(t - \frac{|\mathbf{r} - \mathbf{r}_0|}{c} + \frac{\mathbf{v}t}{v_0} \cdot \frac{\mathbf{r} - \mathbf{r}_0}{|\mathbf{r} - \mathbf{r}_0|} \right) \\
&= g(|\mathbf{r} - \mathbf{r}_0|) \psi_0 \left(t' - \frac{|\mathbf{r} - \mathbf{r}_0|}{v_0} \right),
\end{aligned} \tag{18.19}
$$

where

$$
t' \equiv \left(1 - \frac{\mathbf{v}}{v_0} \cdot \frac{\mathbf{r} - \mathbf{r}_0}{|\mathbf{r} - \mathbf{r}_0|} \right) t. \tag{18.20}
$$

With a sinusoidal source disturbance of angular frequency ω, such as $\psi_s(t) = \psi_0 \cos(\omega t)$, we shall hence observe a disturbance

$$
\psi(t) = \psi_0' \cos(\omega t' - \varphi) = \psi_0' \cos(\omega' t - \varphi), \tag{18.21}
$$

where $\psi_0' = g(|\mathbf{r} - \mathbf{r}_0|) \psi_0$, $\varphi \equiv \omega |\mathbf{r} - \mathbf{r}_0| / v_0$ and

$$
\omega' \equiv \left(1 - \frac{\mathbf{v}}{v_0} \cdot \frac{\mathbf{r} - \mathbf{r}_0}{|\mathbf{r} - \mathbf{r}_0|} \right) \omega, \tag{18.22}
$$

thus reproducing the Doppler shift derived in Section 18.1.1.

If the source is moving at or above the wave-propagation speed v_0, however, we see that it is possible for the emission time-frame to be compressed to a single observation time and the waves emitted over a range of times to be observed at the same instant. If

$$
1 - \frac{\mathbf{v}}{v_0} \cdot \frac{\mathbf{r} - \mathbf{r}_0}{|\mathbf{r} - \mathbf{r}_0|} = 0, \tag{18.23}
$$

i.e. the velocity component $\mathbf{v} \cdot (\mathbf{r} - \mathbf{r}_0)/|\mathbf{r} - \mathbf{r}_0|$ along the observation direction equals the speed of propagation, then a single value of t' will correspond to a range of values of the emission time t. Rather than a continuous disturbance, we may observe a brief pulse of great intensity. This is known as a **shock wave**, and a Huygens depiction of it is shown in Fig. 18.4.

Shock waves are well known as the *sonic boom* of a supersonic aircraft, shown spectacularly in Fig. 18.5, or the crack of a whip or pistol shot. They are observed where the Earth's magnetic field intercepts the *solar wind* of

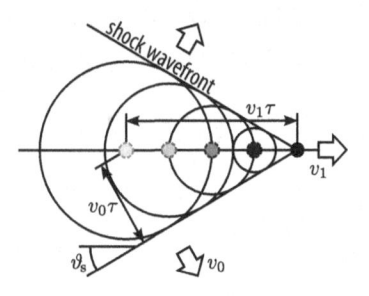

Ripples from a source moving at above the wave-propagation speed may arrive simultaneously to form a shock wave.

A spectacular demonstration of the shock wave from a supersonic F/A-18 Hornet aircraft, causing saturated air to condense into a short-lived cloud.

© US Navy photo by Ensign John Gay aboard USS Constellation in 1999

charged particles from the Sun – or, shown on a larger scale in Fig. 18.6, by the supersonic encounter between a stellar wind and inter-stellar clouds.

18.2.2 Čerenkov radiation

Our analysis in Section 6.4 of refraction at the interface between two media showed how the refracted wave could be regarded as the emission from secondary sources lying along the interface that are triggered by the arrival of the incident wavefront. We could replace this continuous line of momentary sources by a single continuous source that moves along the interface to follow its intersection with the incident wavefront. The refracted wavefront, from this point of view, may be regarded as the shock wave from the single continuous source when its speed along the interface exceeds the wave-propagation speed

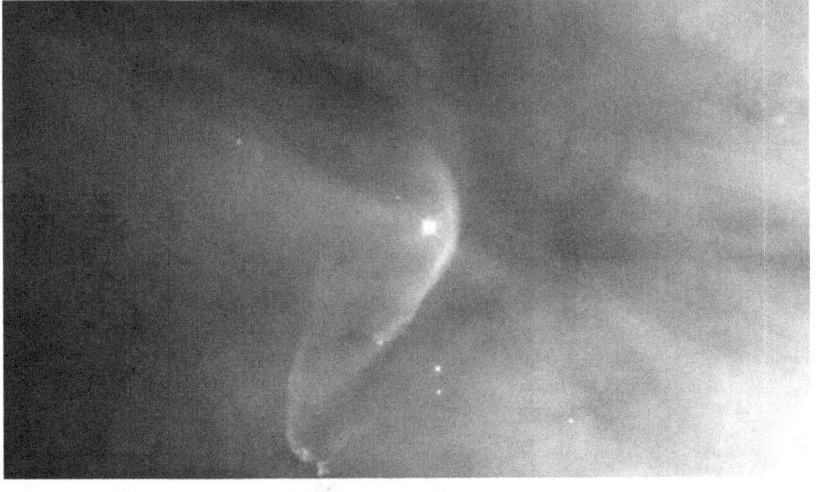

Fig. 18.6 Hubble Telescope image of the bow shock around the young star LL Ori in Orion's Great Nebula. The crescent-shaped wave is formed by the supersonic encounter between the star's stellar wind and plasma flowing from the core of the Nebula, rather as the Earth's own bow shock arises when the solar wind meets its magnetosphere.   Courtesy of NASA and The Hubble Heritage Team (STScI/AURA)
Acknowledgement: C. R. O'Dell (Vanderbilt University) [3]

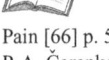

Pain [66] p. 508
P. A. Čerenkov [11]

in the second medium; evanescent waves, as mentioned in Section 11.5, result if the source travels more slowly than the emitted waves.

The same principle applies when real sources replace the imagined Huygens sources above, as in the process of **Čerenkov radiation** when charged particles move at relativistic velocities through refractive media. Since the speed of light in a glass, crystal or even gas is below that in vacuum, it is possible for high-energy particles to exceed it, resulting in the production of electromagnetic shock waves. As above, the condition for Čerenkov radiation is that the component of the particle velocity $\mathbf{v}$ in the emission direction should equal the wave propagation speed c/η in the medium of refractive index η. The angle ϑ between the Čerenkov emission and the particle direction is therefore

$$\vartheta = \cos^{-1}\left(\frac{c}{\eta|\mathbf{v}|}\right). \tag{18.24}$$

Čerenkov radiation accounts for the eerie glow within a nuclear reactor, as shown in Fig. 18.7, and also occurs when high-energy cosmic rays strike the Earth's atmosphere. Its angular dependence is used to determine the energies of, and distinguish between, relativistic products from high-energy particle-physics experiments. It is the basis of the 'Ice Cube' polar neutrino observatory, which collects the Čerenkov radiation emitted from the energetic muons created by the occasional interaction of neutrinos with Antarctic ice.

At particle velocities below but near to the speed of light, the transformations of special relativity cause the radiation from an accelerating charge to be

Fig. 18.7 Čerenkov radiation from high-energy electrons (β-radiation) illuminating the inside of a nuclear reactor at the Plum Brook Reactor Facility.   Courtesy of NASA

strongly directed along its motion. This phenomenon, used in particle-physics detectors, is exploited in *synchrotron radiation sources* and *free-electron lasers*.

18.3   The wake of a ship under way

18.3.1   Kelvin ship waves

The wake left behind a ship, boat or waterfowl as it moves through the water has a quite remarkable property: once the initial turbulence has dissipated, what remains most noticeably is a pair of ripples that form a V-shaped wedge on the surface of the water (Fig. 18.8), and the angle made by these ripples – 19.5° to either side of the vessel's track – is the same, regardless of the speed of the vessel. The explanation, first deduced by Kelvin in 1887, stems entirely from the simple dispersion relations for the deep-water ocean waves that we met in Section 3.3.3. For ripples whose wavelengths $\lambda = 2\pi/k$ are much less than the depth h_0 of the water, we may take the deep-water limit $kh_0 \to \infty$ and simplify equation (3.53) to

$$\omega = v_\mathrm{p}k \to \sqrt{gk}, \qquad (18.25)$$

where g is the acceleration due to gravity. It is easy to show that, in this highly dispersive system and in contrast with the example of Section 13.4.1, the group

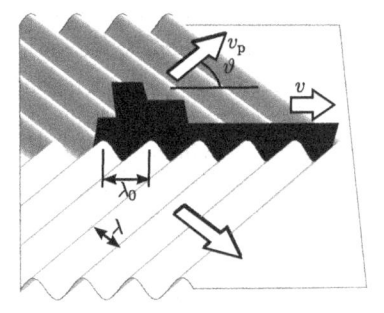

Fig. 18.9   Plane waves propagate away from an imagined vessel whose hull produces a sinusoidal displacement.

velocity v_g is half of the phase velocity v_p:

$$v_p = \frac{\omega}{k} = \sqrt{\frac{g}{k}}, \qquad (18.26a)$$

$$v_g = \frac{\partial \omega}{\partial k} = \frac{1}{2}\sqrt{\frac{g}{k}}. \qquad (18.26b)$$

The propagation of ripples from the moving vessel may be determined by initially imagining a very long vessel, moving with speed v, that produces a sinusoidal displacement (with wavelength λ_0) of the water along its length and hence sets up a sinusoidal motion of the water as depicted in Fig. 18.9. Just as with the phased arrays of Section 9.4, a Huygens construction will show that the resulting disturbance propagates away as a monochromatic plane wave with a wavelength λ and at an angle ϑ to the vessel's heading, where

$$\frac{\lambda}{v_p} = \frac{\lambda_0}{v}, \qquad (18.27a)$$

$$\lambda = \lambda_0 \cos \vartheta, \qquad (18.27b)$$

equation (18.27a) following from equivalent expressions of the wave period $2\pi/\omega$ and equation (18.27b) from simple geometry. It follows that

$$\cos\vartheta = \frac{\lambda}{\lambda_0}$$

$$= \frac{v_p}{v} = \frac{1}{v}\sqrt{\frac{g}{k}}. \qquad (18.28)$$

Equation (18.28) shows that there is therefore a limit to the wavenumber of components that can propagate away from the boat, which is given by setting $\cos\vartheta = 1$,

$$k_{max} = \frac{g}{v^2}, \qquad (18.29)$$

and hence a maximum frequency of $\omega_{max} = g/v$: above this, the phase velocity exceeds the boat speed v and only evanescent waves can result. Below this frequency, however, any sinusoidal component of the original disturbance will propagate away as a plane wave, and on each side of the boat's heading will travel in a single direction ϑ that depends upon the wavelength (and hence wavenumber) of the component according to equation (18.28) above.

A real vessel will displace the water to create a complex disturbance that corresponds to the superposition of many individual sinusoidal components; these, in the propagating wave, travel in frequency-dependent directions according to equation (18.28). We could in principle determine the frequency spectrum by performing a Fourier transform of the actual disturbance: a first approximation might assume this to be a 'top-hat' function the same length as the vessel, giving a frequency spectrum similar to the sinc function of Section 9.2. The three-dimensional nature of the disturbance must be taken into account, and the turbulent motion around the hull will also be rather nonlinear. While detailed measurement or calculation would be needed to determine the detail, the result will usually be a broad continuous spectrum of components that qualitatively show the same characteristic behaviour.

The components within any narrow range of frequencies travel in essentially the same direction, and interfere to form a wavepacket that travels not with the phase velocity but with the group velocity v_g. On combining equations (18.26) and (18.28), we see that the dependences of the phase and group velocities upon the wave-propagation direction will be given by

$$v_p = v\cos\vartheta, \qquad (18.30a)$$

$$v_g = \frac{1}{2}v\cos\vartheta. \qquad (18.30b)$$

These are plotted in Fig. 18.10. As the component wavelength and hence ϑ are varied, each velocity traces out a circle: a geometrical result that greatly aids the next stage of our analysis. The group velocity can never exceed $v/2$.

Figure 18.11 indicates how wavepackets, of wavelengths around a given value, propagate from successive positions of the vessel at intervals τ and

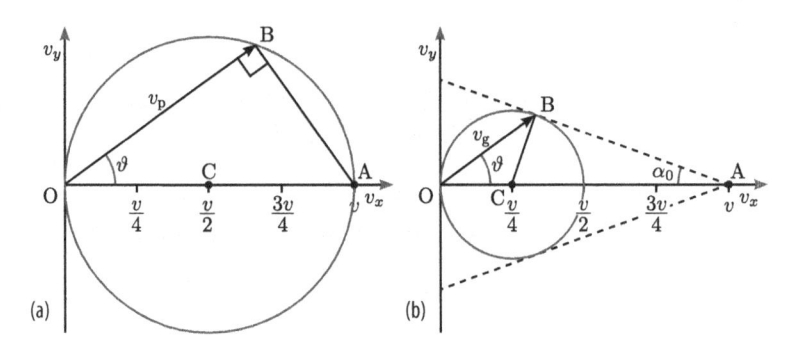

Fig. 18.10 The phase velocity (a) and group velocity (b) shown as functions of the propagation angle ϑ. Waves in each narrow range of frequencies will reinforce each other to form a straight wavepacket parallel to the dashed line that joins their group velocity $\mathbf{v}_g$ to the vessel velocity $\mathbf{v}$. The most advanced of these wavepackets will follow a tangent to the circle.

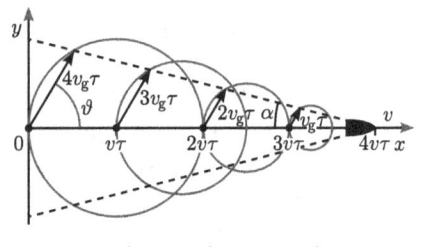

Fig. 18.11 The contributions of successive wavepackets to a plane wavepacket.

combine to produce a plane wavepacket: the distance travelled by each wavepacket from its point of origin will be proportional to the distance travelled by the vessel from that point, so the angle α subtended behind the boat in each case will be the same. The ratio v_g/v, and hence the angle α, varies with wavelength, but we shall be interested in the greatest value that it can take, for this represents the visible extent of the wake.

The maximum angle, α_0, will occur when the plane wavepacket lies along a tangent to the circles of Fig. 18.11. In this case, shown in Fig. 18.10(b), the triangle ABC is right-angled, with lengths $AC = 3v/2$ and $BC = v/2$. It follows that

$$\alpha_0 = \sin^{-1}\left(\frac{1}{3}\right)$$
$$= 19.47°. \tag{18.31}$$

This is not only the leading edge of the wave but, because it corresponds to a tangent to the locus of group velocities, it is the direction in which the group velocity varies least with wavelength and hence in which the greatest range of wavelengths will share similar group velocities. A broad range of wavelengths may therefore contribute to the superposition and, in a manifestation of the bandwidth theorem, form the most sharply defined part of the wavepacket.

Fig. 18.12

Waves propagating in the same direction as a fishing boat off Scotland's Holy Isle have a wavelength approximately equal to the boat length.   © Alison Randle

Within the wavepacket, individual components travel at the phase velocity in a direction ϑ that is not perpendicular to the line of the wavepacket. For the wavelengths of the leading edge, a little geometry gives $\vartheta = \tan^{-1}(1/\sqrt{2}) = 35.3°$. Behind the leading edge, these wavelengths will drift out of phase with each other, and it will be other wavelengths that dominate, with different values of ϑ. The exact shapes of the ripples within the wavepacket are therefore complex, albeit calculable and rather beautiful, curves.

Energy may be carried away from the vessel only by those components with wavenumbers below k_{max}, given in equation (18.29). Whereas the circular ripples from a point disturbance must diminish in amplitude to conserve energy as they spread, the straight wavefronts of a vessel's wake weaken only through processes such as viscous dissipation and changes to the wavepacket shape from residual dispersion, and can remain visible long after the vessel has passed.

W. Thomson (Lord Kelvin) [85]
Billingham & King [7]
pp. 99–109

18.3.2 Hull speed

Our treatment of Kelvin ship waves gives a simple explanation for the relationship between a ship's length and what is known as its **hull speed**, above which the power required for propulsion grows extremely rapidly. From equation (18.28), we see that waves propagating in the same direction as the vessel ($\vartheta = 0$) will have a wavelength

$$\lambda = \frac{2\pi v^2}{g}. \qquad (18.32)$$

At low speeds, the vessel spans a number of peaks and troughs; as it accelerates, however, the wavelength increases. If the wavelength exceeds the length of the vessel itself, the craft will find itself, as in Fig. 18.12, to be always climbing its own *bow wave* – the rising part of the wave that is created around the front of the boat – and, as the speed v is increased further, the effect sharply becomes

Fig. 18.13   Long, narrow rowing shells have a high hull speed and low drag.

New College, Oxford, 2010 Women's 1st VIII © Arthur Barlow

much more pronounced. The hull speed of a vessel of length l may therefore be defined as that at which the wavelength equals the boat length,

Hull speed

$$v_{\text{hull}} = \sqrt{\frac{gl}{2\pi}}, \tag{18.33}$$

and, for a wide range of engine powers, is close to the maximum speed of which the vessel is capable. Travel at higher speeds requires either prodigious power or the use of a 'planing' vessel that rides on the crest of its bow wave, using hydrodynamic forces rather than buoyancy to lift the vessel and hence reducing the volume of water displaced.

Racing yachts and rowing shells, such as that shown in Fig. 18.13, therefore tend to have long, narrow hulls with high hull speeds; where stability is required without the depth and weight of a keel, racing yachts and high-speed ferries may combine two such hulls to form a *catamaran*.

Exercises

18.1 The *Gatsometer* 'speed camera' measures the Doppler shift of a 24-GHz radar signal when it is reflected by a moving vehicle. Assuming initially that the radar beam is horizontal and parallel to the vehicle's motion, and that the reflecting surface is vertical, show that the Doppler frequency shift $\Delta \nu$ will be related to the radar frequency ν_0 through

$$\Delta \nu = 2\nu_0 \frac{v}{c}, \tag{18.34}$$

where v is the vehicle's speed and c is the speed of light, and hence determine the frequency shift caused by a car moving at 30 mph. (You may, of course, assume $v \ll c$.)

18.2 The Acme Corporation's chief inventor has realized that, in practice, speed cameras are placed to the side of and/or above the level of the vehicles that they monitor, and that the radar beam therefore strikes vehicles at an angle (typically 20°) that changes as the vehicle moves past the camera. When a car is close to the camera, the radar set receives principally the reflection from the roof, whereas at greater distances it is the reflection from the tailgate that dominates. Taking typical values for the dimensions involved, estimate the error that such effects might introduce into the measurement of the vehicle's speed, and suggest a vehicle shape that will cause the apparatus to underestimate the speed.

18.3 Calculate the hull speed for a rowing shell of length 20 m, and compare this with the average speed of boats in the annual race between Oxford and Cambridge Universities along the River Thames from Putney to Mortlake – a distance of 6.8 km that, in favourable conditions, is usually completed in about 17 minutes.

19 Radiation from moving charges

Electromagnetic waves have, as light and radio waves, been recurring examples for many of the phenomena that we have met throughout this book, from reflection and refraction to diffraction and interference, and for many of the technological applications, from antireflection coatings to Doppler radar. Using Maxwell's equations of electromagnetism, we can describe in exquisite detail how each of these processes occurs; and, since in characterizing the wave by the electric field strength **E** we refer directly to the force that the wave will exert on a static point-like test charge, we can see quite directly the processes by which electromagnetic waves are detected. The accompanying magnetic field, and its effects and detection, require only small steps further; and even the extension of our classical treatment into a quantum-mechanical description proves to be straightforward. The detailed processes by which moving charges give rise to electromagnetic waves, however, prove to hold many subtleties, and to yield some elegant but somewhat startling results.

Our general approach throughout this book has been to determine the characteristics of wave propagation in each case from the physical mechanisms by which a disturbance at one point affects that at its neighbours. This allows us to write and solve a wave equation for the system, and to determine amongst other properties the phase velocity v_p of the propagating wave. We showed in Section 1.3, however, that wave propagation can be approached in a different order, and that the propagation of a disturbance from an emitter to an observer or receiver may be regarded as a version of the static interaction between the source and detector when the finite propagation speed is taken into account. Electromagnetic and gravitational waves hence correspond to the static Coulomb force between charges and the gravitational force between masses when the apparent position of each distant particle or body is taken to be the *retarded* value at an earlier time to allow for the propagation delay.

We mentioned, however, two deficiencies with our naïve introduction of retardation in Section 1.3. First, although we assumed the principle of special relativity that propagation is limited to the speed of light, c, we overlooked the Lorentz transformations required by the same theory to account for the motions of the charges or bodies. Secondly, we neglected any variation in retardation as the individual masses or charges moved towards or away from the observer. As a result, our initial calculation omitted the magnetic parts of electromagnetic waves and the long-range components of dipole radiation in general.

Although these corrections add little to our understanding of the principles of wave propagation, the results are of far-reaching importance, for, among other examples, they account for the light emitted by atoms and the radiation patterns from radio transmitters. Furthermore, even though in most cases the interactions between the atoms and molecules of a medium may be considered to be instantaneous, and the medium well approximated by a continuous, isotropic solid or fluid, matter is ultimately bound by electromagnetic forces, and their propagation is therefore implicit in virtually every practical example of wave propagation that we have considered.

We therefore end this book by considering in detail the waves radiated by an oscillating electric dipole. A common approach is to match a general solution to the electromagnetic wave equation of Section 15.3.1 to the boundary conditions at the oscillating dipole, and we shall indeed sketch out this method, beginning with a more thorough analysis of the field from a static dipole that forms the near-field limit, and to which retardation may be added in an ad-hoc fashion. We shall then follow the alternative approach discussed above, by considering fully the effects of retardation upon the static Coulomb interaction between distant charges, first through the scalar and vector *Liénard–Wiechert* potentials and then in the retarded electric and magnetic fields. These equivalent but rather different approaches then allow us to write the field radiated by a moving charge and an oscillating electric dipole in full.

Billingham & King [7]
pp. 210–212
Lipson *et al.* [56] Section 5.3

19.1   Solution of the electromagnetic wave equation

The most straightforward, though not necessarily enlightening, method of determining the field radiated by an oscillating dipole is the general approach of solving the electromagnetic wave equation (15.18) to give a general solution in terms, for example, of the spherical harmonics of Section 15.6.2, and fitting this to the boundary conditions by requiring that, at positions close to the dipole where retardation may be neglected, the solution at any time approaches that for a static dipole. A slight modification to this method, which we pursue below, is to determine the corrections that must be made to a naïvely retarded version of the static field to satisfy Maxwell's equations.

19.1.1   Field of a static electric dipole

Before examining the three-dimensional wave emitted by an oscillating electric dipole, we first establish in more detail the static field due to a stationary dipole, comprising positive and negative charges $\pm q_1$ separated by a distance $2a_0$, whose transverse field component we considered in Section 1.3.1. We retrace the derivation of equation (1.2) for a dipole inclined at an angle ϑ to the

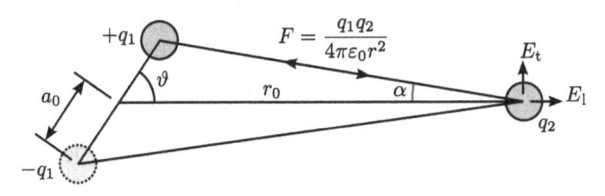

Fig. 19.1 Coulomb interaction between an oscillating dipole (q_1, $-q_1$) and a test charge q_2.

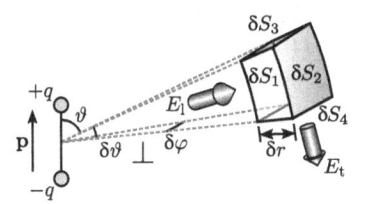

Fig. 19.2 Geometry for application of Gauss's law to the field of a static dipole.

direction of observation, as shown in Fig. 19.1, to determine the transverse and longitudinal components of the electric field experienced by the test charge q_2.

For the transverse field, we find that equation (1.2) is modified only by the introduction of an additional factor $\sin \vartheta$,

$$E_t = \frac{2q_1}{4\pi \varepsilon_0 r_0^3} a \sin \vartheta, \tag{19.1}$$

and similar working yields the longitudinal component of the static field,

$$E_l = \frac{q_1}{4\pi \varepsilon_0 (r_0 - a \cos \vartheta)^2} - \frac{q_1}{4\pi \varepsilon_0 (r_0 + a \cos \vartheta)^2}$$

$$\approx \frac{2q_1}{4\pi \varepsilon_0 r_0^3} 2a \cos \vartheta, \tag{19.2}$$

where we have again assumed $a \ll r_0$ and performed a series expansion, to first order, of the squared term in the denominator. Placing the origin at the centre of the dipole so that $r_0 \to r$, and writing the total field in terms of the transverse and longitudinal components E_t and E_l and unit vectors $\hat{\boldsymbol{\vartheta}}$ and $\hat{\mathbf{r}}$ as $\mathbf{E} \equiv E_t \hat{\boldsymbol{\vartheta}} + E_l \hat{\mathbf{r}}$, we hence obtain the total field due to a static point dipole of dipole moment $p \equiv 2q_1 a$,

Field of a static electric dipole

$$\mathbf{E} = \frac{p}{4\pi \varepsilon_0 r^3} \left(\sin \vartheta \, \hat{\boldsymbol{\vartheta}} + 2 \cos \vartheta \, \hat{\mathbf{r}} \right). \tag{19.3}$$

It is instructive, and helpful for later sections, to check that this field satisfies Gauss's law, equation (15.12a). In vacuum and with no free charges, and written

dS is the product of the area dS of an element of the closed surface, and a unit vector normal to the surface, pointing outwards from the volume enclosed.

in integral form with $d\mathbf{S}$ representing an elemental area, this becomes

$$\int \mathbf{E} \cdot d\mathbf{S} = 0. \tag{19.4}$$

We consider an element of space as shown in Fig. 19.2, with dimensions in polar coordinates δr, $\delta \vartheta$ and $\delta \varphi$, and break the surface integral into terms corresponding to the electric field flux through each of the six faces of the element, whose areas we label $\delta S_{1\ldots6}$, noting that, by symmetry, there will be no components of the field normal to the front and back faces δS_5 and δS_6, which are parallel to the dipole axis. We may therefore write

$$\int \mathbf{E} \cdot d\mathbf{S} \equiv E_1(r + \delta r, \vartheta)\delta S_2 - E_1(r, \vartheta)\delta S_1 + E_t(r, \vartheta + \delta\vartheta)\delta S_4 - E_t(r, \vartheta)\delta S_3.$$

(19.5)

We now substitute the explicit forms for the field components from equations (19.1) and (19.2), and the elemental areas $\delta S_1 = r^2 \sin\vartheta\, \delta\vartheta\, \delta\varphi$, $\delta S_2 = (r + \delta r)^2 \sin\vartheta\, \delta\vartheta\, \delta\varphi$, $\delta S_3 = r \sin\vartheta\, \delta r\, \delta\varphi$ and $\delta S_4 = r\sin(\vartheta + \delta\vartheta)\, \delta r\, \delta\varphi$, to give

$$\int \mathbf{E} \cdot d\mathbf{S} = \frac{p}{4\pi\varepsilon_0}\left[\frac{2\cos\vartheta}{(r+\delta r)^3}(r+\delta r)^2 \sin\vartheta\, \delta\vartheta\, \delta\varphi - \frac{2\cos\vartheta}{r^3}r^2 \sin\vartheta\, \delta\vartheta\, \delta\varphi \right.$$
$$\left. + \frac{\sin(\vartheta + \delta\vartheta)}{r^3}r\sin(\vartheta + \delta\vartheta)\delta r\, \delta\varphi - \frac{\sin\vartheta}{r^3}r\sin\vartheta\, \delta r\, \delta\varphi \right].$$

(19.6)

In the limit of small δr and $\delta\vartheta$ this gives

$$\int \mathbf{E} \cdot d\mathbf{S}$$
$$= \frac{p\,\delta\varphi}{4\pi\varepsilon_0}\left[2\sin\vartheta\cos\vartheta\, \delta\vartheta\left(\frac{1}{r+\delta r} - \frac{1}{r}\right) + \frac{\delta r}{r^2}\left(\sin^2(\vartheta + \delta\vartheta) - \sin^2\vartheta\right) \right]$$
$$\to \frac{p\,\delta\varphi}{4\pi\varepsilon_0}\left[2\sin\vartheta\cos\vartheta\, \delta\vartheta\frac{(-\delta r)}{r^2} + \frac{\delta r}{r^2}2\sin\vartheta\cos\vartheta\, \delta\vartheta \right] = 0, \qquad (19.7)$$

confirming that equation (19.3) satisfies Gauss's law.

19.1.2 Retarded field of a changing electric dipole

Equation (19.3) is a complete result for the field due to a static electric dipole and, when retardation is introduced as in Chapter 1, provides a trial form that is a good approximation in the near field of an oscillating dipole,

$$\mathbf{E} = \frac{p(t - r/c)}{4\pi\varepsilon_0 r^3}\left(\sin\vartheta\,\hat{\boldsymbol{\vartheta}} + 2\cos\vartheta\,\hat{\mathbf{r}}\right). \qquad (19.8)$$

However, if we repeat the exercise above, we find that this retarded field alone is no longer a solution to Gauss's law, for the fields passing through the surfaces δS_1 and δS_2 were created by the oscillating source at slightly different times. The Gaussian integral therefore acquires and retains an additional term

$$\int \mathbf{E} \cdot d\mathbf{S} = \frac{\left[p(t - (r + \delta r)/c) - p(t - r/c)\right]\delta\varphi}{4\pi\varepsilon_0}2\sin\vartheta\cos\vartheta\, \delta\vartheta\frac{1}{r+\delta r}$$
$$\approx \frac{2\cos\vartheta}{4\pi\varepsilon_0 r^3}r^2 \sin\vartheta\, \delta r\, \delta\vartheta\, \delta\varphi\frac{dp(t - r/c)}{dr}$$
$$= -\frac{2\cos\vartheta}{4\pi\varepsilon_0 r^3 c}\delta V\frac{dp}{dt'}, \qquad (19.9)$$

where $t' \equiv t - r/c$, $\delta V \equiv r^2 \sin \vartheta \, \delta r \, \delta \vartheta \, \delta \varphi$ is the volume of the element of integration, and we have again taken the limit of small δr.

The algebra of Section 19.1.1 is more compact, but exactly equivalent, if we work with the differential form of Gauss's law given by equation (15.12a). In spherical polar coordinates (r, ϑ, φ), the three-dimensional electric field vector will be $\mathbf{E} \equiv E_r \hat{\mathbf{r}} + E_\vartheta \hat{\boldsymbol{\vartheta}} + E_\varphi \hat{\boldsymbol{\varphi}}$. In vacuum and the absence of free charges, Gauss's law hence becomes (see the Appendix)

$$\nabla \cdot \mathbf{E} \equiv \frac{1}{r^2} \frac{\partial}{\partial r} \left(r^2 E_r \right) + \frac{1}{r \sin \vartheta} \frac{\partial}{\partial \vartheta} (\sin \vartheta \, E_\vartheta) + \frac{1}{r \sin \vartheta} \frac{\partial}{\partial \varphi} E_\varphi = 0. \quad (19.10)$$

The residual for the trial retarded field of equation (19.8) is given by

$$\nabla \cdot \mathbf{E} = \frac{-2 \cos \vartheta}{4 \pi \varepsilon_0 r^3 c} \frac{\mathrm{d}p}{\mathrm{d}t'}, \quad (19.11)$$

consistently with equation (19.9) above.

We may modify equation (19.8) without changing the field at the dipole by adding terms that vary as higher powers of the radius r and therefore become increasingly negligible as $r \to 0$. Gauss's law is satisfied, for example, if equation (19.8) is modified to provide an improved trial form

$$\mathbf{E} = \frac{p(t')}{4 \pi \varepsilon_0 r^3} \left(\sin \vartheta \, \hat{\boldsymbol{\vartheta}} + 2 \cos \vartheta \, \hat{\mathbf{r}} \right) + \frac{1}{4 \pi \varepsilon_0 r^2 c} \frac{\mathrm{d}p}{\mathrm{d}t'} \sin \vartheta \, \hat{\boldsymbol{\vartheta}}. \quad (19.12)$$

The field must also satisfy the electromagnetic wave equation (15.18), however, requiring two further terms to be added. Using the vector differentials of the Appendix, it is a little messy but straightforward to show that Gauss's law, the wave equation and the boundary condition at the dipole are simultaneously satisfied by

Field radiated by an electric dipole

$$\mathbf{E} = \frac{1}{4 \pi \varepsilon_0 r^3} \left\{ \left(p + \frac{r}{c} \frac{\mathrm{d}p}{\mathrm{d}t'} \right) \left(\sin \vartheta \, \hat{\boldsymbol{\vartheta}} + 2 \cos \vartheta \, \hat{\mathbf{r}} \right) + \left(\frac{r}{c} \right)^2 \frac{\mathrm{d}^2 p}{\mathrm{d}t'^2} \sin \vartheta \, \hat{\boldsymbol{\vartheta}} \right\}.$$

$$(19.13)$$

This is the complete electric field radiated by an oscillating, or more generally varying, dipole. At large distances, the final term dominates and the field becomes purely transverse, with a $1/r$ dependence that, as for a spherical wave, gives the $1/r^2$ variation in intensity required for energy to be conserved.

Maxwell's equations, which yielded the electromagnetic wave equation and propagation speed in vacuum as well as the continuity conditions from which we derived the Fresnel coefficients in Section 11.3, hence allow a rigorous determination of the electric field radiated by a varying dipole. Their further application reveals the accompanying magnetic field, which, according to equation (15.19a) for harmonic waves at large distances, oscillates in phase with the electric field and is perpendicular both to the wave vector and to the electric field.

19.2 Retarded electromagnetic potentials

While the solution of Maxwell's equations subject to boundary conditions at the dipole (and requiring reasonable behaviour at infinity) is a rigorous and relatively straightforward method of determining the field radiated by an oscillating dipole, it offers little insight into the origins of the terms present. Maxwell's equations lack the conceptual simplicity of the Coulomb interaction with which we began, whereby the force between two charges is directed along the line joining them; and, since Einstein's special theory of relativity leads us to regard the speed of light as a fundamental limit upon causal propagation in the Universe, related more to the geometries of space-time than to the mechanisms associated with any particular wave type, there are attractions to an alternative approach in which the speed of light is assumed and Maxwell's equations emerge from, rather than lead to, the retarded Coulomb interaction itself. Provided that we accept – and are willing to deal with – the accompanying transformations of special relativity, this approach proves instructive and fruitful, although the mathematical manipulations and bookkeeping require some care and patience. It is this type of approach, nonetheless, that leads to the more advanced concepts of quantum electrodynamics (QED) and the nature of the Coulomb interaction itself.

In the following pages, we therefore outline two alternative derivations of the radiated field that are based upon careful calculations of the Coulomb interaction that take into account retardation and the transformations of special relativity. While it is more intuitively enlightening to determine the electric field directly, fewer pitfalls are presented by determining it from the corresponding electric potential, which is therefore our first method of approach.

19.2.1 The scalar and vector potentials

It is common that for problems in classical mechanics we may choose to consider either the forces acting upon an object or its energy. The two approaches are essentially equivalent for, if the force $\mathbf{F}$ may be written in terms of the gradient of the potential energy $\mathcal{U}$ as

$$\mathbf{F} = -\nabla \mathcal{U}, \tag{19.14}$$

then the mathematics in one case is the spatial derivative of that in the other. This is certainly true with the conservative forces of electrostatics, when we may refer with comparable utility to the force and potential energy of the Coulomb interaction between two charges. When one of the charges is simply being used to probe the effects of the other, we may equivalently consider the electric field strength $\mathbf{E}$ and electric potential V, which are just the force and potential energy per unit probe charge, and are similarly related by

$$\mathbf{E}(\mathbf{r}) = -\nabla V(\mathbf{r}). \tag{19.15}$$

The *scalar potential* $V(\mathbf{r})$ at position $\mathbf{r}$ of a *static* charge q at position $\mathbf{r}_q$ is therefore the well-known Coulomb potential

$$V(\mathbf{r}) = \frac{1}{4\pi\varepsilon_0}\frac{q}{R}, \tag{19.16}$$

where $R \equiv |\mathbf{R}| \equiv |\mathbf{r} - \mathbf{r}_q|$ is the distance of the measurement position from the charge.

Solutions to Maxwell's equations for dynamic systems of *moving* charges and varying fields require both this scalar potential, which depends upon the charge distribution at any time, and an additional *vector potential* $\mathbf{A}(\mathbf{r}, t)$, which is determined by the distribution of currents that exist when charges move. Electrical currents induce magnetic fields, and varying magnetic fields induce voltages and thereby electric fields, as we have already seen for harmonic fields in Section 15.3.1 and equations (15.19a) and (15.19b). For a *steady* current I along a line element $\delta\mathbf{s}$, the vector potential will be

$$\mathbf{A}(\mathbf{r}, t) = \frac{\mu_0 I}{4\pi R}\,\delta\mathbf{s}. \tag{19.17}$$

The electric field is then given in terms of the scalar and vector potentials by

$$\mathbf{E}(\mathbf{r}, t) = -\nabla V - \frac{\partial \mathbf{A}}{\partial t}. \tag{19.18}$$

For static charges and steady currents, this reduces to equation (19.15), and the vector potential determines only the magnetic field, which is given by

$$\mathbf{B}(\mathbf{r}, t) = \nabla \times \mathbf{A}. \tag{19.19}$$

Before addressing the radiation from a single moving charge, however, we must make some small but significant adjustments to take into account the time delay between the charge and the observer.

19.2.2 Retarded variables

If a charge is moving, and the distance from the charge is great enough for the movement to be appreciable in the time that it takes light (or any causal effect) to propagate from the charge to observer, then we must take into account that at any time t the observer will 'see' the charge to be as it was at an earlier time t', because it will have taken a time R/c for the effect of the charge to reach the observer. We have indeed already done this in an ad-hoc fashion in Section 19.1.2, when we expressed the radiated field at time t and position $\mathbf{r}$ in terms of the dipole moment $p(t - r/c)$ at the origin at the earlier time $t' = t - r/c$, and in our examination of shock waves in Section 18.2.1.

Regrettably, our expression for the retarded time t' involves a circular definition, for the distance r from the moving charge depends upon the time t' at which its position is observed, which depends upon r and so on. Although algebraically messy, r nonetheless remains just the apparent distance of the charge from the observer, which we could imagine determining by reversing time, propagating an imaginary spherical wavefront back from the observer, and noting where and when it encountered the charge as it retraced its path. While such definitions are relatively easily solved, the details depend upon the situation and are unimportant for our general treatment. To avoid cluttering the rest of our derivation, it is therefore helpful to refer to the *retarded* values of the time, position, velocity and so on of the distant charge that are apparent to the observer at a given time t, so that we may, for example, write

$$V(\mathbf{r}, t) = \frac{1}{4\pi\varepsilon_0} \frac{[q]}{[R]}, \tag{19.20}$$

where the square brackets indicate that R takes its retarded value

$$[R] = |[\mathbf{R}]| = |\mathbf{r}(t) - \mathbf{r}_q(t')|. \tag{19.21}$$

Here, $[\mathbf{R}]$ is the retarded value of the vector $\mathbf{R}$ from the charge to the observer, and

$$t' \equiv t - [R]/c. \tag{19.22}$$

For future completeness, equation (19.20) also indicates that the charge q takes its retarded value. Expressions such as equation (19.20) reflect only a change in notation; the retarded values must still be determined, and the implicit way that this is indicated does not reflect any change in the physics of the situation.

There is a further consequence of this retardation that we must consider more closely. We assume for the time being that the charge is moving with a constant velocity $\mathbf{v}$, and therefore that its position at time t' may be written as

$$\mathbf{r}_q(t') = \mathbf{r}_q(t) - (t - t')\mathbf{v}. \tag{19.23}$$

On combining equations (19.21)–(19.23), we therefore obtain

$$[\mathbf{R}] = \mathbf{R}(t) + \frac{[R]}{c}\mathbf{v}, \tag{19.24}$$

where the unretarded relative position is

$$\mathbf{R}(t) \equiv \mathbf{r}(t) - \mathbf{r}_q(t). \tag{19.25}$$

If the component of the velocity $\mathbf{v}$ normal to $\mathbf{R}$ is small in comparison with c,

$$[R] \approx R(t) + \frac{[R]}{c}\mathbf{v} \cdot [\hat{\mathbf{R}}], \tag{19.26}$$

where $R(t)$ again refers to the unretarded distance and $[\hat{\mathbf{R}}]$ is a unit vector in the direction of $[\mathbf{R}]$. Equation (19.26) may now be rearranged to give

$$[R] = \frac{R(t)}{1 - (\mathbf{v} \cdot [\hat{\mathbf{R}}])/c}, \qquad (19.27a)$$

$$t' = t - \frac{R(t)/c}{1 - (\mathbf{v} \cdot [\hat{\mathbf{R}}])/c}. \qquad (19.27b)$$

19.2.3 Temporal compression factor for a moving charge

Equation (19.27b) reveals that what we see in the course of a short period δt is the influence of the charge over a generally different period $\delta t'$. To establish this temporal compression factor, we differentiate the square of the retarded distance $[R]$ with respect to the current time t,

$$\frac{\mathrm{d}}{\mathrm{d}t}[R]^2 = \frac{\mathrm{d}}{\mathrm{d}t}\left([\mathbf{R}] \cdot [\mathbf{R}]\right)$$

$$\Rightarrow \quad 2[R]\frac{\mathrm{d}[R]}{\mathrm{d}t} = 2\frac{\mathrm{d}[\mathbf{R}]}{\mathrm{d}t} \cdot [\mathbf{R}]$$

$$= 2\left(\frac{\mathrm{d}\mathbf{R}(t)}{\mathrm{d}t} + \frac{1}{c}\frac{\mathrm{d}[R]}{\mathrm{d}t}[\mathbf{v}]\right) \cdot [\mathbf{R}], \qquad (19.28)$$

where the final step follows from equation (19.24) and, in the general case, $\mathbf{v}$ takes its retarded value and $\mathbf{R}(t)$ represents the extrapolated current position. On collecting terms in $\mathrm{d}[R]/\mathrm{d}t$,

$$\frac{\mathrm{d}[R]}{\mathrm{d}t}\left(2[R] - \frac{2}{c}[\mathbf{v} \cdot \mathbf{R}]\right) = -2[\mathbf{v} \cdot \mathbf{R}], \qquad (19.29)$$

where we have substituted $\mathrm{d}\mathbf{R}(t)/\mathrm{d}t = -[\mathbf{v}]$, whence

$$\frac{\mathrm{d}[R]}{\mathrm{d}t} = \frac{-[\mathbf{v} \cdot \mathbf{R}]}{[R] - [\mathbf{v} \cdot \mathbf{R}]/c}. \qquad (19.30)$$

Differentiating equation (19.22) with respect to t now gives

$$\frac{\mathrm{d}t'}{\mathrm{d}t} = 1 - \frac{1}{c}\frac{\mathrm{d}[R]}{\mathrm{d}t}$$

$$= 1 + \frac{[\mathbf{v} \cdot \mathbf{R}]/c}{[R] - [\mathbf{v} \cdot \mathbf{R}]/c}$$

$$= \frac{1}{1 - [\mathbf{v} \cdot \hat{\mathbf{R}}]/c}, \qquad (19.31)$$

where $[\mathbf{R}] \equiv [R][\hat{\mathbf{R}}]$. This is another expression of the Doppler effect of equation (18.3) in Section 18.1 (where f' denoted the *observed* frequency).

We have seen throughout this book that, as originally outlined by Huygens, the wave disturbance at a given point is the sum of the contributions from all propagation routes to that point; hence, for a source moving towards the observer, a greater range of contributions can arrive in a given period of time.

It turns out that this effect leads to an enhancement of the observed wave[1] by a factor of dt'/dt, which is commonly written as $1/\kappa$, where the temporal compression factor

Temporal compression factor

$$\kappa \equiv \frac{1}{dt'/dt} = 1 - \frac{[\mathbf{v} \cdot \hat{\mathbf{R}}]}{c}. \qquad (19.32)$$

As the component of the particle velocity along the observation direction, $[\mathbf{v}] \cdot [\hat{\mathbf{R}}]$, approaches c, the factor κ accounts for the concentration of synchrotron radiation in the direction of travel, which, together with a (periodic) transverse impulse that we shall see shortly to be necessary for long-range radiation, is the basis of the *free-electron laser*. The shock wave when $[\mathbf{v} \cdot \hat{\mathbf{R}}] = c$ and $\kappa \to 0$ corresponds to the Čerenkov radiation described in Section 18.2.2.

19.2.4 Retarded potentials for a moving charge

We may now combine our results for the retarded potentials of Section 19.2.2 with the temporal compression of Section 19.2.3, which we introduce as an enhancement factor into equation (19.20), so that the scalar potential a distance R from a charge moving with velocity $\mathbf{v}$ proves to be

$$V(\mathbf{r}, t) = \frac{1}{4\pi\varepsilon_0} \left[\frac{q}{\kappa R} \right]. \qquad (19.33)$$

The same methods may be used to determine the vector potential. Conveniently, for a single charge q moving with velocity $\mathbf{v}$, we may write the current element $I\,\delta\mathbf{s} \equiv q\mathbf{v}$, so, proceeding as for the scalar potential and making the substitution $\mu_0 \equiv 1/(\varepsilon_0 c^2)$, the vector potential of equation (19.17) becomes

$$\mathbf{A}(\mathbf{r}, t) = \frac{1}{4\pi\varepsilon_0 c^2} \left[\frac{q\mathbf{v}}{\kappa R} \right]. \qquad (19.34)$$

Equations (19.33) and (19.34) define what are known as the *Liénard–Wiechert* potentials, which, like the temporal compression factor κ, prove to be of general validity; they are consistent with the principles of special relativity, and may be alternatively derived by Lorentz transformation of the electromagnetic potentials. It was indeed the analysis of such phenomena that prompted Lorentz's work and Einstein's subsequent proposal of the special theory of relativity.

H. A. Lorentz [58]
A. Einstein [19]

The resulting electric field is then given by equation (19.18), yielding

$$\mathbf{E}(\mathbf{r}, t) = -\frac{1}{4\pi\varepsilon_0} \left(\nabla \left[\frac{q}{\kappa R} \right] + \frac{1}{c^2} \frac{\partial}{\partial t} \left[\frac{q\mathbf{v}}{\kappa R} \right] \right), \qquad (19.35)$$

which is relativistically valid for any velocity. The magnetic field is similarly given from equation (19.34) by equation (19.19).

---

[1] The formal proof of this, outlined by Feynman (*Lectures on Physics*, vol. II, 21-5) and described in more detail by, for example, A. R. Janah *et al.* ('On Feynman's formula for the electromagnetic field of an arbitrarily moving charge,' *Am. J. Phys.* **56**, 1036 (1988)), involves writing the propagated field as a Green function and integrating this for the moving source.

For a charge moving towards the observer, equations (19.27a) and (19.32) yield $[\kappa R] \equiv R(t)$, and hence the rather startling result that, despite the propagation delay from the charge to the observer, the scalar and vector potentials for a charge moving with a steady velocity are the same as if the charge were at its *unretarded* position. This proves to be essentially true for any direction of motion: the field *magnitude* depends upon the velocity of the charge relative to the observer but, if the velocity **v** is constant, the electric field always points to or from the *unretarded* position of the charge. For an accelerating charge, the field lines appear to meet at the position that would be reached by continuing at the retarded velocity. The directionality of the compression factor κ accounts for the offset between the retarded position and the effective source of the field lines, and the product $[\kappa R]$ may be re-written as a symmetrical function of the unretarded position as

Panofsky & Phillips [67]
Chapter 19
Lorrain *et al.* [59] Section 16.5

$$[\kappa R] = R(t)\sqrt{1 - \frac{v^2}{c^2}\sin^2\psi}, \tag{19.36}$$

where ψ is the angle between the unretarded position $\mathbf{R}(t)$ and the velocity **v**.

19.2.5 Electric field of a moving charge

Evaluation of equation (19.35) is slightly tricky, for the retarded quantities depend upon t', which is itself a function of the observation position **r** and time t with respect to which we calculate the derivatives. The temporal derivative is easily re-written, from equation (19.31), as

$$\frac{\partial}{\partial t} = \frac{\mathrm{d}t'}{\mathrm{d}t}\frac{\partial}{\partial t'} = \frac{1}{\kappa}\frac{\partial}{\partial t'}. \tag{19.37}$$

The gradient corresponds to the variation with observation position **r** at fixed observation time t, where there is an implicit dependence of the source time t' upon **r**. We separate these dependences upon **r** and t' by writing

$$\nabla \equiv \nabla\big|_{t'} + \left(\nabla\big|_t t'\right)\frac{\partial}{\partial t'}\bigg|_{\mathbf{r}}, \tag{19.38}$$

where the subscript indicates explicitly the constrained property.

We shall require a few further results. From equation (19.22), we may write

$$[R] = c(t - t') \tag{19.39}$$

and hence

$$\frac{\partial[R]}{\partial t'}\bigg|_{\mathbf{r}} = c\left(\frac{\mathrm{d}t}{\mathrm{d}t'} - 1\right)$$

$$= -[\mathbf{v}\cdot\hat{\mathbf{R}}], \tag{19.40}$$

where for the final step we have substituted for dt/dt' from equation (19.31). By application of equation (19.38) we may now determine

$$\nabla[R] = \nabla\big|_{t'}[R] + \left(\nabla\big|_{t}t'\right)\frac{\partial[R]}{\partial t'}\bigg|_{\mathbf{r}}$$

$$= [\hat{\mathbf{R}}] - [\mathbf{v}\cdot\hat{\mathbf{R}}]\nabla\big|_{t}t', \qquad (19.41)$$

where the standard result used for $\nabla\big|_{t'}[R]$ may be shown by considering the explicit form $(\partial/\partial x, \partial/\partial y, \partial/\partial z)^{\mathrm{T}}\cdot(x-x_q, y-y_q, z-z_q)^{\mathrm{T}}$. Equation (19.38) also yields

$$\nabla(c(t-t')) = \nabla\big|_{t'}\left(c(t-t')\right) + \left(\nabla\big|_{t}t'\right)\frac{\partial}{\partial t'}\bigg|_{\mathbf{r}}\left(c(t-t')\right)$$

$$= -c\,\nabla\big|_{t}t'. \qquad (19.42)$$

Equation (19.39) shows equations (19.41) and (19.42) to be equal, so

$$\nabla t' = \left(\nabla\big|_{t}t'\right) = \frac{-[\hat{\mathbf{R}}]}{c - [\mathbf{v}\cdot\hat{\mathbf{R}}]}$$

$$= \frac{-[\hat{\mathbf{R}}]}{c\kappa}. \qquad (19.43)$$

Finally, explicit consideration of the components of $\nabla\big|_{t'}\left([\mathbf{v}]\cdot(\mathbf{r}-\mathbf{r}_q)\right)$ yields

$$\nabla\big|_{t'}[\mathbf{v}\cdot\mathbf{R}] = [\mathbf{v}] \qquad (19.44)$$

and hence, noting from equation (19.21) the sign of $[\mathbf{v}] = -\partial[\mathbf{R}]/\partial t'$,

$$\nabla[\mathbf{v}\cdot\mathbf{R}] = \nabla\big|_{t'}[\mathbf{v}\cdot\mathbf{R}] + \left(\nabla\big|_{t}t'\right)\frac{\partial}{\partial t'}\bigg|_{\mathbf{r}}[\mathbf{v}\cdot\mathbf{R}]$$

$$= [\mathbf{v}] + \left(\nabla\big|_{t}t'\right)\left(\frac{\partial[\mathbf{v}]}{\partial t'}\bigg|_{\mathbf{r}}\cdot[\mathbf{R}] + [\mathbf{v}]\cdot\frac{\partial[\mathbf{R}]}{\partial t'}\bigg|_{\mathbf{r}}\right)$$

$$= [\mathbf{v}] + \nabla t'\left(\left[\frac{d\mathbf{v}}{dt'}\right]\cdot[\mathbf{R}] - v^2\right). \qquad (19.45)$$

Panofsky & Phillips [67]
Chapters 18–20

It is now a matter of careful bookkeeping to substitute equations (19.31), (19.32), (19.40), (19.41), (19.43) and (19.45) into equation (19.35) to give

$$\mathbf{E}(\mathbf{r}, t) = \frac{q}{4\pi\varepsilon_0[\kappa R]^3}\left\{\left(1 - \frac{v^2}{c^2}\right)\left([\mathbf{R}] - \frac{[R\mathbf{v}]}{c}\right)\right.$$

$$\left. + \frac{1}{c^2}[\mathbf{R}]\times\left[\left([\mathbf{R}] - \frac{[R\mathbf{v}]}{c}\right)\times\left[\frac{d\mathbf{v}}{dt'}\right]\right]\right\}. \qquad (19.46)$$

If $r_q \ll r$ and $v \ll c$, we may write

$$\frac{1}{[\kappa R]^3} = \frac{1}{\left\{\left(1 - [\mathbf{v}\cdot\hat{\mathbf{R}}]/c\right)\left(|\mathbf{r} - [\mathbf{r}_q]|\right)\right\}^3}$$

$$\approx \frac{1}{r^3}\left\{1 + 3\left(\frac{[\mathbf{v}]\cdot\hat{\mathbf{r}}}{c} + \frac{[\mathbf{r}_q]\cdot\hat{\mathbf{r}}}{r}\right)\right\}. \qquad (19.47)$$

The dipole may alternatively be represented as a pair of fixed charges of varying magnitude. This simplifies the derivation of the scalar potential, but leaves a trickier calculation of the vector potential to account for the current element that must run between the fixed points. See, for example, Smith [78] Chapter 7.

The field of a dipole, comprising a pair of charges $\pm q$ at $\pm \mathbf{r}_q$, where $\mathbf{p} \equiv 2q\mathbf{r}_q$, may now be determined by adding their individual contributions according to equations (19.46) and (19.47) and taking the limit as $r_q \to 0$, to give

$$\mathbf{E}(\mathbf{r}, t) = \frac{q}{4\pi\varepsilon_0 r^3} \left\{ \left(3[\mathbf{p}] \cdot \hat{\mathbf{r}}\hat{\mathbf{r}} - [\mathbf{p}] \right) + \frac{r}{c} \left(3 \left[\frac{d\mathbf{p}}{dt} \right] \cdot \hat{\mathbf{r}}\hat{\mathbf{r}} - \left[\frac{d\mathbf{p}}{dt} \right] \right) \right.$$
$$\left. + \frac{r^2}{c^2} \hat{\mathbf{r}} \times \left(\hat{\mathbf{r}} \times \left[\frac{d^2\mathbf{p}}{dt^2} \right] \right) \right\}. \tag{19.48}$$

If, as in Section 19.1, the dipole direction $\hat{\mathbf{p}}$ is fixed along the polar axis, we may write $\hat{\mathbf{r}} \times (\hat{\mathbf{r}} \times \hat{\mathbf{p}}) = \sin\vartheta \, \hat{\boldsymbol{\vartheta}}$, where $\hat{\mathbf{r}} \cdot \hat{\mathbf{p}} = \cos\vartheta$, and

$$3\left([\mathbf{p}] \cdot \hat{\mathbf{r}}\right)\hat{\mathbf{r}} - [\mathbf{p}] = 3[p]\cos\vartheta \, \hat{\mathbf{r}} - [p] \left(\cos\vartheta \, \hat{\mathbf{r}} - \sin\vartheta \, \hat{\boldsymbol{\vartheta}} \right)$$
$$= [p] \left(2\cos\vartheta \, \hat{\mathbf{r}} + \sin\vartheta \, \hat{\boldsymbol{\vartheta}} \right), \tag{19.49}$$

and correspondingly for the term in $[d\mathbf{p}/dt]$. With these substitutions, we once again obtain the field of equation (19.13).

19.3 Retarded electromagnetic fields

While it is in many ways more straightforward to work with Maxwell's equations, or the scalar and vector potentials of Section 19.2, it is also possible and perhaps more intuitive to deal from the start with the electrostatic fields themselves. In the inertial rest frame of a uniformly moving source charge, we may assume simply the Coulomb interaction between charges at the source and observer. To determine the force upon, and hence field at, the observer, however, we must transform the coordinates and fields from the frame of the source to that of the observer. The price of this arguably more direct approach is the algebraic complexity of the relativistic Lorentz transformation.

Lorrain *et al.* [59] Section 16.5

As in Section 19.2, we consider a source charge q with position $\mathbf{r}_q(t)$ and constant velocity $\mathbf{v}(t) \equiv d\mathbf{r}_q/dt$ in the frame of the observer. We transform the relative position of the observer to the frame of the source, determine the Coulomb field in that frame, and then transform the field back into the frame of the observer; the electric field upon our test charge is then found to be

$$\mathbf{E}(\mathbf{r}, t) = \frac{1}{4\pi\varepsilon_0} \frac{q}{[\gamma]^2 \left\{ (1 - [\beta]^2) + ([\boldsymbol{\beta}] \cdot \hat{\mathbf{R}})^2 \right\}^{3/2}} \frac{\mathbf{R}}{R^3}, \tag{19.50}$$

where $\beta \equiv v/c$, $\boldsymbol{\beta} \equiv \mathbf{v}/c$ and $\gamma^2 \equiv 1/(1 - \beta^2)$. On writing equation (19.36) as

$$[\kappa R] = R(t) \sqrt{1 - \frac{[v]^2}{c^2} + \left(\frac{[\mathbf{v}] \cdot \hat{\mathbf{R}}(t)}{c} \right)^2}, \tag{19.51}$$

where $[\mathbf{v}] \cdot \hat{\mathbf{R}}(t) = [v]\cos\psi$, we see equation (19.50) to be identical for constant motion to equation (19.46). Note that it is again the *unretarded* position vector that determines the field direction. The electric field therefore resembles that for a static charge, except that its strength varies according to the observation direction: along the direction of motion ($\boldsymbol{\beta} \cdot \hat{\mathbf{R}} = \beta$) the field strength is a factor of γ^2 less than for the same charge when stationary, whereas normal to the direction of motion ($\boldsymbol{\beta} \cdot \hat{\mathbf{R}} = 0$) it is a factor of γ greater.

On repeating our transformations to find the force upon a *moving* test charge q_2 with velocity $\mathbf{u}$ in the frame of the observer, we find a further term, towards or away from the source trajectory, that may be written as $q_2 \mathbf{u} \times \mathbf{B}$, where

$$\mathbf{B}(\mathbf{r}, t) = \frac{1}{4\pi\varepsilon_0} \frac{q}{[\gamma]^2 \left\{ (1 - [\beta]^2) + ([\boldsymbol{\beta}] \cdot \hat{\mathbf{R}})^2 \right\}^{3/2}} \frac{[\mathbf{v}] \times \mathbf{R}}{c^2 R^3}. \tag{19.52}$$

This is the observed magnetic field, which is entirely due to the relativistic transformation of the electrostatic Coulomb interaction, and may be written as

$$\mathbf{B}(\mathbf{r}, t) = \frac{[\mathbf{v}] \times \mathbf{E}(\mathbf{r}, t)}{c^2} = \frac{[\hat{\mathbf{R}}] \times \mathbf{E}(\mathbf{r}, t)}{c}. \tag{19.53}$$

The magnetic field is hence normal to the plane containing the electric field $\mathbf{E}$, the retarded velocity $[\mathbf{v}]$ and the retarded position vector $[\mathbf{R}]$.

19.3.1 The retarded field of an accelerating charge

J. J. Thomson [83] pp. 53*ff*
J. R. Tessman & J. T. Finnell [81]
D. H. Frisch & L. Wilets [30]

The difficulty with the retarded-field approach arises when we allow the source velocity to vary, for we must then perform transformations between *accelerating* frames, requiring *general* relativity. Instead, we adopt the more pictorial analysis originally outlined by J. J. Thomson in 1904.

To determine the field of an accelerating charge, we consider a brief period of acceleration between two spells of uniform motion: the charge is therefore taken to move with a velocity $\mathbf{v}_1$ until time $t = 0$, accelerate with a uniform acceleration $\mathbf{a}$ until $t = \tau$, and then continue with its new velocity $\mathbf{v}_2 = \mathbf{v}_1 + \mathbf{a}\tau$. We have seen in Section 19.2 that the retarded field from periods of uniform motion will appear to originate at the current or extrapolated position of that motion, so we can begin to construct the electric field apparent at a later time $t > \tau$, as shown in Fig. 19.3. The initial motion, as far as point A, defines the field beyond a circle of radius ct about A, in which region it appears to originate at the point C that the charge would have reached had there been no acceleration. The later period of uniform motion defines the field within a circle of radius $c(t - \tau)$ about the point B where the acceleration ceased, within which the field appears to come from the current position D. For simplicity, we consider here only acceleration parallel to the initial velocity, and may therefore refer henceforth to scalar values; it is always possible to transform to an inertial frame in which this will be the case.

We now deduce the field in the region between the two circles, in which the retarded field was generated between points A and B while the charge was accelerating during the period from $t = 0$ to $t = \tau$. Assuming that there are no

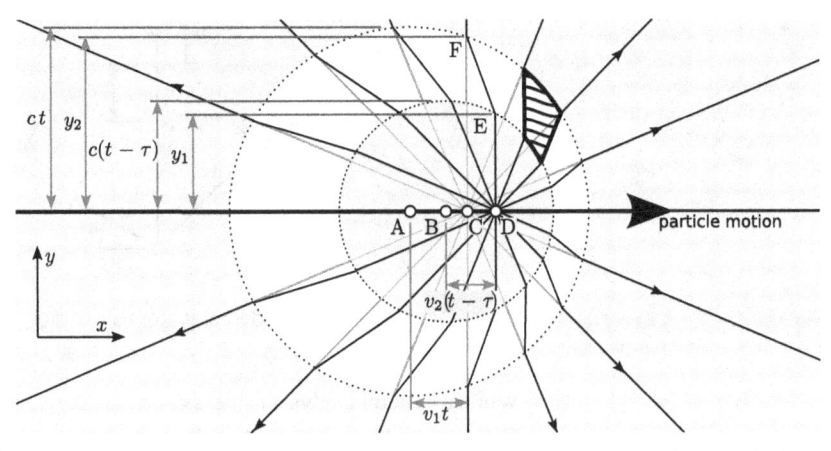

Fig. 19.3 Radiation from an accelerating charge (after J. J. Thomson). The particle is taken to have travelled along the x axis with speed v_1 until reaching point A at time $t = 0$, thereafter undergoing an acceleration a until reaching point B at time $t = \tau$. The figure is a snapshot at time t, when the particle has reached point D; at its original speed, it would only have reached C.

free charges, we may infer from Gauss's law that no field line can end in the region of interest, and we may therefore join the field lines in the two known regions to determine the field between them, assuming that, if τ and/or a are sufficiently small, then the field lines in this region too will everywhere be straight.

Although this pictorial method may appear to invest rather more significance than is common in the spacing of field lines, which are usually drawn schematically, it is nonetheless sound. We may divide the region between the two circles into segments that are bounded by sections of the circles and pairs of field lines, such as the segment shown hatched in Fig. 19.3. Gauss's law requires that no net field flux must enter the region and, if the field lines are spaced to achieve a regular division of the flux, this means that the fluxes crossing the two circular edges will be balanced, and therefore that the components across the straight edges must also cancel each other out. While this could be achieved for various field directions between the circles, those shown are the only ones that give a radial field along the symmetry axis. They are also the only ones to satisfy Faraday's law that the integral of the azimuthal field component around any closed path that has reflection symmetry about the particle trajectory (such as one following the middle of the unshaded region) must be zero.

Applying Gauss's law to a thin 'pillbox' G enclosing an arc of one of the circles in Fig. 19.4 also establishes a continuity condition that the radial field components must be equal inside and outside the circle. The radial component of the field between the circles, measured with respect to the current *retarded* position of the charge, will therefore be the same as if it were moving with a constant velocity equal to its current value, and is therefore unaffected by the acceleration, which contributes only a transverse component to the field.

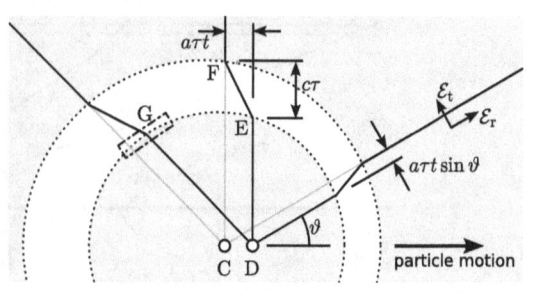

Fig. 19.4 Radiation at an angle ϑ to the trajectory of an accelerated charge.

While a general analysis is messy, we may derive a useful result by determin-ing the transverse component of the field that propagates in a direction normal to the trajectory of the accelerated particle, as shown in Fig. 19.4. The reflection symmetry of the field of a moving charge, apparent in equation (19.13), about a plane through the charge normal to the trajectory, means that those field lines that before and after the acceleration lie perpendicular to the trajectory will be connected. Elementary geometry shows that points C and D will be separated by a distance $a\tau(t - \tau/2)$ and that, for v_1, $v_2 \ll c$, the vertical distance to be filled by the field line will be approximately $c\tau$. The inserted field line will therefore lie at an angle $\tan^{-1}(a(t - \tau/2)/c)$ to the normal, and the transverse field component will therefore be this fraction of the radial component due to the Coulomb term. Writing the distance from the source $r = ct$ and taking the limit as $\tau \to 0$, this fraction may be written as ar/c^2. Figure 19.4 also shows that, for slowly moving charges for which the circles are roughly concentric and the field lines parallel, the acceleration-dependent term varies as $\sin \vartheta$. The transverse field component, varying with the inverse square of the radial distance, will therefore be

$$E = \frac{q}{4\pi\varepsilon_0} \frac{1}{c^2} \left[\frac{a}{r}\right] \sin \vartheta, \qquad (19.54)$$

in agreement with the acceleration-dependent term of equations (19.12) and (19.46). A full calculation proves this to be a general result, and finds that, as indicated, both the acceleration and the radial distance take their retarded values.

Our derivation in Section 19.2 of the field of an accelerated charge proceeded simply by retarding, with appropriate correction factors, the electrostatic poten-tial. It is instructive to consider the difficulty in applying this method directly to the electrostatic field, which stems from the relationship, given by equation (19.15), between the field and the potential. Whereas for uniform motion the fields in the particle's rest frame are steady, acceleration of the charge always introduces a time dependence and, as we found in Section 19.2.5, the spatial derivatives in such cases acquire a dependence upon the temporal evolution as well. Where the Coulomb potential varies only gently with distance, changes in the charge position can create strong gradients in the direction of propagation. Indeed, we might consider a moving charge to be represented by a series of 'snapshots', each static but present only for an instant: the potential propagating from each snapshot would have extremely sharp radial gradients, which could well exceed the static fields created by such charges.

19.4 Radiation from moving charges

We have explored three approaches to the determination of the fields of moving charges. The solution of Maxwell's equations, and consequently of the electromagnetic wave equation, provided a straightforward method when we imposed the boundary condition that, sufficiently near to the source charges for retardation to be neglected, the solution should approach that for a static distribution. The perhaps more fundamental starting point of Coulomb's law then allowed us an alternative approach, whereby a careful consideration of the effects of retardation led to the Liénard–Wiechert potentials, from which the fields could be directly determined provided that, for the dynamic case, the static, scalar potential was augmented by the addition of a vector potential. Finally, the relativistic transformation of the electrostatic force between charges allowed the derivation of the fields for the case of steady motion, although further reasoning was required for the acceleration-dependent component. The three methods happily yield equivalent results.

Although some care is clearly needed whichever method of analysis we adopt, the eventual result is satisfyingly elegant. Expressed in terms of the retarded position of the source as apparent to the observer, the electric field of a single moving charge comprises three terms: the retarded static Coulomb force, which varies as $1/[R]^2$ and hence dominates the near field; a velocity-dependent term related to the temporal compression factor κ; and an acceleration-dependent term, known as the *radiation field*, that for an accelerating charge is dominant at large distances. Although the derivation is a little troublesome, it proves possible to write the electric field $\mathbf{E}(\mathbf{R}, t)$ in a particularly elegant form as a function of the unit retarded relative position vector $\hat{\mathbf{R}}(t)$, its magnitude and its derivatives – that is, in terms of the retarded position, velocity and acceleration of the charged particle – as

O. Heaviside [39] p. 174
Feynman [23] Vol. II, p. 21-1
A. R. Janah *et al.* [49]

$$\mathbf{E}(\mathbf{R}, t) = \frac{q}{4\pi \varepsilon_0} \left(\underbrace{\left[\frac{\hat{\mathbf{R}}}{R^2} \right]}_{\text{static}} + \underbrace{\frac{[R]}{c} \frac{\mathrm{d}}{\mathrm{d}t} \left[\frac{\hat{\mathbf{R}}}{R^2} \right]}_{\text{predictive}} + \underbrace{\frac{1}{c^2} \frac{\mathrm{d}^2}{\mathrm{d}t^2} \left[\hat{\mathbf{R}} \right]}_{\text{radiative}} \right). \qquad (19.55)$$

Field radiated by a moving charge

The simple form of equation (19.55), and its dependence throughout upon the retarded position of the charge, reinforce the contention that the observed field is fundamentally no more than the Coulomb interaction between the source charge and a test charge at the point of observation, and differs only through the effects of retardation and coordinate transformation. The accompanying magnetic field is then given directly by equation (19.53), and is seen to be a correction to account for the relativistic transformation of the Coulomb force when experienced by a charge undergoing motion in the frame of the observer.

We have seen in Section 19.2.4, and used in Section 19.3, that the field of a uniformly moving charge appears to diverge from the unretarded position, and more generally points to or from the position extrapolated to the current time from the retarded position and velocity; and we saw that the temporal compression factor κ provided the corresponding correction to the potential. We may therefore consider the velocity-dependent term to be a 'predictive' correction to the retarded position to that which, at a uniform rate of change, it would have at the time of observation, for, combined with the static term, it may be written in the form

$$\mathbf{X} + \Delta t \, \frac{d\mathbf{X}}{dt}, \tag{19.56}$$

where $\mathbf{X} \equiv [\hat{\mathbf{R}}/R^2]$ and $\Delta t \equiv [R]/c$ is the retardation time from the source to the observer.

Given the result for a single charge, the field radiated by a dipole may be written in terms of the retarded dipole moment $\mathbf{p}(t)$ and its derivatives. For a dipole of fixed alignment $p(t)\hat{\mathbf{p}}$, the electric field may be written as in equation (19.13) as

$$\mathbf{E} = \frac{1}{4\pi\varepsilon_0 r^3} \underbrace{\left\{ \left(p + \frac{r}{c} \frac{dp}{dt'} \right) \left(2\cos\vartheta \, \hat{\mathbf{r}} + \sin\vartheta \, \hat{\boldsymbol{\vartheta}} \right)}_{\text{static} + \text{predictive}} + \underbrace{\left(\frac{r}{c} \right)^2 \frac{d^2 p}{dt'^2} \sin\vartheta \, \hat{\boldsymbol{\vartheta}} \right\}}_{\text{radiative}}.$$

$$\tag{19.57}$$

Like the field of a single charge from which it is obtained, the dipole field exhibits terms corresponding to the static field, a component proportional to the first temporal derivative of the retarded dipole moment and the 'radiation' term that depends upon the second derivative; the first two terms retain the shape of the static dipole field, and all terms show the vector spherical harmonics of Section 15.6. Once more, the electric field of equation (19.57) is accompanied by a magnetic field that may be obtained from it by the application of equation (19.53).

The first and second (static and predictive) terms in equations (19.55) and (19.57) prove to be near-field components, akin to an evanescent field that, once established, transmits no energy away from the source. Their energies, integrated over the full solid angle, decrease with the distance from the source, and their time-averaged Poynting vectors are zero. The integrated energy of the third term is in contrast the same at all distances, and this term indeed accounts for the net propagation of energy from the accelerating charge. This radiated field is a transverse wave, polarized in the plane of the dipole and strongest in the direction normal to the dipole axis.

Equation (19.55) shows this radiated field to be determined only by the acceleration of the unit vector $\hat{\mathbf{R}}$, and hence by the direction from which the propagating disturbance arrives, irrespective of its distance. While the source is clearly necessary for the radiation of such a disturbance, this feature perhaps encourages us to adopt a degree of conceptual detachment and to consider the effect of the propagating wave in some isolation.

Exercises

The following exercises, which mainly concern definitions and distinctions, may be used as revision of the principles of wave propagation and its analysis that have been encountered throughout this book.

19.1 What is meant by a *wave motion*? Give examples that are *transverse*, *longitudinal* and *neither*.

19.2 How does the displacement of one element of a *stretched string* influence the motion of adjacent points? How does the displacement of one element of an *air column* influence adjacent regions?

19.3 Which physical properties of the media in each case determine the speeds of wave propagation on strings, of sound in fluids and of shallow-water waves?

19.4 Explain the distinction and connection between *travelling* and *standing* waves.

19.5 Why are *sinusoidal* wave motions so often considered? How are they related to *complex exponential* waves?

19.6 Explain with examples how the *amplitude*, *energy density* and *intensity* (or power) of a wave motion are related.

19.7 What is meant by *dispersion*? What are the *dispersion relations*?

19.8 How may *phasors* represent complex exponential or sinusoidal wave motions? How may they be combined to indicate the total disturbance of a *superposition*?

19.9 Outline the *Huygens description* of wave propagation.

19.10 State *Fermat's principle of least time* for wave propagation. Draw a diagram showing several possible paths from a source A in air to a point B in a plane-edged block of glass. Explain why the ray path does not in general form a straight line between the two points.

19.11 How may a lens be used to transform between angle and position?

19.12 What is meant by wave *interference*? Explain how it results in fringes when light passes through a double slit.

19.13 What is meant by *diffraction*? Outline the conditions necessary for the observation of *Fraunhofer* diffraction. Give examples of practical situations in which Fraunhofer diffraction may be observed.

19.14 Explain, with examples, what is meant by the *continuity conditions* for wave motion at the interface between two media. How does a discontinuity in the characteristics of a medium affect the motion of a wave propagating through it?

19.15 Explain, with examples, what is meant by a *boundary condition* for wave motion. What rôle do the boundary conditions play in determining the pitch and sound of musical instruments?

19.16 State the boundary condition that must be met at a point where the string described by equation (2.18) is fixed. Hence find the real *standing-wave* solutions to the wave equation, and determine the allowed oscillation frequencies, when such a string of length l is fixed at its ends.

19.17 Show that the diffusion equation (10.23) may be solved by the damped sinusoidal wave

$$\Theta(x, t) = \Theta_0 + \Theta_1 \cos(kx - \omega t) \exp(-kx) \qquad (19.58)$$

and find how the parameter k depends upon the angular frequency ω.

19.18 What is meant by the *linearity* of a wave equation? Why is it significant?

19.19 What is meant by the *phase velocity* and *group velocity* of a wave motion? How do they depend upon the dispersion relations?

19.20 Explain, with examples, what is meant by the *frequency spectrum* of a time-dependent wave signal $f(t)$.

19.21 Explain the principles of *Fourier synthesis* and *Fourier analysis*.

19.22 Outline what is meant by the *Fourier transform*, and how it may be defined mathematically.

19.23 Describe what is meant by the *convolution* of two functions, and how the *convolution theorem* may be used to determine the diffraction patterns of regular arrays of a basic pattern.

19.24 Explain, with examples, what is meant by an *operator* for wave motions.

19.25 Describe how the *mean* frequency, and its *standard deviation*, may be determined for a wave motion, both from its expression as a function of time and from its frequency spectrum.

19.26 Outline the *bandwidth theorem*, and explain its significance for both classical and quantum mechanical wave motions.

19.27 Describe the *Doppler effect*, and discuss several different methods by which it may be theoretically derived.

19.28 Explain what is meant by a *shock wave*, and determine the angle between the wavefront of a shock wave and the path of its source in terms of the source and wave speeds v and v_0.

Appendix: **Vector mathematics**

A.1 Cartesian coordinates

Although we live on a globe and have long been fascinated by the heavens, we are usually most familiar with the representation of spatial positions in terms of rectilinear, Cartesian coordinates, (x, y, z). In this representation, a vector field $\mathbf{E}(x, y, z)$ is described by its components along and dependence upon the orthogonal x, y and z axes in the directions of the unit vectors $\hat{\mathbf{i}}$, $\hat{\mathbf{j}}$ and $\hat{\mathbf{k}}$ as

$$\mathbf{E} \equiv E_x \hat{\mathbf{i}} + E_y \hat{\mathbf{j}} + E_z \hat{\mathbf{k}} \equiv \begin{pmatrix} E_x \\ E_y \\ E_z \end{pmatrix}, \tag{A1.1}$$

where E_x, E_y and E_z are all scalar functions of the coordinates x, y and z.

A.1.1 Scalar product

The scalar ('dot') product of two vectors $\mathbf{A}$ and $\mathbf{B}$ is given by

$$\mathbf{A} \cdot \mathbf{B} \equiv A_x B_x + A_y B_y + A_z B_z \tag{A1.2a}$$

$$\equiv AB \cos \vartheta, \tag{A1.2b}$$

where A and B are the magnitudes of the vectors $\mathbf{A}$ and $\mathbf{B}$ and ϑ is the angle between them.

A.1.2 Vector product

The vector product of two vectors $\mathbf{A}$ and $\mathbf{B}$ is given by the determinant

$$\mathbf{A} \times \mathbf{B} \equiv \begin{vmatrix} \hat{\mathbf{i}} & \hat{\mathbf{j}} & \hat{\mathbf{k}} \\ A_x & A_y & A_z \\ B_x & B_y & B_z \end{vmatrix} \equiv \begin{pmatrix} A_y B_z - A_z B_y \\ A_z B_x - A_x B_z \\ A_x B_y - A_y B_x \end{pmatrix}. \tag{A1.3}$$

A.1.3 Gradient operator

The gradient or differential operator ∇ is written in Cartesian coordinates as

$$\nabla \equiv \begin{pmatrix} \partial / \partial x \\ \partial / \partial y \\ \partial / \partial z \end{pmatrix}. \tag{A1.4}$$

A.1.4 Scalar differentials

The gradient of a scalar field $\psi(x, y, z)$ is given in Cartesian coordinates by

$$\nabla \psi \equiv \hat{\mathbf{i}} \frac{\partial \psi}{\partial x} + \hat{\mathbf{j}} \frac{\partial \psi}{\partial y} + \hat{\mathbf{k}} \frac{\partial \psi}{\partial z} \,. \tag{A1.5}$$

The scalar Laplacian $\nabla^2 \equiv \nabla \cdot \nabla$ is then given by

P.-S. Laplace [54]

$$\nabla^2 \psi \equiv \frac{\partial^2 \psi}{\partial x^2} + \frac{\partial^2 \psi}{\partial y^2} + \frac{\partial^2 \psi}{\partial z^2} \,. \tag{A1.6}$$

A.1.5 Vector differentials

The divergence of a vector field $\mathbf{E}(x, y, z)$ may be written as

$$\nabla \cdot \mathbf{E} \equiv \frac{\partial E_x}{\partial x} + \frac{\partial E_y}{\partial y} + \frac{\partial E_z}{\partial z} \,. \tag{A1.7}$$

The curl is most clearly written in terms of a determinant,

$$\nabla \times \boldsymbol{E} \equiv \begin{vmatrix} \hat{\mathbf{i}} & \hat{\mathbf{j}} & \hat{\mathbf{k}} \\ \partial/\partial x & \partial/\partial y & \partial/\partial z \\ E_x & E_y & E_z \end{vmatrix} \equiv \begin{pmatrix} \partial E_z/\partial y - \partial E_y/\partial z \\ \partial E_x/\partial z - \partial E_z/\partial x \\ \partial E_y/\partial x - \partial E_x/\partial y \end{pmatrix}. \tag{A1.8}$$

The Laplacian ∇^2 takes a straightforward form

$$\nabla^2 \mathbf{E} \equiv \begin{pmatrix} \partial^2 E_x/\partial x^2 + \partial^2 E_x/\partial y^2 + \partial^2 E_x/\partial z^2 \\ \partial^2 E_y/\partial x^2 + \partial^2 E_y/\partial y^2 + \partial^2 E_y/\partial z^2 \\ \partial^2 E_z/\partial x^2 + \partial^2 E_z/\partial y^2 + \partial^2 E_z/\partial z^2 \end{pmatrix}. \tag{A1.9}$$

A.2 Spherical polar coordinates

In spherical polar coordinates (r, ϑ, φ), where r is the radial coordinate, ϑ the colatitude and ϕ the longitude, a vector field $\mathbf{E}(r, \vartheta, \varphi)$ may be written in terms of its radial and azimuthal components as

$$\mathbf{E} \equiv E_r \hat{\mathbf{r}} + E_\vartheta \hat{\boldsymbol{\vartheta}} + E_\varphi \hat{\boldsymbol{\varphi}} \equiv \begin{pmatrix} E_r \\ E_\vartheta \\ E_\varphi \end{pmatrix}, \tag{A2.1}$$

where E_r, E_ϑ and E_φ are all scalar functions of r, ϑ and φ.

A.2.1 Scalar differentials

The gradient of a scalar field $\psi(r, \vartheta, \varphi)$ is given in spherical polar coordinates by

$$\nabla \psi \equiv \hat{\boldsymbol{r}}\frac{\partial \psi}{\partial r} + \hat{\boldsymbol{\vartheta}}\frac{1}{r}\frac{\partial \psi}{\partial \vartheta} + \hat{\boldsymbol{\varphi}}\frac{1}{r\sin\vartheta}\frac{\partial \psi}{\partial \varphi}. \tag{A2.2}$$

The scalar Laplacian ∇^2 is then given by

$$\frac{1}{r^2}\frac{\partial}{\partial r}\left(r^2\frac{\partial \psi}{\partial r}\right) + \frac{1}{r^2\sin\vartheta}\frac{\partial}{\partial \vartheta}\left(\sin\vartheta\frac{\partial \psi}{\partial \vartheta}\right) + \frac{1}{r^2\sin^2\vartheta}\frac{\partial^2 \psi}{\partial \varphi^2}. \tag{A2.3}$$

A.2.2 Vector differentials

The divergence of a vector field $\boldsymbol{E}(r, \vartheta, \varphi)$ may be written as

$$\nabla \cdot \boldsymbol{E} \equiv \frac{1}{r^2}\left\{\sin\vartheta\frac{\partial}{\partial r}\left(r^2 E_r\right) + r\frac{\partial}{\partial \vartheta}(\sin\vartheta\, E_\vartheta) + r\frac{\partial E_\varphi}{\partial \varphi}\right\}. \tag{A2.4}$$

The curl is most clearly written in terms of a determinant,

$$\nabla \times \boldsymbol{E} \equiv \frac{1}{r^2\sin\vartheta}\begin{vmatrix} \hat{\boldsymbol{r}} & r\hat{\boldsymbol{\vartheta}} & r\sin\vartheta\,\hat{\boldsymbol{\varphi}} \\ \partial/\partial r & \partial/\partial\vartheta & \partial/\partial\varphi \\ E_r & rE_\vartheta & r\sin\vartheta\, E_\varphi \end{vmatrix}. \tag{A2.5}$$

The Laplacian ∇^2 may be written, using the vector identity

$$\nabla^2 \boldsymbol{E} \equiv \nabla(\nabla \cdot \boldsymbol{E}) - \nabla \times (\nabla \times \boldsymbol{E}), \tag{A2.6}$$

as

$$\nabla^2 \boldsymbol{E} = \begin{pmatrix} \nabla^2 E_r - \dfrac{2}{r^2}E_r - \dfrac{2}{r^2}\dfrac{\partial E_\vartheta}{\partial \vartheta} - \dfrac{2\cos\vartheta}{r^2\sin\vartheta}E_\vartheta - \dfrac{2}{r^2\sin\vartheta}\dfrac{\partial E_\varphi}{\partial \varphi} \\[2mm] \nabla^2 E_\vartheta - \dfrac{1}{r^2\sin^2\vartheta}E_\vartheta + \dfrac{2}{r^2}\dfrac{\partial E_r}{\partial \vartheta} - \dfrac{2\cos\vartheta}{r^2\sin^2\vartheta}\dfrac{\partial E_\varphi}{\partial \varphi} \\[2mm] \nabla^2 E_\varphi - \dfrac{1}{r^2\sin^2\vartheta}E_\varphi + \dfrac{2}{r^2\sin\vartheta}\dfrac{\partial E_r}{\partial \varphi} + \dfrac{2\cos\vartheta}{r^2\sin^2\vartheta}\dfrac{\partial E_\vartheta}{\partial \varphi} \end{pmatrix}. \tag{A2.7}$$

Arfken & Weber [1] Section 2.5
Boas [8] Section 10.9

References

[1] Arfken, G. B., Weber, H.-J., and Harris, F. 2005. *Mathematical Methods for Physicists*. Sixth edn. New York: Academic Press.

[2] Atal, B. S., and Schroeder, M. R. 1966. *Apparent sound source translator*. US Patent 3,236,949.

[3] Bally, J., O'Dell, C. R., and McCaughrean, M. J. 2000. Disks, microjets, windblown bubbles, and outflows in the Orion Nebula. *Astronomical Journal*, **119**, 2919–2959.

[4] Benade, A. H. 1960. The physics of woodwinds. *Scientific American*, **203**(4), 145–154.

[5] Benade, A. H. 1973. The physics of brasses. *Scientific American*, **229**(1), 24–35.

[6] Benade, A. H. 1992. *Horns, Strings and Harmony*. New York: Dover.

[7] Billingham, J., and King, A. C. 2000. *Wave Motion*. Cambridge: Cambridge University Press.

[8] Boas, M. L. 2005. *Mathematical Methods in the Physical Sciences*. Third edn. Chichester: Wiley.

[9] Boyle, R. 1664. *Experiments and Considerations Touching Colours*. London: Henry Herringman. Available from `http://infomotions.com/etexts/gutenberg/dirs/1/4/5/0/14504/14504.htm`.

[10] Bragg, W. L. 1922. The diffraction of X-rays by crystals. Nobel lecture. Available from `http://nobelprize.org/nobel_prizes/physics/laureates/1915/wl-bragg-lecture.pdf`.

[11] Čerenkov, P. A. 1958. Radiation of particles moving at a velocity exceeding that of light, and some of the possibilities for their use in experimental physics. Nobel lecture. Available from `http://www.nobelprize.org/nobel_prizes/physics/laureates/1958/cerenkov-lecture.html`.

[12] Chanaud, R. C. 1970. Aerodynamic whistles. *Scientific American*, **222**(1), 40–46.

[13] *Concise Oxford Dictionary of Current English*. 1990. Eighth edn. Oxford: Oxford University Press.

[14] Coulomb, C.-A. de. 1785. Premier mémoire sur l'électricité et le magnétisme. *Histoire de l'Académie Royale des Sciences*, pp. 569–577. Available from `http://books.google.co.uk/books?id=by5EAAAAcAAJ&pg=PA569`.

[15] Coulson, C. A., and Jeffrey, A. 1977. *Waves*. London: Longman.

[16] Crawford, Jr., F. S. 1965. *Waves. Berkeley Physics Course*. New York: McGraw-Hill.

[17] Dholakia, K., Spalding, G., and MacDonald, M. 2002. Optical tweezers: the next generation. *Physics World*, **15**, 31–35.

[18] Doppler, C. A. 1842. Ueber das farbige Licht der Doppelsterne und einiger anderer Gestirne des Himmels. *Abhandlungen der königlichen böhmischen*

Gesellschaft der Wissenschaften zu Prag, **2**, 465–482. Available from `http://www.archive.org/details/ueberdasfarbige100doppuoft`.

[19] Einstein, A. Zur Elektrodynamik bewegter Körper. *Annalen der Physik*, **322**(10), 891–921.

[20] Farkas, I., Helbing, D., and Vicsek, T. 2002. Mexican waves in an excitable medium. *Nature*, **419**, 131–132. Further details are given in the website `http://angel.elte.hu/wave/`.

[21] Fermat, P. de. 1657. Letter to Marin Cureau de la Chambre. Available from `http://wlym.com/~animations/fermat/16570800Fermat todelaChambre.pdf`.

[22] Feynman, R. P. 1990. *QED – The Strange Theory of Light and Matter*. London: Penguin.

[23] Feynman, R. P., Leighton, R. B., and Sands, M. 1971. *Lectures on Physics*. Reading, MA: Addison-Wesley.

[24] Fleisch, D. 2008. *A Student's Guide to Maxwell's Equations*. Cambridge: Cambridge University Press.

[25] Fletcher, N. H., and Thwaites, S. 1983. The physics of organ pipes. *Scientific American*, **248**(1), 94–103.

[26] Fourier, J. B. J. 1822. *Théorie analytique de la chaleur*. Paris: Firmin Didot, père et fils. Available from `http://www.archive.org/stream/thorieanalytiqu00fourgoog`.

[27] Franklin, B., Brownrigg, W., and Mr Farish. 1774. On the stilling of waves by means of oil. *Philosophical Transactions of the Royal Society of London*, **64**, 445–460. Available from `http://rstl.royalsocietypublishing.org/content/64.toc`.

[28] Franklin, R. E., and Gosling, R. G. 1953. Molecular configuration in sodium thymonucleate. *Nature*, **171**, 740–741.

[29] French, A. P. 1971. *Vibrations and Waves*. London: Chapman and Hall.

[30] Frisch, D. H., and Wilets, L. 1956. Development of the Maxwell–Lorentz equations from special relativity and Gauss's law. *American Journal of Physics*, **24**(8), 574–579.

[31] Gabor, D. 1948. The new microscopic principle. *Nature*, **161**, 777–778.

[32] Gabor, D. 1971. Holography, 1948–1971. Nobel lecture. Available from `http://nobelprize.org/nobel_prizes/physics/laureates/1971/gabor-lecture.pdf`.

[33] Goos, H. F. G., and Hänchen, H. 1947. Ein neuer und fundamentaler Versuch zur Totalreflexion. *Annalen der Physik*, **436**(7–8), 333–346.

[34] Gouy, L. G. 1890a. Sur la propagation anomale des ondes. *Comptes rendus hebdomadaires des séances de l'Académie des Sciences de Paris*, **111**, 33–35. Available from `http://gallica.bnf.fr/ark:/12148/bpt6k3067d.image`.

[35] Gouy, L. G. 1890b. Sur une propriété nouvelle des ondes lumineuses. *Comptes rendus hebdomadaires des séances de l'Académie des Sciences de Paris*, **110**, 1251–1253. Available from `http://gallica.bnf.fr/ark:/12148/bpt6k30663.image`.

[36] Grier, D. G. 2003. A revolution in optical manipulation. *Nature*, **424**, 810–816.

[37] Grimaldi, F. M. 1665. *Physico-mathesis de Lumine, Coloribus, et Iride*. Bologna: Hæredis Victorij Benatij. Available from `http://fermi.imss.fi.it/rd/bdv?/bdviewer/bid=300682\#`.

[38] Hänsch, T. W., and Schawlow, A. L. 1975. Cooling of gases by laser radiation. *Optics Communications*, **13**, 68–69.

[39] Heaviside, O. 1912. *Electromagnetic Theory*, vol. III. London: "The Electrician" Printing and Publishing Company. Available from `http://openlibrary.org/books/OL7145891M/Electromagnetic_theory`.

[40] Hecht, E. 2002. *Optics*. Fourth edn. San Francisco, CA: Addison-Wesley.

[41] Helene, O., and Yamashita, M. T. 2006. Understanding the tsunami with a simple model. *European Journal of Physics*, **27**, 855–863.

[42] Helmholtz, H. 1954. *On the Sensations of Tone*. New York: Dover.

[43] Hooke, R. 1665. *Micrographia*. London: John Martyn and James Allestry. Available from `http://www.gutenberg.org/files/15491/15491-h/15491-h.htm\#obsIX`.

[44] Horowitz, P., and Hill, W. 1989. *The Art of Electronics*. Second edn. Cambridge: Cambridge University Press.

[45] Houghton, J. T. 2002. *The Physics of Atmospheres*. Third edn. Cambridge: Cambridge University Press.

[46] Huygens, C. 1690. *Traité de la Lumière*. Leiden: Pieter van der Aa. Available from `http://gallica.bnf.fr/ark:/12148/bpt6k5659616j`. Translated as *Treatise on Light* by S. P. Thompson, Macmillan, London, 1912; reprinted by various publishers. Available from `http://www.gutenberg.org/ebooks/14725`.

[47] Jackson, J. D. 1998. *Classical Electrodynamics*. Third edn. New York: John Wiley & Sons.

[48] James, J. F. 2002. *A Student's Guide to Fourier Transforms*. Second edn. Cambridge: Cambridge University Press.

[49] Janah, A. R., Padmanabhan, T., and Singh, T. P. 1988. On Feynman's formula for the electromagnetic field of an arbitrarily moving charge. *American Journal of Physics*, **56**(11), 1036–1038.

[50] Jeans, Sir J. H. 1968. *Science and Music*. New York: Dover.

[51] Jones, R. C. 1941. A new calculus for the treatment of optical systems. *Journal of the Optical Society of America*, **31**(7), 488–493.

[52] Khintchine, A. Ya. 1934. Korrelationstheorie der stationären stochastischen Prozesse. *Mathematische Annalen*, **109**, 604–615.

[53] Kirchhoff, G. R. 1882. Zur Theorie der Lichtstrahlen. *Sitzungsberichte der Königlich Preußischen Akademie der Wissenschaften zu Berlin*, **II**, 641–669.

[54] Laplace, P.-S. 1799. *Mécanique céleste*, vol. 1. Paris: Crapelet. Available from `http://www.archive.org/stream/traitdemcani01lapl#page/136/mode/2up`, Chapter II, p. 136.

[55] Lay, T., Kanamori, H., Ammon, C. J. *et al.* 2005. The Great Sumatra–Andaman Earthquake of 26 December 2004. *Science*, **308**, 1127–1133.

[56] Lipson, A., Lipson, S. G., and Lipson, H. 2011. *Optical Physics*. Fourth edn. Cambridge: Cambridge University Press.

[57] Longhurst, R. S. 1974. *Geometrical and Physical Optics*. Third edn. London: Longman.

[58] Lorentz, H. A. 1895. *Versuch einer Theorie der electrischen und optischen Erscheinungen in bewegten Körpern*. Leiden: E. J. Brill. Reprinted by Adament Media Corporation, 2001.

[59] Lorrain, P., Corson, D. R., and Lorrain, F. 1988. *Electromagnetic Fields and Waves*. Third edn. San Francisco, CA: W. H. Freeman.

[60] Main, I. G. 1993. *Vibrations and Waves in Physics*. Third edn. Cambridge: Cambridge University Press.

[61] Margaritondo, G. 2005. Explaining the physics of tsunamis to undergraduate and non-physics students. *European Journal of Physics*, **26**, 401–407. See *corrigendum* in *European Journal of Physics*, **28**, 779 (2007).

[62] Matthews, P. T. 1974. *Introduction to Quantum Mechanics*. Third edn. New York: McGraw-Hill.

[63] Michelson, A. A. 1881. The relative motion of the Earth and of the luminiferous ether. *American Journal of Science*, **22**, 120–129. Available from `http://www.archive.org/stream/americanjournal62unkngoog\#page/n142/`.

[64] Michelson, A. A., and Morley, E. W. 1887. On the relative motion of the Earth and the luminiferous ether. *American Journal of Science*, **34**, 333–345. Available from `http://www.aip.org/history/gap/PDF/michelson.pdf`.

[65] Newton, I. 1704. *Opticks*. London: Sam Smith and Benjamin Walford. Available from `http://www.rarebookroom.org/Control/nwtopt/index.html`.

[66] Pain, H. J. 2005. *Vibrations and Waves*. Sixth edn. Chichester: John Wiley and Sons.

[67] Panofsky, W. K. H., and Phillips, M. N. 2005. *Classical Electricity and Magnetism*. Second (revised) edn. New York: Dover.

[68] Pedrotti, F. L., Pedrotti, L. M., and Pedrotti, L. S. 2006. *Introduction to Optics*. Third edn. San Francisco, CA: Addison-Wesley.

[69] Pliny the Elder. 77 A.D. The wonders of fountains and rivers, Chapter 106 of *Naturalis Historia*, vol. II. Available from `http://www.perseus.tufts.edu/hopper/text?doc=Perseus:text:1999.02.0138`. Translated as *Natural History* by John Bostock, Taylor and Francis, London (1855), Available from `http://www.perseus.tufts.edu/hopper/text?doc=Perseus:text:1999.02.0137`.

[70] Poincaré, H. 1885. Sur les courbes définies par les équations différentielles (3ème partie). *Journal de mathématiques pures et appliquées*, **4**, 167–244.

[71] Poynting, J. H. 1884. On the transfer of energy in the electromagnetic field. *Philosophical Transactions of the Royal Society of London*, **175**, 343–361.

[72] Pretor-Pinney, G. 2010. *The Wavewatcher's Companion*. London: Bloomsbury.

[73] Ramsauer, C. W. 1921. Über den Wirkungsquerschnitt der Gasmoleküle gegenüber langsamen Elektronen. *Annalen der Physik*, **369**(6), 513–540.

[74] Rossby, C.-G. A., and collaborators. 1939. Relation between variations in the intensity of the zonal circulation of the atmosphere and the displacements of the semi-permanent centers of action. *Journal of Marine Research*, **2**(1), 38–55.

[75] Rossing, T. D. 1982. The physics of kettle drums. *Scientific American*, **247**(5), 172–178.

[76] Russell, J. S. 1844. Report on waves. *Report of the British Association for the Advancement of Science*, **14**(September), 311–390. Available from `http://books.google.com/books?id=994EAAAAYAAJ`.

[77] Schrödinger, E. 1926. Quantisierung als Eigenwertproblem. *Annalen der Physik*, **384**(4), 361–376.

[78] Smith, G. S. 1997. *An Introduction to Classical Electromagnetic Radiation*. First edn. Cambridge: Cambridge University Press.

[79] Smith, W. H. F., Scharroo, R., Titov, V. V., Arcas, D., and Arbic, B. K. 2005. Satellite altimeters measure tsunami. *Oceanography*, **18**(2), 11–13. See also `http://www.noaanews.noaa.gov/stories2005/s2365.htm` (10 January 2005).

[80] Stokes, G. G. 1852. *Transactions of the Cambridge Philosophical Society*, **9**, 399–416.

[81] Tessman, J. R., and Finnell Jr., J. T. 1967. Electric field of an accelerating charge. *American Journal of Physics*, **95**(6), 523–527.

[82] Teufel, J. D., Donner, T., Li, D. *et al.* 2011. Sideband cooling of micromechanical motion to the quantum ground state. *Nature*, **475**, 359–363.

[83] Thomson, Sir J. J. 1904. *Electricity and Matter*. London: Archibald Constable. Available from `http://www.archive.org/stream/electricityandma00thomiala`.

[84] Thomson, Sir W. (Lord Kelvin). 1879. On gravitational oscillations of rotating water. *Proceedings of the Royal Society of Edinburgh*, **10**, 92–100.

[85] Thomson, Sir W. (Lord Kelvin). 1887. On ship waves. *Proceedings of the Institution of Mechanical Engineers*, **38**, 409–434.

[86] Titov, V., Rabinovich, A. B., Mofjeld, H. O., Thomson, R. E., and González, F. I. 2005. The global reach of the 26 December 2004 Sumatra tsunami. *Science*, **309**, 2045–2048.

[87] Townsend, J. S. E., and Bailey, V. A. 1921. The motion of electrons in gases. *Philosophical Magazine*, Series 6, **42**(252), 873–891.

[88] von Laue, M. 1915. Concerning the detection of X-ray interferences. Nobel lecture. Available from `http://nobelprize.org/nobel_prizes/physics/laureates/1914/laue-lecture.pdf`.

[89] Website: *LIGO: Laser Interferometric Gravitational-Wave Observatory*. `http://www.ligo.caltech.edu`.

[90] Website: *LISA: Laser Interferometer Space Antenna*. `http://sci.esa.int/home/lisa`.

[91] Website: *GEO600: the German–British gravitational wave detector*. `http://www.geo600.org`.

[92] Westcott, W. J. 1970. *Bells and Their Music*. New York: G. P. Putnam. Available from `https://www.msu.edu/~carillon/batmbook/`.

[93] Wood, A. 1913. *The Physical Basis of Music*. Cambridge: Cambridge University Press.

[94] Young, T. 1804. The Bakerian Lecture. Experiments and calculations relative to physical optics. *Philosophical Transactions of the Royal Society of London*, **94**, 1–16. Available from `http://rstl.royalsocietypublishing.org/content/94.toc`.

Index

aberrations, 88, 122
accelerating charge
 electromagnetic field, 274, 276, 277, 279
acousto-optic modulation, 214
adiabatic expansion, 41, 128
æther, 8, 71
Airy, Sir George Biddell
 diffraction pattern for circular aperture, 108, 112
Ampère, André-Marie
 circuital law, 27, 29, 142
amplitude, 48, 73
 complex, 63
 interference by division of, 100
 modulation, 204, 233
 per unit frequency, 181
 reflection and transmission coefficients, 136, 138, 146
angular frequency, 48
angular momentum
 quantum operator, 239
anisotropy, 176
antireflection coating, 145
apodization, 118
approximation
 paraxial, 83, 86, 87, 96, 122
 ray, 85
 thin lens, 86, 87, 91
Arago, François
 Poisson's bright spot, 112
Argand, Jean-Robert
 diagram, 64, 67, 74
argument, 225
astigmatism, 104

Babinet, Montmorency
 principle, 112, 120
bandwidth theorem, 235, 259
beamsplitter, 101–104, 153
beats, 174
bell, 168
Bernoulli, Daniel
 equation, 35
Bessel, Friedrich
 function, 109

binomial expansion, 43, 51, 98
 Taylor series, 179
birefringence, 220
black-body radiation, 243
blooming, 143
Boltzmann, Ludwig Eduard
 constant, 244
 Maxwell–Boltzmann distribution, 251
boundary conditions, 136, 158
 capillary waves, 165
 cyclic, 167
 electromagnetic waves, 165
 ocean waves, 166
 sound waves, 161
 thermal waves, 166
 water waves, 165
bow wave, 260, 261
Boyle, Robert
 observation of interference, 97
Bragg, William Lawrence and William Henry
 diffraction, 107, 197
brass instrument, 59
breakers (surf waves), 40
Brewster, Sir David
 angle, 147, 222
Brunt–Väisälä frequency, 44

cable, coaxial, 23, 25
 capacitance, 26
 inductance, 27
 wave equation, 27
camera, 88
capacitance
 coaxial cable, 24, 26
capillary wave
 boundary conditions, 165
 deep water, 39
 shallow water, 38
capillary–gravity waves, 39
cardioid response, 219
Cartesian coordinates, 283
catamaran, 261
Čerenkov, Pavel Alekseyevich
 radiation, 254, 272

chaos, 173
characteristic impedance, 140, 145
 electromagnetic, 142
 of free space, Z_0, 140
chemical waves, 10
chop (short sea waves), 40
chord, musical, 160
clarinet, 59
clothoid, 74, 109
cloud, lenticular, 41
coaxial cable, 23, 25–27
cochleoid, 111
coherence
 longitudinal (temporal), 101
 transverse (spatial), 100
collapse of quantum wavefunction, 242, 243
complex harmonic waves, 63
compressibility, 40, 132
compression factor
 moving charge, 272
Compton, Arthur Holly
 effect, 214
Concorde, 173
conjugate variables, 92, 187, 236
constructive interference, 98, 174
continuity conditions, 135, 139, 140, 151
 for electromagnetic waves, 141
conventions
 real and virtual images, 88
converging lens, 91
convolution, 198, 251
 theorem, 197, 199, 204
 proof, 200
coordinates
 Cartesian, 283
Coriolis, Gaspard-Gustave
 force, 44
Cornu, Marie Alfred
 spiral, 74, 109
correlation, 199, 204
Coulomb, Charles-Augustin de
 law of electrostatic attraction, 4, 265, 268
critical angle, 147

cross-talk compensation, 99
crystal, photonic, 212
curl, vector
    Cartesian coordinates, 284
    spherical polar coordinates, 285
cyclic boundary conditions, 167

damped oscillations, 67
de Broglie, Louis-Victor-
        Pierre-Raymond
    quantum wavefunction, 214
δ-function
    Dirac, 112, 188, 191
    Kronecker, 61
destructive interference, 98, 174
diffraction
    Bragg, 107
    crystal, 212
    electron, 213
    Fraunhofer, 92, 106, 107, 109, 115
        complex apertures, 202
        definition, 121
        Fourier analysis, 201
    Fresnel, 106, 109
    grating, 101, 112, 148
    order, 115
    single-slit, 107
    three dimensions, 212
    X-ray, 213
diffusion equation, 63, 130
dipole, electric, 4
    electrostatic (Coulomb) field, 4, 264
    radiation, 4, 218, 266, 275, 279
Dirac, Paul Adrien Maurice
    δ-function, 112, 188, 191
dispersion, 47, 49, 65, 103, 115, 132,
        176, 197
    normal, 178
    relation, 65, 177, 179
dissipation, 47, 63, 65, 195
disturbance, 3
divergence, vector
    Cartesian coordinates, 284
    spherical polar coordinates, 285
diverging lens, 91
DNA (deoxyribonucleic acid)
    X-ray diffraction pattern, 213
Doppler, Christian Andreas
    effect, 247, 253, 271
        line-broadening, 251
        moving observer, 249
        moving source, 248
        quantum-mechanical, 250
        relativistic, 249
double glazing, 157
driven boundaries, 167
drum skin, 9

eigenfunctions, 59
eigenmodes, 59
eikonal equation, 83
Einstein, Albert
    special relativity, 178, 263, 268, 272,
        275
electric dipole moment, 218, 265, 269,
        280
electric field
    coaxial cable, 26
    continuity conditions, 142
    moving charge, 273–277, 279
    plane electromagnetic wave, 211
    retarded, 266, 275
    strength, 5
electromagnetic potential
    retarded, 268, 272
electromagnetic wave, 4, 8, 28, 76
    boundary conditions, 165
    continuity conditions, 141
    longitudinal, 9
electron diffraction, 213
energy
    conservation of, 141
    kinetic
        guitar string, 51
        ocean wave, 54
        quantum particle, 153, 239, 250
        sound, 132
    of wave motion, 50
    potential
        guitar string, 51
        ocean wave, 54
        quantum particle, 154, 239
        sound, 132
    quantum operator, 238
energy density, 52
    electromagnetic waves, 211, 229
    ocean waves, 53
    sound, 132
ENSO (El Niño Southern Oscillation),
        44
equation
    partial differential, 15
    wave, 14
étalon
    Fabry–Perot, 101
Euler, Leonhard
    spiral, 74, 109
evanescent wave, 36, 148, 166, 255, 258,
        280
    applications, 154
    characterization of, 150
    conundrums, 155
    energy flow in, 150
    motion of, 149
expectation value, 227, 237

eye, 88, 93
    resolution, 123

Fabry, Charles
    Fabry–Perot étalon, 101
Faraday, Michael
    law of induction, 24, 28, 141
Fender Stratocaster, 170
Fermat, Pierre de
    principle of least time, 81, 92
Feynman, Richard Phillips, 30, 106,
        272
    quantum electrodynamics, 71
fibre, optical, 154
field strength, electric, 5
flame front, 10
flexural waves, 168
flute, 59
flutter, 132
focal length, 89, 91
focal plane, 92
focal point, 87
$4f$ arrangement, 123
four-vector, 175
Fourier, Jean Baptiste Joseph
    analysis, 172, 181
        conjugate variables, 187
        Fraunhofer diffraction, 201
    integral, 181
    principle, 180, 191
    series, 180, 182
    synthesis, 180
    transform, 233
        alternative forms, 186
        complex, 187
        in multiple dimensions, 211
        inverse, 186
        spectroscopy, 203
Franklin, Benjamin
    observation of effect of oil on water,
        40
Fraunhofer, Joseph von, 112
    diffraction, 92, 106, 107, 109, 115,
        148, 197
        complex apertures, 202
        definition, 121
        Fourier analysis, 201
free-electron laser, 256, 272
frequency, 48
    angular, 48
Fresnel, Augustin-Jean, 73, 112
    diffraction, 76, 106, 109, 148
    equations (reflection/transmission
        coefficients), 136, 138, 140, 146,
        219
    integral, 74, 109
function, 225

fundamental frequency, 160
Fylingdales, 116

Gabor, Dennis, 119
γ (ratio of specific heats), 41, 128
Gauss, Carl Friedrich
  Gaussian function, 189, 190, 198, 205,
    230, 235, 236, 251
  lens formula, 91
geometrical optics, 85
glider, 41
Goos, Hermann Fritz Gustav
  Goos–Hänchen effect, 151
Gouy, Louis Georges
  phase shift, 77, 111
gradient operator
  Cartesian coordinates, 283
  spherical polar coordinates, 285
grating, diffraction, 101, 112, 148
  spectrometer, 203
gravitational wave, 6, 7, 218
  longitudinal, 9
gravity wave, 30
  compressible fluids, 40
  deep water, 34, 154
  shallow water, 30
Grimaldi, Francesco Maria
  observation of diffraction, 97
group velocity, 176–178
guitar string, 9, 17, 20, 50, 52, 57, 59, 65,
    158
  boundary conditions, 159
  frayed, 134

Hänchen, Hilda
  Goos–Hänchen effect, 151
hairy-ball theorem, 217
Hamiltonian operator, 239
harmonic, 59, 160
  musical, 160
harmonic oscillator, 244
harmony, 160
hearing
  frequency resolution of, 205, 236
  intensity range of, 132
heart, 10
Heisenberg, Werner Karl
  uncertainty principle, 188, 238
holography, 119, 120, 214
Hooke, Robert
  law of elasticity, 126
  observation of interference, 97
hull speed, 260
Huygens, Christiaan, 71
  model of wave propagation, 8, 67, 71,
    96, 107, 116, 148, 149, 247, 271
  free space, 71

reflection, 77, 78
refraction, 78
spherical wavefront, 71
subtleties, 75
hydroelectric power, 36

ideal gas, 41, 127, 169
  constant, 41, 127
image, 84, 87, 88, 90, 92, 93, 119
  plane, 106, 115, 122
  real, 88
  virtual, 89
impedance
  characteristic, 140
  complex electrical, 232
index, refractive, 5, 79
inductance
  coaxial cable, 27
  per unit length, 25
induction, Faraday's law of, 24
intensity, 52, 73, 132, 138, 192, 193, 211
  spectral, 193–195, 243
interference, 72, 96
  by division of amplitude, 100
  by division of wavefront, 100
  constructive, 98, 174
  destructive, 98, 174
  filter, 101
  thin film, 143
irrotational flow, 34
isotropic media, 85, 211, 220

jet engine, 9
jetstream, 45
Jones, R. Clark
  calculus, 222
  matrix, 221
  vector, 221

Kelvin, Lord (William Thomson)
  ship wave (wedge), 50, 256
  waves in fluids, 44
Kirchhoff, Gustav Robert, 73, 96, 112
  diffraction, 76
Korteweg–de-Vries equation, 70
Kronecker, Leopold
  δ-function, 61

Laplace, Pierre-Simon
  scalar Laplacian
    Cartesian coordinates, 208, 284
    spherical polar coordinates, 216,
      285
  vector Laplacian
    Cartesian coordinates, 284
    spherical polar coordinates, 285
lapse rate, 43

laser
  free-electron, 256, 272
  ring cavity, 169
least action, principle of, 82
least time, Fermat's principle of, 81, 92
Legendre, Adrien-Marie
  associated polynomials, 216
lens, 90
  converging, 91
  diverging, 91
  focal length, 89, 91
  thin, 90
lensmaker's formula, 91
lenticular cloud, 41
Liénard–Wiechert potentials, 272
linearity, 19, 47, 61, 63, 172
longitudinal coherence, 101
longitudinal waves, 9, 51, 125, 154, 161,
    216, 218, 265
Lorentz, Hendrik Antoon
  transformation, 6, 249, 263, 272, 275

mackerel, Atlantic (*Scomber scombrus*),
    157
magnesium fluoride, 143
magnetic field
  coaxial cable, 24, 27
  continuity conditions, 142
  moving charge, 276, 279
  plane electromagnetic wave, 211
  solar wind, 253
  from vector potential, 269
Maxwell, James Clerk
  equations, 141, 210, 268
  Maxwell–Boltzmann distribution, 251
measurement
  quantum
    non-commutativity, 244
measurement, quantum, 241, 243
medium, definition, 3
Mexican wave, 3
Michelson, Albert Abraham
  interferometer, 101, 203
  stellar interferometer, 100
microscope, 94
mirage, Huygens' description, 79
mirror, spherical, 91
mode, 59
  eigen-, 59
  normal, 59
  order, 59
  proper, 59
modulus of elasticity, 126, 127
momentum
  conservation of, 141
  quantum operator, 238
mountain lee wave, 9, 41

moving charge
  compression factor, 272
  electromagnetic field, 274–277, 279
multipole radiation, 218
musical instrument
  bell, 168
  drum, 208
  guitar, 159
  organ, 161–164
  stringed, 159
  wind, 161–164
musical stave, 160

Navier, Claude-Louis
  Navier–Stokes equation, 35
Newton, Sir Isaac
  rings, 97
Niña, la, 45
Niño, el, 44
nodes, 59, 99, 109, 161–165, 167, 168,
  241
non-commutativity, 244
nonlinear systems
  wave equations, 172
  wave propagation in, 14
normal dispersion, 178
normal modes, 59, 61, 172

object
  real, 89
  virtual, 89
ocean wave, 30
  boundary conditions, 166
  deep water, 34, 154
  elliptical motion, 34
  energy density, 53
  power, 55
  shallow water, 30
Ohm, Georg Simon
  law of resistance, 165, 233
oil
  on troubled waters, 40
ola, la, 3
opal, 213
operator, 225
  angular momentum, 239
  expectation value, 227
  frequency, 226
  Hamiltonian, 239
  momentum, 238
  quantum-mechanical, 238
  uncertainty, 229
  wavenumber, 226
optical fibre, 154
optical information processing,
  202
optical tweezers, 121

order
  diffraction, 115
  of mode, 59
organ pipe, 161–164
orthogonality, 61, 181, 193
oscillating dipole
  radiation, 266, 275
overtone, 59, 160

paraxial approximation, 83, 86, 87, 96,
  122
partial derivative, 14
partial differential equation, 15
period, 48
Perot, Alfred
  Fabry–Perot étalon, 101
phase, 48
phase-contrast microscopy, 202
phase velocity, 49, 176, 177
phased array
  radar, 116
  sonar, 118
phasors, 67, 73, 97, 108, 109
phonon, 141, 214, 215, 243
photon, 141
photon scanning tunnelling microscopy,
  155
photonic crystal, 212
pianoforte, 59
pipe
  open, 162
  stopped, 163, 186
    conical, 164
plane of incidence, 146
plane wave, 209
plasmon, 141, 154
Pliny the Elder
  observation of effect of oil on water,
  40
Poincaré, Jules Henri
  –Heinz Hopf, theorem, 218
Poisson, Siméon Denis
  bright spot, 112
  ratio, 127
polariton, 141, 154
polarization, 146, 219
  circular, 222
  elliptical, 221
  plane (linear), 222
  random, 223
polaron, 141
position–angle transducer, lens as, 92
power, 193
  ocean waves, 55
  transmitted by wave motion, 52
Poynting, John Henry
  vector, 76, 211

propagation, definition, 3
proper modes, 59
pseudo-momentum, 214
pulse, blood flow, 9, 127

quadrupole radiation, 7, 218
quantum electrodynamics (QED), 71,
  268
quantum harmonic oscillator, 244
quantum measurement, 241
quantum mechanics, 141
  operators, 238
quantum wavefunction, 240
  collapse, 242
  measurement, 241, 243
  uncertainty, 244
quarter-wavelength coating, 144

radar, phased array, 116
radiation
  Čerenkov, 254
  dipole, 4, 218, 266, 275, 279
  in three dimensions, 216
  moving charge, 274, 276, 277,
  279
  quadrupole, 218
  synchrotron, 256, 272
Ramsauer, Carl Wilhelm
  Ramsauer–Townsend effect, 145
ray, 81, 85
  approximation, 85
reaction–diffusion wave, 10
real image, 88
real object, 89
reciprocal lattice, 197, 213
reflection
  by multiple interfaces, 143
  dielectric interface, 136, 138, 146
  Huygens' description, 77, 78
  total internal, 146, 148
    frustrated, 152, 154, 155
refraction
  acoustic, 80
  at a spherical surface, 86
  Huygens' description, 78, 254
  wind gradient, 80
refractive index, 5, 79
relativity, special theory of, 5, 263
resolution
  of an imaging system, 123
  of hearing, 236
retardation, 5, 263, 269
retarded field, electromagnetic, 266, 275,
  276
retarded potential, electromagnetic, 268,
  272
road traffic, 127

Rossby, Carl-Gustaf Arvid
  ocean waves, 44

sailplane, 41
sand-bar, 90
scalar (dot) product, 283
scalar potential, 269
scanning near-field optical microscopy
      (SNOM/NSOM), 155
Schrödinger, Erwin R. J. A.
  wave equation, 63, 177, 239
Scott Russell, John
  wave of translation (soliton), 70
seismology, 10
separation of variables, 57, 67, 131, 208,
      216
Severn, River, 70
ship
  hull speed, 260
  wake, 50, 256
shock wave, 173, 252, 253, 272
sign convention
  imaging, 89, 91
sinc function, 108, 110, 164, 206,
      236
single-slit diffraction, 107
sinusoidal waves, 47, 63
  in dispersive systems, 50
Snell (Snellius), Willebrord van Royen
  law of refraction, 79, 82, 85, 92
solar wind, 253
soliton, 70
sonar, 90
  phased array, 118, 124
sonic boom, 253
sound, 9, 125
  barrier, 132
  boundary conditions, 161
  speed in air, 127
spatial coherence, 100
spatial light modulator, 120
spectacles, 93
spectral intensity, 195, 243
spectrometer, 192
  grating, 203
spectroscopists' wavenumber, 48
spectrum, 181, 192
  amplitude, 192
  phase, 192
  power, 192
speed of light in vacuum, 29
spherical harmonics
  scalar, 216
  vector, 218
spherical mirror, 91
spherical wave, 164, 216
spin wave, 9

square wave, Fourier components, 182,
      195
standard deviation, 229, 235
standing wave, 47, 57, 63, 159
stave, musical, 160
stealth aircraft, 145
stereo sound, true reproduction of, 99
Stokes, George Gabriel
  Navier–Stokes equation, 35
  parameters, 223
  vector, 223
string
  boundary conditions, 159
  frayed, 134
  waves on, 9, 17, 20, 50, 52, 57, 59, 65,
      158
superluminal waves, 58, 156, 171, 178
superposition, 19, 61, 63, 72, 74, 96,
      103, 172–176, 180–182, 190,
      194, 218, 222
surface acoustic wave, 9
surface tension, 37
synchrotron radiation, 256, 272

television aerial, 116
temporal coherence, 101
temporal compression factor, 272
thermal wave, 10, 63, 129, 180, 195
  boundary conditions, 166
thin-film interference, 143
thin lens, 90
  approximation, 86, 87, 91
total internal reflection, 146, 148
  frustrated, 152, 154, 155
Townsend, John Sealy Edward
  Ramsauer–Townsend effect, 145
trade winds, 44
transmission
  dielectric interface, 136, 138, 146
transverse coherence, 100
transverse waves, 9, 125
  three dimensions, 217
travelling wave, 13, 63
  sinusoidal, 48
triangular wave, Fourier components,
      184
trombone, 59
trumpet, 59
tsunami, 53, 55, 173
tuning fork, 168
tunnelling, 153

ultrasound imaging, 90
ultraviolet catastrophe, 243
uncertainty, 229, 235, 237
  principle, 244
  quantum measurement, 244

variables, separation of, 57, 208
variance, 229
vector potential, 269
vector product, 283
velocities, confusion between, 32
very-long-baseline interferometry
      (VLBI), 119
violin, 59
virial theorem, 52
virtual image, 89
virtual object, 89

wake, ship, 50, 256
water wave, 9
  boundary conditions, 165
wave
  capillary, 37–39
    boundary conditions, 165
  capillary–gravity, 39
  chemical, 10
  definition, 3, 6
  drum skin, 9
  electromagnetic, 4, 8, 9, 28, 76
    boundary conditions, 165
    intensity, 194
    magnetic field, 267
    three-dimensional, 210
  evanescent, 148, 255
    characterization of, 150
    conundrums, 155
    energy flow in, 150
    motion of, 149
  of fear, 2, 10
  flexure, 168
  gravitational, 6, 7, 9, 218
  gravity, 30, 34, 40, 41, 154
  guitar string, 9
  Kelvin, 44
  longitudinal, 9, 51, 125, 154, 161, 216,
    218, 265
  Mexican, 3
  mountain lee, 9, 41
  multiple dimensions, 207
  ocean, 30, 34, 154
    boundary conditions, 166
    elliptical motion, 34
  plane, 209
  power transmitted by, 52
  of protest, 2, 10
  quickly moving source, 252
  reaction–diffusion, 10
  Rossby, 44
  shock, 173, 252, 253
  slowly moving source, 247
  sound, 9
    boundary conditions, 161
  spherical, 164, 216

wave (*cont.*)
  spin, 9
  square, 182
  standing, 47, 63
  on a string, 17, 20, 50, 52, 57, 59, 65,
      158
  superluminal, 156, 178
  surface acoustic, 9
  thermal, 10, 63, 129, 180, 195
    boundary conditions, 166
  transverse, 9, 125
    three dimensions, 217
  travelling, 13, 63
    sinusoidal, 48
  triangular, 184
  water, 9
    boundary conditions, 165
wave equation, 14
  coaxial cable, 27
  complex solutions, 63

dispersive
  sinusoidal solutions, 49
drum skin, 208
electromagnetic, 29
  three-dimensional, 210
nonlinear, 172
Schrödinger, 239
sinusoidal solutions, 47, 63
thermal diffusion, 130
wave power, 36
wave–particle duality, 214, 237
wavefront, 49
  division, 100
  interference by division of, 100
  reconstruction (holography),
      119
wavefunction
  definition, 4
  quantum, 10, 238, 240
    collapse, 242, 243

cyclic boundary condition, 168
  measurement, 241, 243
  uncertainty, 244
wavelength, 48
wavenumber, 48
  spectroscopists', 48
wavepacket, 173–176, 203,
    235
  Gaussian, 190, 230, 236
wavevector, 152, 209
weather, 44
Wiener–Khintchine theorem, 204
wind instrument, 59, 125,
    161–164

X-ray diffraction, 213

Young, Thomas
  double-slit experiment, 97
  modulus of elasticity, 126, 127